Nonlinear semigroups and differential equations in Banach spaces

Nonlinear semigroups and differential equations in Banach spaces

Viorel Barbu

Editura Academiei
București România

Noordhoff International Publishing
Leyden The Netherlands

1976

ISBN: 90 286 0205 4
Noordhoff International Publishing
A division of A.W. Sijthoff International Publishing Company B.V. Leyden, The Netherlands

Revised and enlarged translation of the Romanian edition of "Semigrupuri de contracţii neliniare în spaţii Banach", first published in 1974 by Editura Academiei Române, Calea Victoriei 125, Bucureşti, România.

Printed in Romania

772152

Contents

Chapter IV

NONLINEAR DIFFERENTIAL EQUATIONS IN HILBERT SPACES

Chapter V

SECOND ORDER NONLINEAR DIFFERENTIAL EQUATIONS

Preface to the first edition

This book is concerned with nonlinear semigroups of contractions in Banach spaces and their application to the existence theory for differential equations associated with nonlinear dissipative operators. The study of nonlinear semigroups resulted from the examination of nonlinear parabolic equations and from various nonlinear boundary value problems. The first work done by Y. Komura stimulated much further work and interest in this subject. Thus a series of studies was begun and then continued by T. Kato, M. G. Crandall, A. Pazy, H. Brézis and others, who made important contributions to the development of the theory. The theory as developed below is a generalisation of the Hille-Yosida theory for one-parameter semigroups of linear operators and is a collection of diversified results unified more or less loosely by their methods of approach. This theory is also closely related to the theory of nonlinear monotone operators.

Of course not all aspects of this theory could be covered in our exposition, and many important contributions to the subject have been excluded for the sake of brevity. We have attempted to present the basic results to the reader and to orient him toward some of the applications.

This book is intended to be self-contained. The reader is assumed to have only a basic knowledge of functional analysis, function theory and partial differential equations. Some of the necessary prerequisites for the reading of this book are summarized, with or without proof, in Chapter I.

The list of references given at the end of the book is not intended to be a complete bibliography on the subject, but it should nevertheless be detailed enough to enable further independent work to be done. Each chapter is concluded with a detailed section entitled "Bibliographical Notes" containing references to the principal results treated, as well as information on important topics related to, but not included in the body of the text.

The present book is based on lectures given by the author at a seminar on nonlinear semigroups and differential equations at the University of Jassy in 1969 – 1972.

I wish to thank Professor C. Corduneanu who suggested that I write this book and provided the first impulse for work on the subject of this book.

VIOREL BARBU

Jassy, September 1972

Preface to the English edition

The English edition differs from the Romanian edition in that several alterations of the text have been made. Thus, Chapter I, Section 3 in Chapter III, Section 4 in Chapter IV and Section 2 in Chapter V have been entirely rewritten for this new edition. Some new sections pertaining to Riccati equations, semi-linear second order elliptic equations in L^1, nonlinear variational inequations, integro-differential equations and two-point boundary value problems were inserted.

The errors in the first edition were corrected through the kindness of many friends.

VIOREL BARBU

Jassy, January 1975

Chapter I

Preliminaries

The purpose of this chapter is to provide some fundamental results for the convenience of later references. To keep the chapter within reasonable length, some basic theorems (for example the renorming theorem and the "trace" theorem for Sobolev spaces) are stated without proof. For other results, which can be easily found in text books or monographs, only a sketch of the proof and adequate references are given.

§ 1. Metric properties of normed spaces

1.1. Duality mappings

In this and the following chapters X will be a real normed space and X^* will denote its dual. The value of $x^* \in X^*$ at $x \in X$ will be denoted by either (x, x^*) or (x^*, x). The norm of X will be denoted by $\|.\|$ and the norm of X^*, which will always be the dual norm of X^*, will be denoted $\|.\|_*$. If there is no danger of confusion we shall omit the star from the norm $\|.\|_*$ in X^* and denote both the norm in X and its dual norm in X^* by $\|.\|$.

We shall use the symbol "lim" or $\rightarrow$ to indicate the strong convergence and "w-lim" or $\rightharpoonup$ for the *weak convergence* on X (respectively *weak-star convergence* on X^*).

The space X endowed with the weak topology will be denoted by X_w. Similarly, X_w^* will denote the space X^* with weak-star topology.

With each $x \in X$ we associate the set

$$F(x) = \{x^* \in X^*; (x, x^*) = \|x\|^2 = \|x^*\|^2\}. \tag{1.1}$$

Using the Hahn-Banach theorem it is immediately clear that $F(x) \neq \emptyset$ for any $x \in X$.

DEFINITION 1.1. The multivalued operator $F: X \to X^*$ is called *the duality mapping* of X.

It turns out that the properties of F are closely related to the differentiability of the norm $\|\cdot\|$ in X.

Let $\Delta(x_0, y, t) = t^{-1}(\|x_0 + ty\| - \|x_0\|)$ be the difference quotient of $\|x\|$ at x_0 in the direction y. Since the function $t \to \|x_0 + ty\|$ is convex we easily deduce that $t \to \Delta(x_0, y, t)$ is monotonic increasing for $t > 0$ and $\Delta(x_0, y, t) \geqslant -\|y\|$ for all $t > 0$. This implies the existence of the right derivative $\Delta'_+(x_0, y) = \lim_{t \to 0^+} \Delta(x_0, y, t)$ of $\|x_0 + ty\|$ at $t = 0$. As $\Delta(x_0, y, -t) = -\Delta(x_0, -y, t)$ we deduce that $\Delta(x_0, y, t)$ is also monotone increasing and bounded above for negative t. Thus the left derivative $\Delta'_-(x_0, y) = \lim_{t \to 0^-} \Delta(x_0, y, t)$ exists and we have

$$\Delta'_-(x_0, y) = -\Delta'_+(x_0, -y).$$

Finally, we obtain the inequality

$$\Delta(x_0, y, -t) \leqslant \Delta'_-(x_0, y) \leqslant \Delta'_+(x_0, y) \leqslant \Delta(x_0, y, t), \quad t > 0. \tag{1.2}$$

If $\Delta'_-(x_0, y) = \Delta'_+(x_0, y)$, the common value $\Delta'(x_0, y)$ is the Gâteaux derivative of $\|x\|$ at $x = x_0$.

PROPOSITION 1.1. *For every $x_0 \in X$, $F(x_0)$ is the set of all $x_0^* \in X^*$ such that*

$$\Delta'_-(x_0, y)\,\|x_0\| \leqslant (y, x_0^*) \leqslant \Delta'_+(x_0, y)\,\|x_0\|, \qquad \forall\, y \in X. \tag{1.3}$$

Proof. Let x_0^* satisfy (1.3). From (1.2) it follows that

$$|(y, x_0^*)| \leqslant \|x_0\|\,\|y\|, \qquad \forall\, y \in X \tag{1.4}$$

because $\Delta(x_0, y, -t) \geqslant -\|y\|$ and $\Delta'_+(x_0, y) \leqslant \|y\|$.

In (1.2) we take $y = x_0$. Thus by (1.3) we obtain $(x_0, x_0^*) = \|x_0\|^2$, and this, together with (1.4), implies that $x_0^* \in F(x_0)$.

Conversely let $x_0^* \in F(x_0)$. By (1.1) we have

$$\|x_0\|^2 + t(y, x_0^*) = (x_0 + ty, x_0^*) \leqslant \|x_0\|\,\|x_0 + ty\|, \quad t > 0.$$

From this we obtain

$$(y, x_0^*) \leqslant t^{-1}(\|x_0 + ty\| - \|x_0\|)\,\|x_0\|, \qquad t > 0,\ y \in X.$$

Similarly, we find that

$$(y, x_0^*) \geqslant t^{-1}(\|x_0\| - \|x_0 + ty\|)\,\|x_0\|, \qquad \forall\, t > 0,\ y \in X.$$

Thus (1.3) has been established.

Let $< , >_s : X \times X \to R^1 =]-\infty, +\infty[$ be the function defined by

$$<y, x>_s = \sup\{(y, x^*);\ x^* \in F(x)\}. \tag{1.5}$$

Since $F(x)$ is a weak-star compact subset of X^*, it is clear that for every pair $(x, y) \in X \times X$, $\langle y, x \rangle_s < \infty$ and the supremum defining $\langle y, x \rangle_s$ is attained.

PROPOSITION 1.2. *Let* $x, y \in X$. *Then*

(i) $\langle \alpha y, \beta x \rangle_s = \alpha\beta \langle y, x \rangle_s$ *for* $\alpha, \beta \geqslant 0$

(ii) $\langle \alpha x + y, x \rangle_s = \alpha\|x\|^2 + \langle y, x \rangle_s$ *for* $\alpha \in R^1$

(iii) $\langle\ ,\ \rangle_s : X \times X \to R^1$ *is upper-semicontinuous.*

Proof. The proof of (i) and (ii) is not difficult and therefore we shall omit it. To prove (iii) we must show that $x_n \to x_0$ and $y_n \to y_0$ in X imply that

$$\limsup_{n\to\infty} \langle y_n, x_n \rangle_s \leqslant \langle y_0, x_0 \rangle_s. \tag{1.6}$$

Since $F(x_n)$ is weak-star compact, there exists $\eta_n^* \in F(x_n)$ such that $\langle y_n, x_n \rangle_s = (y_n, \eta_n^*)$. We may assume, without loss of generality, that $\lim\limits_{n\to\infty} \langle y_n, x_n \rangle_s$ exists. Let η_0^* be any weak-star limit point of $\{\eta_n^*\}$. Then we have

$$(x_0, \eta_0^*) = \lim_{n\to\infty} (x_n, \eta_n^*) = \lim_{n\to\infty} \|x_n\|^2 = \|x_0\|^2$$

and

$$\|\eta_0^*\| \leqslant \liminf_{n\to\infty} \|\eta_n^*\| = \lim_{n\to\infty} \|x_n\| = \|x_0\|.$$

This shows that $\eta_0^* \in F(x_0)$. Moreover, since $y_n \to y_0$, we obtain

$$\lim_{n\to\infty} \langle y_n, x_n \rangle_s = \lim_{n\to\infty} (y_n, \eta_n^*) = (y_0, \eta_0^*) \leqslant \langle y_0, x_0 \rangle_s$$

as claimed.

1.2. Strictly convex normed spaces

A normed space X is called *strictly convex* if the unity ball S of X is strictly convex, i.e. the boundary ∂S of S contains no line segment. Other equivalent conditions for strict convexity may be found in G. Köthe [1].

PROPOSITION 1.3. If X^* *is strictly convex then the duality mapping* $F: X \to X^*$ *is single-valued and demicontinuous* (i.e. *continuous from* X to X_w^*).

Proof. Obviously, for every $x \in X$, $F(x)$ is a closed convex subset of X^*. Since $F(x) \subset \partial S(0, \|x\|)$ ($S(0, \|x\|)$ denotes a ball with center in origin and having a radius $\|x\|$) we conclude that $F(x)$ consists of a single point. Let $\{x_n\} \subset X$ be strongly convergent to x_0 and let x_0^* be a weak star-cluster point of $F(x_n)$.

It is easy to see that $(x_0, x_0^*) = \|x_0\|^2$. Since $\|x_0^*\| \leqslant \lim_{n\to\infty} \|F(x_n)\| = \|x_0\|$ we deduce that $x_0^* \in F(x_0)$, which concludes the proof. The following theorem is due to E. Asplund [1], [2].

THEOREM 1.1. *Let X be a reflexive Banach space with norm* $\|\cdot\|$. *Then there exists an equivalent norm* $\|\cdot\|_0$ *on X such that X is strictly convex under this norm and* X^* *is strictly convex under the dual norm* $\|\cdot\|_0^*$.

We have omitted the proof of this important theorem since it involves special considerations and results rarely used in books on functional analysis.

1.3. Uniformly convex Banach spaces

A linear normed space X is said to be *uniformly convex* if for each ε with $0 < \varepsilon \leqslant 2$ there exists $\delta(\varepsilon) > 0$ such that $\|x\| \leqslant 1$, $\|y\| \leqslant 1$ and $\|x - y\| \geqslant \varepsilon$ implies $\|x + y\| \leqslant 2(1 - \delta(\varepsilon))$.

In applications the following criterion for strong convergence is very useful.

PROPOSITION 1.4. *Let X be a uniformly convex Banach space and let* $\{x_n\} \subset X$ *be weakly convergent to x. Suppose that*

$$\limsup_{n\to\infty} \|x_n\| \leqslant \|x\|. \tag{1.7}$$

Then $\{x_n\}$ *converges strongly to x in X.*

Proof. It is well known that (1.7) implies that

$$\lim_{n\to\infty} \|x_n\| = \|x\|. \tag{1.8}$$

We set $y_n = x_n/\|x\|^{-1}$ and $y_0 = x/\|x\|^{-1}$. Assume that $\{y_n\}$ does not converge to y_0 in the strong topology of X.

Therefore, there exists $\varepsilon > 0$ such that $\|y_{n'} - y_0\| \geqslant \varepsilon$ where $\{y_{n'}\}$ is a subsequence of $\{y_n\}$. We argue from this to a contradiction. Since X is uniformly convex, there is $\delta > 0$, such that $\|y_{n'} + y_0\| \leqslant 2(1-\delta)$. As y_n converges weakly to y_0, this implies that $\|y_0\| \leqslant 1 - \delta$. This concludes the proof.

The following criterion for reflexivity is owed to D. Milman (for the proof see K. Yosida [1]):

THEOREM 1.2. *Every uniformly convex Banach space is reflexive.*

We have already seen that the continuity properties of the duality mapping are closely related to the convexity of the dual norm. In this connection proposition 1.3 can be precised as follows (T. Kato [2]):

PROPOSITION 1.5. *If the dual* X^* *of a real Banach space X is uniformly convex, then the duality mapping* $F\colon X \to X^*$ *is uniformly continuous on every bounded subset of X.*

Proof. Suppose that the sequences $\{u_n\}$ and $\{v_n\}$ exist in X such that $\|u_n\| \leqslant M$, $\|u_n - v_n\| \to 0$ as $n \to \infty$ and $\|F(u_n) - F(v_n)\| \geqslant \varepsilon > 0$ for all n. We argue from this to a contradiction. Let $x_n = u_n/\|u_n\|$ and $y_n = v_n/\|v_n\|$. Without any loss of generality we may assume that for all n, $\|u_n\| > \alpha$ and $\|v_n\| > \alpha$ where $\alpha > 0$. Then, as is easily verified,

$$\|x_n - y_n\| \to 0 \quad \text{for} \quad n \to +\infty.$$

Furthermore, we have

$$(x_n, F(x_n) + F(y_n)) = \|x_n\|^2 + \|y_n\|^2 + (x_n - y_n, \;\; F(y_n)) \geqslant 2 - \|x_n - y_n\|$$

which implies that $\|F(x_n) + F(y_n)\|/2 \geqslant 1 - 2^{-1}\|x_n - y_n\|$. Since $\|F(x_n)\| = \|F(y_n)\| = 1$ and X^* is uniformly convex, we can deduce that

$$F(x_n) - F(y_n) \to 0 \quad \text{in} \quad X^* \text{ as } n \to \infty.$$

On the other hand, we have

$$F(u_n) - F(v_n) = \|u_n\|(F(x_n) - F(y_n)) + (\|u_n\| - \|v_n\|)\, F(y_n)$$

so that trivially

$$\lim_{n\to\infty} (F(u_n) - F(v_n)) = 0 \quad \text{in } X^*.$$

The contradiction we have arrived at concludes the proof.

§2. Vectorial functions defined on real intervals

2.1. Absolutely continuous vectorial functions

Let $[0, T]$ be a fixed real interval $(0 < T < +\infty)$ and let X be a real Banach space.

An X-valued function $x(t)$ defined on $[0, T]$ is said to be absolutely continuous on $[0, T]$ if for each $\varepsilon > 0$, there exists $\delta(\varepsilon) > 0$ such that $\sum_n \|x(\beta_n) - x(\alpha_n)\| \leqslant \varepsilon$ whenever $\sum_n |\beta_n - \alpha_n| \leqslant \delta(\varepsilon)$ and $]\alpha_n, \beta_n[\cap]\alpha_m, \beta_m[= \emptyset$ for $m \neq n$. Here $]\alpha_n, \beta_n[$ are arbitrary intervals contained in $[0, T]$.

It is well known that every real-valued absolutely continuous function $x(t)$ on a real interval $[0, T]$ is almost everywhere differentiable on $]0, T[$ and it is expressed as the indefinite integral of the derivative.

The following simple example shows that this fails when $x(t)$ is absolutely continuous from $[0, T]$ to a general Banach space X. Let $X = L^1(0, T)$ be the space of all real-valued summable functions on $]0, T[$ and $x(t) = \chi[0, t]: [0, T] \to X$. Here $\chi[0, t]$ denotes the characteristic function of the interval $[0, t]$. It is easy to see that $x(t)$ is absolutely continuous on $[0, T]$. However, $x(t)$ has no differentiability point on $[0, T]$. In fact, if $x(t)$ was differentiable at $t = t_0$, this would imply that the function $f(t) = (x(t), x^*)$ would be differentiable for every $x^* \in L^\infty(0, T) = (L^1(0, T))^*$. In particular the function $f(t) = \int_0^t h(s)\mathrm{d}s$ would be differentiable at $t = t_0$ for any $h \in L^\infty(0, T)$. Let h be defined as follows

$$h(s) = \begin{cases} 1 & \text{if} \quad s < t_0 \\ -1 & \text{if} \quad s > t_0 \end{cases}$$

Then f is given by

$$f(t) = \begin{cases} t & \text{if} \quad t < t_0 \\ 2t_0 - t & \text{if} \quad t \geqslant t_0 \end{cases}$$

and clearly it is not differentiable at $t = t_0$.

However, we can prove the following theorem (Y. Komura [1]).

THEOREM 2.1. *Let X be a reflexive Banach space. Then every X-valued, absolutely continuous function $x(t)$ on $[0, T]$, is a.e. differentiable on $]0, T[$ and*

$$x(t) = x(0) + \int_0^t (\mathrm{d}x/\mathrm{d}s)(s)\,\mathrm{d}s, \qquad 0 \leqslant t \leqslant T. \tag{2.1}$$

Proof. Since $x(t)$ is continuous on $[0, T]$, $\{x(t);\ 0 \leqslant t \leqslant T\}$ is a compact subset of X. Thus the strong closure of the subspace spanned by $\{x(t); 0 \leqslant t \leqslant T\}$ is a separable Banach space. We may therefore assume that X is a separable Banach space. Because the function $x(t)$ is absolutely continuous it is of bounded variation on $[0, T]$. Let $V(x, t)$ denote the variation of x on the interval $[0, t]$. Since $t \to V(x, t)$ is monotonic nondecreasing on $[0, T]$, the inequality

$$\|x(t + h) - x(t)\| \leqslant V(x, t + h) - V(x, t), \qquad 0 \leqslant t \leqslant t + h \leqslant T$$

implies that

$$\int_0^{T-h} \left\| \frac{x(t + h) - x(t)}{h} \right\| \mathrm{d}t \leqslant V(x, T), \qquad \forall\ h > 0. \tag{2.2}$$

From (2.2) it follows using a standard argument that

$$\limsup_{h \to 0} \left\| \frac{x(t + h) - x(t)}{h} \right\| < \infty \qquad \text{a.e. } t \in]0, T[. \tag{2.3}$$

Because the space X is reflexive and separable, X^* is also separable. Let $\{x_n^*\}$ be strongly dense in X^*. Obviously, for each n there exists a subset N_n of $[0, T]$

such that the function $t \to (x(t), x_n^*)$ is differentiable at every point of $]0,T[\setminus N_n$ and meas. $N_n=0$ (in all cases where the notion of measure intervenes, Lebesgue measure is understood unless the contrary is stated expressly.). Denote by N_0 the set of all points $t \in]0, T[$ such that $\limsup_{h\to\infty} \left\| \frac{x(t+h)-x(t)}{h} \right\| = \infty$ and let $N = \bigcup_{n=0} N_n$. From (2.3) we have that meas $N=0$. For every $t \in]0, T[\setminus N$, $\left\| \frac{x(t+h)-x(t)}{h} \right\|$ are bounded as $h\to 0$. As $\lim_{h\to\infty} \left(\frac{x(t+h)-x(t)}{h}, x_n^* \right)$ exists for each $t \in]0, T[\setminus N$ we deduce that

$$\lim_{h\to 0} \left(\frac{x(t+h)-x(t)}{h}, x^* \right) = (\dot{x}(t), x^*)$$

for every $t \in]0, T[\setminus N$ and $x^* \in X^*$. Thus we have shown that $x(t)$ is a.e. weakly differentiable on $]0, T[$. We have denoted by $\dot{x}$ the weak derivative of x. Being weakly measurable, $\dot{x}$ is strongly measurable by Pettis's theorem.

As $\frac{x(t+h)-x(t)}{h}$ converges weakly to $\dot{x}(t)$ we have

$$\|\dot{x}(t)\| \leqslant \liminf_{h\to 0} \left\| \frac{x(t+h)-x(t)}{h} \right\|$$

for all $t \in]0, T[\setminus N$. Estimate (2.2) and Fatou's Lemma then imply that $\dot{x} \in L^1(0, T-h; X)$ for all $0 < h < T$. The symbol $L^1(a, b; X)$ will denote the usual Banach space of all X-valued measurable functions $u(t)$ defined a.e. on $]a, b[$ such that $\|u(t)\|$ is Lebesgue integrable over $]a, b[$.

On the other hand we have

$$(x(t)-x(0), x^*) = \int_0^t (\dot{x}(s), x^*)\,ds, \qquad \forall\ t \in [0, T].$$

Hence

$$x(t) = x(0) + \int_0^t \dot{x}(s)ds, \qquad \forall\, t \in [0, T].$$

This implies that $\lim_{h\to 0} \frac{x(t+h)-x(t)}{h} = \dot{x}(t)$, a.e. $t \in]0, T[$. In other words, $\frac{dx}{dt} \in L^1(0, T; X)$ and equality (2.1) holds.

2.2. Vectorial distributions and $W^{k,p}$ spaces

Let $[0, T]$ be a fixed real interval and let $\mathscr{D}(0, T)$ denote the space of all real-valued functions defined in $[0, T]$ which are infinitely differentiable on $[0, T]$ and have compact support in $]0, T[$. The space $\mathscr{D}(0, T)$ is topologized as the strict inductive limit of $\mathscr{D}_K(0, T)$'s where K ranges over all compact subsets of $]0, T[$ and $\mathscr{D}_K(0, T) = \{\varphi \in D(0, T);\ \operatorname{supp} \varphi \subset K\}$ (see K. Yosida [1], Ch. I). If X is a Banach space, we denote by $\mathscr{D}'(0, T; X)$ the space of all linear continuous operators from $\mathscr{D}(0, T)$ to X. An element u of $\mathscr{D}'(0, T; X)$ is called an *X-valued distribution* on $]0, T[$.
If $u \in \mathscr{D}'(0, T; X)$, then

$$D^k u(\varphi) = (-1)^k u(D^k \varphi / \mathrm{d}t^k) \qquad \forall\ \varphi \in \mathscr{D}(0, T) \tag{2.4}$$

defines another *vectorial distribution* $D^k u \in \mathscr{D}'(0, T; X)$, k being a positive integer. The distribution $D^k u$ is called the *derivative of order* k of u.

From the theory of vectorial distributions, only certain elementary results will be used in the present book. For a detailed treatment of this subject we refer to the book of L. Schwartz [2].

If $1 \leqslant p \leqslant \infty$, we denote by $L^p(0, T; X)$ the space of all X-valued strongly measurable functions $x(t)$ defined a.e. on $]0, T[$ such that $\|x(t)\|^p$ is Lebesgue integrable over $]0, T[$. It is well known that $L^p(0, T; X)$ is a Banach space with the norm defined by

$$\|x\|_p = \left(\int_0^T \|x(t)\|^p \mathrm{d}t \right)^{1/p}, \qquad 1 \leqslant p \leqslant \infty$$

with the usual modification in the case $p = \infty$. If X is reflexive, then $L^p(0, T; X)$; $1 < p < \infty$ is reflexive too and its dual space coincides with $L^{p'}(0, T; X^*)$, $1/p + 1/p' = 1$ (see e.g. R. E. Edwards [1]). Background material for the theory of L^p-spaces of vectorial functions relevant for this book can be found in Dunford and Schwarz [1], K. Yosida [1], and M. Nicolescu [1]. We note that to every $u \in L^1(0, T; X)$ a distribution (denoted again u) corresponds uniquely defined by

$$u(\varphi) = \int u(t)\, \varphi(t)\, \mathrm{d}t, \qquad \forall\ \varphi \in \mathscr{D}(0, T).$$

Thus $L^1(0, T; X)$ can be regarded as a linear subspace of $\mathscr{D}'(0, T; X)$. In all which follows a function $u \in L^1(0, T; X)$ will be identified with the corresponding vectorial distribution.

Let k be a positive integer and $1 \leqslant p \leqslant \infty$. The symbol $W^{k,p}(0, T; X)$ denotes the space of all vectorial distributions $u \in \mathscr{D}'(0, T; X)$ with the property that

$$D^j u \in L^p(0, T; X) \quad \text{for} \quad j = 0, 1, \ldots, k \tag{2.5}$$

where D^j is the derivative in the sense of distributions (see (2.4)).

We denote by $A^{k,p}(0, T; X)$ the space of all absolutely continuous functions u from $[0, T]$ to X whose derivatives $\mathrm{d}^j u/\mathrm{d}t^j$ (defined almost everywhere) are absolutely continuous for $j = 1, 2, \ldots, k-1$ and belong to $L^p(0, T; X)$ for $j = 0, 1, \ldots, k$. In particular, $A^{1,p}(0, T; X)$ consists of all absolutely continuous functions $u: [0, T] \to X$ with the property that function $t \to \frac{\mathrm{d}u}{\mathrm{d}t}(t)$ exists a.e. on $]0, T[$ and belongs to $L^p(0, T; X)$. If X is reflexive, then theorem 2.1 implies that function $u: [0, T] \to X$ belongs to $A^{1,p}(0, T; X)$ if and only if there is $g \in L^p(0, T; X)$ such that $u(t) = u(0) + \int_0^t g(s)\,\mathrm{d}s$ for every $t \in [0, T]$.

It turns out that the space $W^{k,p}(0, T; X)$ can be identified with $A^{k,p}(0, T; X)$. More precisely, we have the following result.

THEOREM 2.2. *Let X be a Banach space and let* $u \in L^p(0, T; X)$, $1 \leqslant p \leqslant \infty$. *Then the following conditions are equivalent:*

(i) $u \in W^{k,p}(0, T; X)$

(ii) *There is* $u_1 \in A^{k,p}(0, T; X)$ *such that* $u(t) = u_1(t)$ a.e. on $]0, T[$.

Proof. We shall consider only the special case where $k = 1$. The general result is obtained by induction with respect to k. First we assume that condition (i) is satisfied i.e., $u \in L^p(0, T; X)$ and $Du \in L^p(0, T; X)$. We then define the *regularization* u_n of u:

$$u_n(t) = \int_0^T \rho_n(t-s)\, u(s)\,\mathrm{d}s \tag{2.6}$$

where $\{\rho_n\} \subset \mathscr{D}(R^1)$, $\rho_n \geqslant 0$, $\int_0^T \rho_n(t)\,\mathrm{d}t = 1$ $\rho_n(t) = \rho_n(-t)$ and $\operatorname{supp} \rho_n \subset \left]-\frac{1}{n}, \frac{1}{n}\right[$.

It is well known that u_n is infinitely differentiable on $[0, T]$ and $\lim_{n\to\infty} u_n = u$ in the topology of $L^p(0, T; X)$ (see e.g. K. Yosida [1], Ch. I). Let $\varphi \in \mathscr{D}(0, T)$ be such that $\operatorname{supp}\varphi \subset \left]\frac{1}{n}, T - \frac{1}{n}\right[$. Then the regularization φ_n of φ has the support contained in $]0, T[$ and a simple calculation involving Eq. (2.6) yields

$$\int_0^T \frac{\mathrm{d}u_n}{\mathrm{d}t}(t)\varphi(t)\,\mathrm{d}t = -\int_0^T u_n(t)\frac{\mathrm{d}}{\mathrm{d}t}\varphi(t)\,\mathrm{d}t = -\int_0^T u(t)\frac{\mathrm{d}\varphi_n}{\mathrm{d}t}(t)\mathrm{d}t.$$

In other words

$$\int_0^T \frac{\mathrm{d}u_n}{\mathrm{d}t}(t)\,\varphi(t)\,\mathrm{d}t = \int_0^T (Du)(t)\varphi_n(t)\,\mathrm{d}t = \int_0^T (Du)_n(t)\varphi(t)\,\mathrm{d}t$$

which implies that $\frac{\mathrm{d}u_n}{\mathrm{d}t}(t) = (Du)_n(t)$ a.e $t \in \left]\frac{1}{n}, T - \frac{1}{n}\right[$. Let Γ be the set of all $t \in]0, T[$ with the property that $u_n(t) \to u(t)$.

Thus the equality: $u_n(t) - u_n(s) = \int_s^t \frac{du_n}{d\tau}(\tau)\, d\tau,\ t, s \in \left]\frac{1}{n},\ T - \frac{1}{n}\right[$ implies that

$$u(t) - u(s) = \int_s^t (Du)(\tau)\, d\tau \quad \text{for all } t, s \in \Gamma \tag{2.7}$$

because $(Du)_n \to Du$ in $L^p(0, T; X)$. In particular, equality (2.7) implies that $\lim_{t \to 0} u(t) = u_1(0)$ exists and

$$u(t) = u_1(0) + \int_0^t (Du)(\tau)\, d\tau \qquad \text{for all } t \in \Gamma.$$

So we have proven that $u(t) = u_1(t)$ a.e. $t \in]0, T[$ where $u_1 \in A^{1,p}(0, T; X)$ is defined by

$$u_1(t) = u_1(0) + \int_0^t (Du)(\tau)\, d\tau, \qquad 0 \leqslant t \leqslant T.$$

Conversely, assume that $u \in A^{1,p}(0, T; X)$. We have to prove that $Du \in L^p(0, T; X)$. By definition we have

$$Du(\varphi) = -\int_0^T u(t) \frac{d\varphi}{dt}(t)\, dt = -\lim_{h \to 0} \frac{1}{h} \int_0^T u(t)(\varphi(t) - \varphi(t - h))\, dt$$

$$= -\lim_{h \to 0} \frac{1}{h} \int_0^{T-h} (u(t) - u(t+h))\varphi(t)\, dt - \lim_{h \to 0} \frac{1}{h} \int_{T-h}^T u(t)\,\varphi(t)\, dt$$

$$+ \lim_{h \to 0} \frac{1}{h} \int_0^h u(t)\,\varphi(t - h)\, dt, \qquad \text{for all } \varphi \in D(0, T). \tag{2.8}$$

We note that

$$\int_0^T u(t)\,(\varphi(t) - \varphi(t-h))\, dt = \int_0^{T-h} (u(t) - u(t+h))\,\varphi(t)\, dt$$

for all sufficiently small h's. Thus as $h \to 0$, Eq. (2.8) yields

$$Du(\varphi) = \int_0^T \frac{du}{dt}(t)\,\varphi(t)\, dt, \qquad \forall \varphi \in \mathscr{D}(0, T)$$

which shows that $Du \in L^p(0, T; X)$ and $(Du)(t) = \frac{du}{dt}(t)$ a.e. $t \in]0, T[$. This completes the proof of Theorem 2.2.

PROPOSITION 2.1. *Let X be reflexive and let $f \in L^p(0, T; X)$ with $1 \leqslant p \leqslant \infty$. Then the following two conditions are equivalent*

(i) *There exists* $f_1 \in W^{1,p}(0, T; X)$ *such that* $f(t) = f_1(t)$, *a.e. on* $]0, T[$.

(ii) $\displaystyle\int_0^{T-h} \|f(t+h) - f(t)\|^p \mathrm{d}t \leqslant Ch^p$ *for every* $h \in]0, T[$.

Proof. (i) $\Rightarrow$ (ii). By Theorem 2.2 we have

$$f(t+h) - f(t) = \int_t^{t+h} \frac{\mathrm{d}f_1}{\mathrm{d}s}(s)\,\mathrm{d}s, \qquad \text{a.e. } t, t+h \in]0, T[$$

and therefore

$$\int_0^{T-h} \|f(t+h) - f(t)\|^p \mathrm{d}t \leqslant h^{p-1} \int_0^{T-h} \mathrm{d}t \int_t^{t+h} \left\| \frac{\mathrm{d}f_1}{\mathrm{d}s} \right\|^p \mathrm{d}s.$$

Since $\displaystyle\lim_{h \to 0} \int_0^T \mathrm{d}t \frac{1}{h} \int_t^{t+h} \left\| \frac{\mathrm{d}f_1}{\mathrm{d}s} \right\|^p \mathrm{d}s = \int_0^T \left\| \frac{\mathrm{d}f_1}{\mathrm{d}s} \right\|^p \mathrm{d}s$, we obtain (ii) with some suitably chosen constant C. (ii)$\Rightarrow$(i). Let f_n be the regularization of f. We observe that (ii) implies that $\left\{ \dfrac{\mathrm{d}f_n}{\mathrm{d}t} \right\}$ is bounded in $L^p(0, T; X)$. Since $f_n \to f$ in $L^p(0, T; X)$ and $\dfrac{\mathrm{d}f_n}{\mathrm{d}t} \to \dfrac{\mathrm{d}f}{\mathrm{d}t}$ in $\mathscr{D}'(0, T; X)$ as $n \to \infty$, we infer that $\dfrac{\mathrm{d}f}{\mathrm{d}t} \in L^p(0, T; X)$ and therefore $f \in W^{1,p}(0, T; X)$ follows.

2.3. Sobolev spaces

We assume a familiarity with the concepts and fundamental results in the theory of scalar distributions defined in an Euclidean space R^n. However, we recall for easy reference some basic notations and definitions.

Let Ω be an open subset of R^n. Denote by $\mathscr{D}(\Omega)$ the space of all real-valued infinitely differentiable functions defined in Ω and with compact support in Ω. By $C^k(\Omega)$, $0 \leqslant k \leqslant \infty$, we denote the set of all real-valued functions defined in Ω which have continuous partial derivatives of an order up to and including k. Define on $\mathscr{D}(\Omega)$ the usual topology (see e.g. L. Schwartz [1], K. Yosida [1]). Denote by $\mathscr{D}'(\Omega)$ the space of all scalar *distributions* defined on Ω, i.e. the dual space of $\mathscr{D}(\Omega)$. We use the multi-index notation

$$D^\alpha u(x) = D_1^{\alpha_1} D_2^{\alpha_2} \ldots D_n^{\alpha_n} u(x), \quad x \in \Omega, \alpha = (\alpha_1, \ldots, \alpha_n)$$

where $D_i = \partial/\partial x_i$, $i = 1, \ldots, n$. The distribution $D^\alpha u$ defined by

$$D^\alpha u(\varphi) = (-1)^{|\alpha|} u(D^\alpha \varphi), \qquad \forall \varphi \in D(\Omega)$$

is called the derivative of order α of $u \in \mathscr{D}'(\Omega)$. Here $|\alpha| = \alpha_1 + \alpha_2 + \ldots + \alpha_n$.

Let $L^p(\Omega)$, $1 \leqslant p \leqslant \infty$, denote the usual Banach space of (equivalence classes of) Lebesgue measurable functions from Ω to R^1 under the norm

$$\| u \|_p = \left(\int_\Omega | u(x) |^p \mathrm{d}x \right)^{1/p} \quad \text{if} \quad 1 \leqslant p < +\infty$$

$$\| u \|_\infty = \operatorname*{ess\,sup}_{x \in \Omega} | u(x) |.$$

For $1 \leqslant p \leqslant \infty$ and $k \geqslant 1$, we denote by $W^{k,p}(\Omega)$ the set of all functions $u(x)$ defined in Ω such that u and its distributional derivatives $D^\alpha u$ of order $|\alpha| \leqslant k$ all belong to $L^p(\Omega)$. $W^{k,p}(\Omega)$ is a Banach space under the norm

$$\| u \|_{k,p} = \sum_{|\alpha| \leqslant k} \| D^\alpha u \|_p. \tag{2.9}$$

In particular $W^{k,2}(\Omega)$ is a Hilbert space by the scalar product

$$(u, v)_k = \sum_{|\alpha| \leqslant k} \int_\Omega D^\alpha u(x) D^\alpha v(x) \mathrm{d}x.$$

Let $C_0^k(\Omega)$ denote the space of all functions $u \in C^k(\Omega)$ with compact support in Ω. Completion of the space $C_0^k(\Omega)$ normed by (2.9) will be denoted by $W_0^{k,p}(\Omega)$. For simplicity we shall denote $W^{k,2}(\Omega) = H^k(\Omega)$. The space $W_0^{k,2}(\Omega)$ will similarly be denoted by $H_0^k(\Omega)$. Finally, denote by $W^{-k,p'}(\Omega)$, $1 \leqslant p' \leqslant \infty$ the set of all distributions $u \in \mathscr{D}'(\Omega)$ which can be represented as

$$u = \sum_{|\alpha| \leqslant k} D^\alpha f_\alpha \qquad f_\alpha \in L^{p'}(\Omega).$$

Now we state without proof some important theorems concerning Sobolev spaces $W_0^{k,p}(\Omega)$ and $W^{k,p}(\Omega)$ (for the proof see e.g. N. Dunford and J. Schwartz [2], Ch. XIV).

THEOREM 2.3. *The dual space of* $W_0^{k,p}(\Omega)$ *coincides with the space* $W^{-k,p'}(\Omega)$; $1 \leqslant p \leqslant \infty$, $1/p + 1/p' = 1$.

THEOREM 2.4. *Assume that* Ω *is a bounded and open subset of* R^n *with a sufficiently smooth boundary.*

(i) *If* $1/q > 1/p - k/n$ *where* $1 \leqslant p \leqslant \infty$, $1 \leqslant q \leqslant \infty$ *and* $k \geqslant 1$, *then* $W^{k,p}(\Omega) \subset L^q(\Omega)$. *Moreover, if* $p < \infty$ *and* $q < \infty$, *then the injection mapping of* $W^{k,p}(\Omega)$ *into* $L^q(\Omega)$ *is completely continuous.*

(ii) *If* $m < k - n/p$ *then* $W^{k,p}(\Omega) \subset C^m(\bar{\Omega})$.

For every $s \geqslant 0$ define,

$$H^s(R^n) = \{ u \in L^2(R^n);\ (1 + |\xi|^2)^{s/2} \hat{u}(\xi) \in L^2(R^n) \}$$

where $\hat{u}$ denotes the Fourier transform of u. If $k=s$ is a positive integer, then using Parseval's formula, one can easily deduce that $H^s(R^n) = H^k(\Omega)$, where $\Omega = R^n$.

Now we suppose that Ω is a bounded open subset of R^n with a sufficiently smooth boundary. More precisely, it will be assumed that the boundary Γ is an $n-1$ *dimensional manifold of class* C^∞.

This allows us to define Sobolev space $H^s(\Gamma)$ for any real s. Let $L^2(\Gamma)$ be the space of all square-integrable functions on Γ with respect to measure $d\sigma$. Let $\{\chi_i\}$ be a family of *local charts* on Γ and let $\{\theta_i\}$ be *a finite partition of unity* subordinate to it. For any $u \in L^2(\Gamma)$ we have

$$u = \sum_i \theta_i u.$$

Let $u_i = \chi_i \circ \theta_i u$. We say that $u \in H^s(\Gamma)$ if $u_i \in H^s(R^{n-1})$ for all i. $H^s(\Gamma)$ is a Hilbert space with the norm defined by

$$\| u \|^2_{H^s(\Gamma)} = \sum_i \| u_i \|^2_{H^s(R^{n-1})}$$

and it easily follows that $H^s(\Gamma)$ as well as its norm do not depend on the chosen family of local charts and the subordinated partition of unity. If $s < 0$ we set $H^s(\Gamma) = (H^{-s}(\Gamma))'$.

Let $\mathscr{D}(\bar{\Omega})$ be the space of all infinitely differentiable functions on R^n with compact support in $\bar{\Omega}$. For any $u \in \mathscr{D}(\bar{\Omega})$ we can define the *outward normal derivative* of order j to Γ, $\partial^j u/\partial n^j$.

It turns out that the mapping $u \to \{u|_\Gamma, \partial u/\partial n, \ldots \partial^{k-1} u/\partial n^{k-1}\}$ can be extended by continuity to all u in $H^k(\Omega)$. More precisely, we have the following important theorem:

THEOREM 2.5. *The mapping*

$$u \to \{\partial^j u/\partial n^j;\quad j = 0, 1, \ldots \mu\}$$

from $\mathscr{D}(\bar{\Omega})$ to $(\mathscr{D}(\Gamma))^{\mu+1}$ *extends to a linear continuous operator* $u \to \{\partial^j u/\partial n^j;$ $j = 0, \ldots \mu\}$ *from* $H^k(\Omega)$ *onto* $\prod_{j=0}^{\mu} H^{k-j-1/2}(\Gamma)$ *where* μ *is the largest integer such that* $\mu < k - \dfrac{1}{2}$.

The proof of this theorem can be found in J. L. Lions and E. Magenes [1], Ch. I. Also (see L. Hörmander [1], Ch. II). In particular, the theorem above shows that for each $u \in H^k(\Omega)$, $\partial^j u/\partial n^j$ is well defined and belongs to $H^{k-j-1/2}(\Gamma)$. One can also deduce that the space $H^k_0(\Omega)$ coincides with the kernel of the mapping $u \to \{\partial^j u/\partial n^j; j = 0,1, \ldots \mu\}$. In other words,

$$H^k_0(\Omega) = \{u \in H^k(\Omega);\ \partial^j u/\partial n^j = 0,\ j = 0, 1, \ldots, k-1\}.$$

§ 3. Semigroups of continuous linear operators

In this section we shall review some facts about semigroups of class (C_0) in Banach spaces. (For a fuller exposition see E. Hille and R.S. Philips [1], K. Yosida [1], and S. G. Krein [1].)

3.1. Semigroups of class (C_0). Hille-Yosida theorem

Let X be a Banach space. Denote by $L(X, X)$ the Banach algebra of all linear continuous operators from X into itself. The norm in $L(X, X)$ will be denoted by $\|\cdot\|_{L(X, X)}$.

DEFINITION 3.1. One parameter family of continuous linear operators $\{S(t); t \geqslant 0\} \subset L(X, X)$ satisfying the conditions

$$S(t)S(s) = S(t+s) \qquad \text{for } t, s \geqslant 0 \tag{3.1}$$

$$S(0) = I \tag{3.2}$$

$$\lim_{t \to 0} S(t)x = x \qquad \text{for each } x \in X \tag{3.3}$$

is called a *semigroup of class* (C_0).

Here I denotes the identity operator in X. From the semigroup property (3.1), one easily deduces that (3.3) implies that the function $t \to S(t)x$ is continuous on $[0, \infty[$ for each $x \in X$. Moreover, condition (3.3) is equivalent to

$$w\text{-}\lim_{t \to 0} S(t)x = x \qquad \text{for every } x \in X, \tag{3.4}$$

where *w-lim* denotes the limit in the weak topology of X. We note also that a semigroup $S(t)$ of class (C_0) satisfies the growth condition

$$\| S(t) \|_{L(X, X)} \leqslant M \exp(\beta t) \qquad \text{for } 0 \leqslant t < \infty \tag{3.5}$$

with some constants $M > 0$ and β. (Proof of these facts can be found in the works cited above).

If, in particular $M = 1$ and $\beta = 0$, i.e.,

$$\|S(t)\|_{L(X, X)} \leqslant 1 \qquad \text{for } t \geqslant 0,$$

then the semigroup $S(t)$ is called a *contraction semigroup of class* (C_0).

DEFINITION 3.2. Let $S(t)$ be a semigroup of class (C_0) defined on X. The *infinitesimal generator* A of $S(t)$ is the linear operator defined by

$$Ax = \lim_{h\to 0} (S(h)x - x)/h, \qquad \text{for} \quad x \in D(A) \tag{3.6}$$

where $D(A) = \{x \in X;\ \lim_{h\to 0} (S(h)x - x)/h \text{ exists in } X\}$.

PROPOSITION 3.1. *The infinitesimal generator A is a closed linear and densely defined operator in X. If $x \in D(A)$ then $S(t)x \in C^1(0, \infty; X)$ and*

$$\frac{\mathrm{d}}{\mathrm{d}t} S(t)x = AS(t)x = S(t)Ax \ \textit{on}\ [0, \infty[. \tag{3.7}$$

Proof. First we shall prove that $D(A)$ is a dense subset of X. Let x be an arbitrary element of X. For every $t > 0$ define

$$x_t = t^{-1}\int_0^t S(s)x\,\mathrm{d}s.$$

Condition (3.3) clearly implies that $x_t \to x$ as $t \to 0$. On the other hand, we observe that $x_t \in D(A)$ for all $t > 0$. In fact, $(S(h)x_t - x_t)/h$ may be written as

$$\frac{1}{t}\left[\frac{1}{h}\int_t^{t+h} S(s)x\mathrm{d}s - \frac{1}{h}\int_0^h S(s)x\mathrm{d}s\right].$$

Hence

$$\lim_{h\to 0}\frac{S(h)\,x_t - x_t}{h} = \frac{S(t)x - x}{t} \qquad \text{for every } t > 0,$$

as claimed.

Next we shall prove Eq. (3.7). Let x be arbitrary but fixed in $D(A)$. By the semigroup property (3.1) it follows that

$$\frac{S(h) - I}{h} S(t)x = S(t)\,\frac{(Sh)\,x - x}{h} \qquad \text{for all } t, h > 0.$$

Since $\lim_{h\to 0}\dfrac{S(h)x - x}{h} = Ax$ we deduce that $S(t)\,x \in D(A)$ for all $t \geqslant 0$. We have thus proved that the right derivative $\dfrac{\mathrm{d}^+}{\mathrm{d}t} S(t)x$ exists for each $x \in D(A)$ and

$$\frac{\mathrm{d}^+}{\mathrm{d}t} S(t)x = AS(t)x = S(t)Ax \qquad \text{for every } t \geqslant 0.$$

To prove that the left derivative also exists and is equal to the right derivative, we proceed as follows. Let u^* be an arbitrary element of X^*. Then, the scalar function $t \to (S(t)x, u^*)$ is continuous in $t \geqslant 0$ and its right derivative which is equal to $(AS(t)x, u^*) = (S(t)Ax, u^*)$ is also continuous in t. Thus, according to a well-known result due to Dini (see e.g. M. Nicolescu [1]), we conclude that $t \to (S(t)\,x, u^*)$ is differentiable on $[0, \infty[$. Hence

$$(S(t)x - S(s)x, u^*) = \int_s^t (S(\tau)\,Ax, u^*)\,\mathrm{d}\tau, \text{ for } t, s \geqslant 0.$$

Since $u \in X^*$ was arbitrary, we have

$$S(t)x - S(s)x = \int_s^t S(\tau)\,Ax\,\mathrm{d}\tau, \qquad \text{for all } t, s \geqslant 0.$$

Because function $t \to S(t)Ax$ is continuous on $[0, \infty[$, it follows that $S(t)x$ is differentiable in t and $\dfrac{\mathrm{d}}{\mathrm{d}t} S(t)x = S(t)Ax$ for all $t \geqslant 0$ and $x \in D(A)$.

Now we shall prove that A is closed. For this purpose, take any $\{x_n\} \subset D(A)$ such that $x_n \to x_0$ and $Ax_n \to y_0$ in the strong topology of X. We see, by

$$S(t)x_n - x_n = \int_0^t S(s)Ax_n\mathrm{d}s, \qquad t \geqslant 0$$

that $S(t)x_0 - x_0 = \int_0^t S(s)y_0\mathrm{d}s$ for all $t \geqslant 0$. This clearly implies that $y_0 = Ax_0$.

THEOREM 3.1. (Hille and Yosida) *Let A be a densely defined linear operator with domain and range both in a complex (or real) Banach space X. Then A is the infinitesimal generator of a uniquely determined semigroup of class (C_0) satisfying*

$$\| S(t) \|_{L(X,X)} \leqslant M \exp \omega t, \qquad t \geqslant 0 \tag{3.8}$$

where $M > 0$ and $\omega \in R^1$, if and only if $(\lambda I - A)^{-1} \in L(X, X)$ and

$$\| (\lambda I - A)^{-n} \|_{L(X,X)} \leqslant M/(\mathrm{Re}\,\lambda - \omega)^n, \; n = 1,2, \ldots \tag{3.9}$$

for all complex (or real) λ such that Re $\lambda > \omega$.

Proof. Suppose that A is the infinitesimal generator of a semigroup $S(t)$ satisfying condition (3.8) with some real ω and positive M.

Define,

$$R(\lambda)x = \int_0^\infty \exp(-\lambda t)\, S(t)x\mathrm{d}t, \qquad x \in X. \tag{3.10}$$

Denote by C the complex λ-plane. By (3.8) it follows that $R(\lambda) \in L(X, X)$ for every λ in the right half-plane $\{\lambda \in C; Re\lambda > \omega\}$. On the other hand, by differentiating (3.10) n-times with respect to λ, we have

$$\frac{d^n}{d\lambda^n} R(\lambda)x = \int_0^\infty \exp(-\lambda t)(-t)^n S(t)x\,dt$$

and so, for any $x \in X$ and $Re\,\lambda > \omega$

$$\left\| \frac{d^n}{d\lambda^n} R(\lambda)x \right\| \leqslant Mn!\,\|x\|(Re\lambda - \omega)^{n+1}. \tag{3.11}$$

On the other hand, from Proposition 3.1 it follows that $R(\lambda)\,x \in D(A)$ for any $x \in X$, $Re\lambda > \omega$ and

$$AR(\lambda)x = \int_0^\infty \exp(-\lambda t)\frac{d}{dt} S(t)x\,dt = -x + \lambda R(\lambda)x$$

because A is closed. This shows that $\{\lambda \in C; Re > \omega\}$ is in the resolvent set $\rho(A)$ of A, and we have

$$R(\lambda) = (\lambda I - A)^{-1} \qquad \text{for all } \lambda \in C,\ Re\,\lambda > \omega.$$

From the resolvent equation

$$(\mu I - A)^{-1} - (\lambda I - A)^{-1} = (\lambda - \mu)(\lambda I - A)^{-1}(\mu I - A)^{-1}$$

it follows that

$$\frac{d^n}{d\lambda^n}(\lambda I - A)^{-1} = (-1)^n n!\,(\lambda I - A)^{-n-1}, \qquad n = 0,1,2,\ldots$$

Comparing this with estimate (3.11) we obtain condition (3.8).

We shall now prove that if A satisfies condition (3.9) then it generates a semigroup of class (C_0) on X. We shall consider only the special case $\omega = 0$. The general result is obtained by appropriately translating the special case we are considering. Thus for any $n = 1, 2, \ldots$ the operator

$$J_n = (I - n^{-1}A)^{-1}$$

belongs to $L(X, X)$. Define also

$$A_n = n\,(J_n - I), \qquad n = 1, 2, \ldots$$

We note that for every $x \in D(A)$, $AJ_n x = J_n Ax = A_n x$. Moreover, by condition (3.8) we have $\|J_n x\| \leqslant M\,\|x\|$ for any $x \in X$ and $n = 1, 2, \ldots$

This implies that

$$\|A_n x\| \leqslant M\|Ax\| \text{ for all } x \in D(A), \qquad n = 1, \ldots \tag{3.12}$$

and

$$\lim_{n\to\infty} J_n x = x \quad \text{for all } x \in \overline{D(A)} = X. \tag{3.13}$$

Since $A_n \in L(X, X)$, for each n the approximative equation

$$\frac{\mathrm{d}u_n}{\mathrm{d}t}(t) = A_n u_n(t), \qquad t \geqslant 0 \tag{3.14}$$

$$u_n(0) = x, \qquad x \in X$$

has a unique solution $u_n \in C^1(0, \infty; X)$.

We put $S_n(t)x = u_n(t)$. It is easy to see that the semigroup $S_n(t)$ can be expressed as

$$S_n(t)x = \exp(tA_n)x = \mathrm{e}^{-nt} \sum_{k=0}^{\infty} \frac{(nt)^k J_n^k x}{\mathrm{k}!}.$$

Thus by simple calculation involving condition (3.8), one obtains

$$\|S_n(t)\|_{L(X,X)} \leqslant M, \quad \text{for } t \geqslant 0, \ n = 1, 2, \ldots \tag{3.15}$$

On the other hand, from the approximate equation (3.14) it follows that

$$\frac{\mathrm{d}}{\mathrm{d}s} S_n(t-s)S_m(s)x = S_n(t-s)(A_m - A_n)S_m(s)x, \quad 0 < s < t.$$

Consequently

$$\|S_n(t)x - S_m(t)x\| \leqslant \int_0^t \|S_n(t-s)\|_{L(X,X)} \|S_m(s)\|_{L(X,X)} \|A_n x - A_m x\| \, \mathrm{d}s$$

$$\leqslant M^2 t \|A_n x - A_m x\|. \tag{3.16}$$

Let $x \in D(A)$. Then from (3.13) it follows that $A_n x \to Ax$ as $n \to \infty$. From (3.16) we then deduce that $S_n(t)x$ converges uniformly on every compact interval of $[0, \infty[$ and for each $x \in D(A)$. We put

$$S(t)x = \lim_{n\to\infty} S_n(t)x, \quad \text{for } t \geqslant 0. \tag{3.17}$$

Since $D(A)$ is dense in X, (3.17) extends to all $x \in X$. One easily deduces from this that $S(t)$ is a semigroup of class (C_0) on X. Moreover, the equality

$$S_n(t)x - x = \int_0^t S_n(s)A_n x \, \mathrm{d}s, \qquad t \geqslant 0$$

implies that

$$S(t)x - x = \int_0^t S(s)Ax\,\mathrm{d}s, \quad \text{for } t \geqslant 0 \text{ and } x \in D(A). \tag{3.18}$$

Let $\tilde{A}$ denote the infinitesimal generator of $S(t)$. By Eq. (3.18) it follows that $D(A) \subset D(\tilde{A})$ and $\tilde{A}x = Ax$ for all $x \in D(A)$.

On the other hand, from the first part of this theorem it follows that $(I - n^{-1}\tilde{A})^{-1} \in L(X, X)$ for $n = 1, \ldots$ This implies that $(I - n^{-1}A)^{-1} = (I - n^{-1}\tilde{A})^{-1}$ for all n, and $D(A) = D(\tilde{A})$. This shows that $A = \tilde{A}$. To prove the uniqueness of semigroup $S(t)$ we proceed as follows. Let $\tilde{S}(t)$ be a semigroup of class (C_0) whose infinitesimal generator is again A. We construct the semigroup $S_n(t)$. As before one has

$$\frac{\mathrm{d}}{\mathrm{d}s} S_n(t-s)\tilde{S}(s) = S_n(t-s)(A_n - A)\tilde{S}(s)x, \qquad 0 < s < t$$

for all $x \in D(A)$, because S_n and A_n are commutative with $\tilde{S}$. Then, for $x \in D(A)$, we obtain, as in condition (3.16)

$$\|S_n(t)x - \tilde{S}(t)x\| \leqslant \int_0^t \|S_n(t-s)\|_{L(X,X)} \|\tilde{S}(s)\|_{L(X,X)} \|A_n x - Ax\|\,\mathrm{d}s.$$

Since $\lim_{n\to\infty} A_n x = Ax$, we find that $S_n(t)x \to \tilde{S}(t)x$ for all $x \in D(A)$ and $t \geqslant 0$. Since $D(A)$ is dense in X, it follows that $\tilde{S}(t) = S(t)$ for any $t \geqslant 0$. This completes the proof of Theorem 3.1.

COROLLARY 3.1. *Let A be a linear and densely defined operator with both its domain and range in a Banach space X. Then A generates a contraction semigroup of class (C_0) if and only if $\lambda \in \rho(A)$ for all $\lambda > 0$ and*

$$\|(\lambda I - A)^{-1}\|_{L(X,X)} \leqslant 1/\lambda, \quad \textit{for every } \lambda > 0. \tag{3.19}$$

3.2. Analytic semigroups

Let $S(t) : [0, \infty[\to L(X, X)$ be a semigroup of class (C_0) and let A be its infinitesimal generator.

DEFINITION 3.3. The semigroup $S(t)$ is said to be *differentiable* if for all $t > 0$, $\quad S(t)X \subset D(A)$.

From Proposition 3.1 it easily follows that if $S(t)$ is a differentiable semigroup, then, for any $x \in X$, $S(t)x$ is infinitely differentiable in $t > 0$ and we have

$$S^{(n)}(t)x = (S'(t/n))^n x, \quad \text{for all } t > 0 \tag{3.20}$$

where $S'(t)x = \dfrac{\mathrm{d}}{\mathrm{d}t} S(t)x$, and $S^{(n)}(t)x = \dfrac{\mathrm{d}}{\mathrm{d}t} S^{(n-1)}(t)x$.
Moreover, for any n and $t > 0$, $S^{(n)}(t) \in L(X, X)$.

DEFINITION 3.4. The differentiable semigroup $S(t)$ is said to be *analytic* if

$$\sup_{0<t\leqslant 1} t \, \| S'(t) \|_{L(X,\,X)} < \infty. \tag{3.21}$$

It follows from Eq. (3.20) that if $S(t)$ is *analytic*, then for any $x \in X$ the function $t \to S(t)x$ is analytic on $]0, \infty[$. Moreover, the semigroup $S(t)$ admits a *holomorphic extension* $S(\lambda)$ given locally by

$$S(\lambda)x = \sum_{n=0}^{\infty} (\lambda - t)^n S^{(n)}(t)x/n!, \ \forall x \in X$$

in a certain sector $\{\lambda; |\arg \lambda| \leqslant M\}$ of a complex plane.

The infinitesimal generators of analytic semigroups are characterized by the following theorem (for proof see K. Yosida [1]).

THEOREM 3.2. *Let A be a densely defined linear operator in a Banach space X. Then A generates an analytic semigroup on X, if and only if, there is $M > 0$ such that*

$$\|(\lambda \mathrm{I} - A)^{-1}\|_{L(X,\,X)} \leqslant M/|\lambda| \tag{3.22}$$

for all complex λ with $\mathrm{Re}\,\lambda > \omega$.

3.3. Nonhomogeneous linear differential equations

Consider the Cauchy problem

$$\frac{\mathrm{d}u}{\mathrm{d}t} = Au + f(t), \qquad t \geqslant 0 \tag{3.23}$$

$$u(0) = x$$

where A is the infinitesimal generator of a semigroup $S(t)$ of class (C_0) on X, x is a given element of X and $f(t)$ is a given function defined on $]0, \infty[$ with values in X.

As we have already seen, if $x \in D(A)$ and $f \equiv 0$, then the above problem has a unique differentiable solution $u(t)$ given by

$$u(t) = S(t)x, \qquad t \geqslant 0.$$

Suppose that $f(t)$ is continuous on $[0, \infty[$ and that problem (3.23) admits a solution $u(t)$ continuous on $[0, \infty[$ and continuously differentiable on $]0, \infty[$. Then, by simple calculation one finds that $u(t)$ can be represented as

$$u(t) = S(t)x + \int_0^t S(t-s)f(s)\,\mathrm{d}s, \qquad t \geqslant 0. \tag{3.24}$$

The above formula can be regarded as the *variation constants formula* for Eq. (3.23). For every continuous (or summable) function $f(t) : [0, T] \to X$, this formula defines a continuous function $u(t) : [0, T] \to X$ which can be interpreted as a *generalized solution* (*mild solution* in other terminology) of the Cauchy problem (3.23). A function $u(t) : [0, T] \to X$ which is continuously differentiable and satisfies Eq. (3.23) for all $t \in [0, T]$ is called a *strong solution* of this equation.

PROPOSITION 3.2. *Assume that $x \in D(A)$ and one of the following two conditions holds.*

(i) *$f(t) \in D(A)$ for all $t \in [0, T]$, $f(t)$ and $Af(t)$ are continuous in t on* $[0, T]$.

(ii) *$f(t)$ is continuously differentiable on* $[0, T]$.

Then function $u(t)$ given by Eq. (3.24) *is a strong solution of problem* (3.23).

Proof. First assume that condition (i) holds. Let $u(t)$ be given by formula (3.24). Then we have

$$\frac{u(t+h) - u(t)}{h} = S(t)\frac{(S(h)\,x - x)}{h} +$$

$$+ \frac{1}{h}\int_0^t (S(t+h-s) - S(t-s))f(s)\,\mathrm{d}s + \frac{1}{h}\int_t^{t+h} S(t+h-s)f(s)\,\mathrm{d}s$$

for all $t, h > 0$ such that $t + h < T$. Since $\lim_{h\to 0} \dfrac{S(t+h-s) - S(t-s)}{h}\,y = Ay$ for all $y \in D(A)$ and $Af(t) \in C(0, T; X)$, the above equality yields

$$\frac{\mathrm{d}^+}{\mathrm{d}t}u(t) = S(t)Ax + \int_0^t S(t-s)Af(s)\,\mathrm{d}s + f(t).$$

Hence

$$\frac{\mathrm{d}^+}{\mathrm{d}t}u(t) = Au(t) + f(t) \quad \text{for all } 0 \leqslant t \leqslant T,$$

because A is closed and commutes with $S(t)$ on $D(A)$. Since the right derivative of $u(t)$ is continuous on $[0, T]$, we deduce that $u(t)$ is continuously differentiable on $[0, T]$. Obviously $u(t)$ satisfies Eq. (3.23) on $[0, T]$.

Suppose now that condition (ii) holds. This case may be treated in much the same manner as the preceding case, observing that $\dfrac{u(t+h)-u(t)}{h}$ can be written as

$$S(t)\,\frac{(S(h)x-x)}{h}+\int_0^t S(s)\,\frac{f(t+h-s)-f(t-s)}{h}\,\mathrm{d}s$$

$$+\frac{1}{h}\int_t^{t+h} S(s)f(t+h-s)\,\mathrm{d}s.$$

Remark. If instead of condition (i), respectively condition (ii), we assume that

(*j*) $f(t)\in D(A)$ a.e. $t\in\,]0, T[$, and f, $Af\in L^1(0, T, X)$ respectively

(*jj*) $f\in W^{1,1}(0, T; X)$,

then the generalized solution $u(t)$ of Eq. (3.23) belongs to $W^{1,1}(0, T; X)$ and satisfies Eq. (3.23) a.e. on$]\,0, T[$.

If in addition, A is the infinitesimal generator of an analytic semigroup and x is arbitrary in X, then for any $0<\delta<T$ the function $u(t)$ belongs to $W^{1,1}(\delta, T; X)$.

Chapter II

Nonlinear operators in Banach spaces

§ 1. Maximal monotone operators

1.1. Definitions and fundamental concepts

If X and Y are linear spaces then $X \times Y$ will denote their cartesian product space. An element of the product space $X \times Y$ will be written in the form $[x, y]$ for $x \in Y$ and $y \in Y$. A *multivalued* operator A from X to Y will be viewed as a subset of $X \times Y$. If $A \subset X \times Y$, we define

$$Ax = \{y \in Y;\ [x, y] \in A\};\ D(A) = \{x \in X;\ Ax \neq \emptyset\}. \tag{1.1}$$

$$R(A) = \bigcup_{x \in D(A)} Ax;\ A^{-1} = \{[y, x)];\ [x, y] \in A\}. \tag{1.2}$$

If $A, B \subset X \times Y$, and λ is real, one sets

$$\lambda A = \{[x, \lambda y];\ [x, y] \in A\}, \tag{1.3}$$

$$A + B = \{[x, y + z];\ [x, y] \in A;\ [x, z] \in B\}, \tag{1.4}$$

$$AB = \{[x, z];\ [x, y] \in B;\ [y, z] \in A, \text{ for some } y \in Y\}. \tag{1.5}$$

In what follows the operators from X to Y will not be distinguished from their graphs in $X \times Y$. If A is *single-valued*, Ax will denote either the value of A at x or the set defined in (1.1).

Throughout this chapter, X will be a real Banach space and X^* will denote its dual space. Notations for norms, convergence, duality mapping and scalar product will be the same as in Chapter I, Section 1.1. In particular the value of $x^* \in X^*$ at $x \in X$, will be denoted by either (x, x^*) or (x^*, x).

DEFINITION 1.1. A set $A \subset X \times X^*$ is called *monotone* if

$$(x_1 - x_2, y_1 - y_2) \geqslant 0$$

for any $[x_i, y_i] \in A$, $i = 1, 2$.

A monotone subset of $X \times X^*$ is said to be *maximal monotone* if it is not properly contained in any other monotone subset of $X \times X^*$.

If A is a single-valued operator from X to X^*, then the monotonicity, condition becomes

$$(x_1 - x_2, Ax_1 - Ax_2) \geqslant 0, \qquad \text{for all } x_1, x_2 \in D(A).$$

DEFINITION 1.2. A subset $A \subset X \times X^*$ is said to be *locally bounded* at $x_0 \in X$ if there exists a neighbourhood V of x_0 such that $A(V) = \{Ax; x \in D(A) \cap V\}$ is a bounded subset of X^*. The operator $A : X \to X^*$ is said to be *bounded* if it maps every bounded subset of X into a bounded subset of X^*.

DEFINITION 1.3. Let A be a single-valued operator defined from X to X^* such that $D(A) = X$.

(i) A is said to be *hemicontinuous* on X if

$$w\text{-}\lim_{t \to 0} A(x + ty) = Ax \tag{1.6}$$

for any x and y in X.

(ii) A is said to be *demicontinuous* on X if it is strongly-weakly continuous from X to X^*, i.e.,

$$w\text{-}\lim_{n \to \infty} Ax_n = Ax \tag{1.7}$$

for any sequence $\{x_n\}$ strongly convergent to x in X.

DEFINITION 1.4. The (multivalued) operator $A : X \to X^*$ is called *coercive* if

$$\lim_{n \to \infty} (x_n, x_n^*)/\|x_n\| = \infty \tag{1.8}$$

for all $[x_n, x_n^*] \in A$ such that $\lim_{n \to \infty} \|x_n\| = \infty$.

We shall begin the study of nonlinear monotone operators with the following important extension theorem.

THEOREM 1.1. *Let X be a reflexive Banach space and let K be a closed convex subset of X. Let A and B be two monotone subsets of $X \times X^*$ such that*

(a) *$D(A) \subset K$, $0 \in D(A)$*

(b) *B is a single-valued, hemicontinuous, bounded and coercive operator from $D(B) = X$ to X^*.*

Then, there exists $x \in K$ *such that*

$$(u - x, Bx + v) \geqslant 0, \textit{ for all } [u, v] \in A. \tag{1.9}$$

For this proof we need the following lemma.

LEMMA 1.1. *Let E be a finite dimensional Banach space and let B be a monotone and hemicontinuous operator from E to E*. Then B is continuous on E.*

Proof. First we shall prove that B is bounded on every bounded subset of E. If otherwise, there exists a sequence $\{x_n\} \subset E$ such that $\|Bx_n\| \to \infty$ and $x_n \to x_0$. Since operator B is monotone, we have

$$(x_n - x, Bx_n - Bx) \geqslant 0, \qquad \forall\, x \in E.$$

Hence

$$\left(x_n - x, \frac{Bx_n}{\|Bx_n\|} - \frac{Bx}{\|Bx_n\|}\right) \geqslant 0, \qquad \forall\, x \in E. \tag{1.10}$$

Without any loss of generality we may assume that $\lim_{n \to \infty} Bx_n/\|Bx_n\| = y_0$. Then by (1.10) it follows that

$$(x_0 - x, y_0) \geqslant 0 \qquad \forall\, x \in E.$$

Hence $y_0 = 0$. This is a contradiction, and so B must be bounded. Now we shall prove the continuity of B. To this purpose we shall consider an arbitrary sequence $\{x_n\} \subset E$ which converges to x_0. Let y_0 be a cluster point of $\{Bx_n\}$. Using again the monotonicity of B, we get

$$(x_0 - x, y_0 - Bx) \geqslant 0, \text{ for all } x \in E. \tag{1.11}$$

In (1.11) we take $x = x_t = tu + (1 - t)x_0$ where $0 \leqslant t \leqslant 1$ and u is arbitrary in E. Since B is hemicontinuous, it follows that

$$(x_0 - u, y_0 - Bx_0) \geqslant 0$$

and so by the arbitrariness of u, we see that $y_0 = Bx_0$.
The next lemma is the main step in proving Theorem 1.1.

LEMMA 1.2. *Let E be a finite dimensional Banach space and let K be a convex and compact subset of E. Let A be a monotone subset of E × E* such that D(A) ⊂ K and B a single-valued monotone and continuous operator from E to E*.*

Then, there exists $x \in K$ *such that*

$$(u - x, Bx + v) \geqslant 0, \qquad \forall\, [u, v] \in A. \tag{1.12}$$

Proof. Let $T : E \to K$ be the multivalued operator defined by

$$Ty = \{x \in K;\ (u - x,\ By + v) \geqslant 0, \quad \forall\, [u, v] \in A\}.$$

First we shall prove that $D(T) \supset K$, i.e., $Ty \neq \emptyset$ for any $y \in K$. To this end consider y to be arbitrary but fixed in K and put $f = By$. We have to show that the system

$$(u - x,\ f + v) \geqslant 0, \qquad \forall\, [u, v] \in A, \tag{1.13}$$

has at least one solution, $x \in K$.

For every $[u, v] \in A$, denote by $K_{u, v}$ the closed subset of K defined by

$$K_{u, v} = \{x \in K;\ (u - x, f + v) \geqslant 0\}.$$

Then system (1.13) may be written as

$$x \in \bigcap_{[u, v] \in A} K_{u, v}.$$

Since K is compact and $K_{u, v}$ are closed, it suffices to show that for every finite sequence $[u_i, v_i] \in A$ $i = 1, 2, \ldots, n$, $\bigcap_{i=1}^{n} K_{u_i, v_i} \neq \emptyset$. In other words, it suffices to prove that the following system of inequations

$$(u_i - x, f + v_i) \geqslant 0, \qquad [u_i, v_i] \in A,\ i = 1, 2, \ldots, n \tag{1.14}$$

has at least one solution, $x \in K$. Denote by P_n the set of all probabilistic vectors $\lambda = (\lambda_1, \ldots, \lambda_n)$ in R^n, i.e. $0 \leqslant \lambda_i \leqslant 1$, $i = 1, 2, \ldots, n$ and $\sum_{i=1}^{n} \lambda_i = 1$. Let $H(\lambda, \mu) : P_n \times P_n \to R^1$ be the function defined by

$$H(\lambda, \mu) = \sum_{i=1}^{n} \mu_i \left(f + v_i,\ \sum_{j=1}^{n} \lambda_j u_j - u_i \right).$$

The function H being biliniar on $P_n \times P_n$, according to a well-known minimax theorem, there exists $(\lambda_0, \mu_0) \in P_n \times P_n$ such that

$$H(\mu, \lambda_0) \leqslant H(\mu_0, \lambda_0) \leqslant H(\mu_0, \lambda) \tag{1.15}$$

for all λ, μ in P_n.

On the other hand, by a simple calculation one finds that

$$H(\lambda, \lambda) = \sum_{i, j=1} \lambda_i \lambda_j (v_i, u_j - u_i), \qquad \forall\, \lambda \in P_n.$$

Hence

$$H(\lambda, \lambda) \leqslant 0, \qquad \forall \lambda \in P_n$$

because A is monotone. The latter inequality combined with (1.15) yields

$$H(\mu, \lambda_0) \leqslant 0, \text{ for any } \mu \in P_n.$$

In particular, for $\mu = \mu_i = (\delta_1^i, \ldots, \delta_n^i)$ (δ_j^i denotes Kronecker's symbol) this inequality implies that

$$(u_i - \sum_{j=1} \lambda_j^0 u_j, f + v_i) \geqslant 0.$$

Thus we have shown that $x = \sum_{j=1} \lambda_j^0 u_j$ is a solution of inequality (1.14). So we proved that for any $y \in K$, $Ty \neq \emptyset$. Furthermore, for every $y \in K$, Ty is a convex compact subset of K. It is also obvious that the multivalued operator T is upper-semicontinuous on E, i.e. $y_n \to y$ and $z_n \in Ty_n \to z$ imply that $z \in Ty$. Thus in virtue of the well-known fixed-point theorem of Kakutani, there is at least one $x \in K$ such that $x \in Tx$. This concludes the proof.

Proof of Theorem 1.1. Let Λ be the family of all finite dimensional subspaces X_α of X, ordered by inclusion. For $X_\alpha \in \Lambda$, let $j_\alpha : X_\alpha \to X$ denote the inclusion mapping of X_α into X, and $j_\alpha^* : X^* \to X_\alpha^*$ the dual projection mapping of X^* onto X_α^*. The operators $A_\alpha = j_\alpha^* A j_\alpha$ and $B_\alpha = j_\alpha^* B j_\alpha$ then map X_α into X_α^* and are monotone in $X_\alpha \times X_\alpha^*$. Since B is hemicontinuous from X to X^* and j_α^* is continuous from the weak topology on X^* to the (unique) topology on X_α^*, from Lemma 1.1 it follows that B_α is continuous from X_α to X_α^*. We shall apply Lemma 1.2 where $E = X_\alpha$, $A = A_\alpha$, $B = B_\alpha$ and $K = K_\alpha = K \cap X_\alpha$. For each $X_\alpha \in \Lambda$, there exists $x_\alpha \in K \cap X_\alpha$ such that

$$(u - x, B_\alpha x_\alpha + v) \geqslant 0 \qquad \forall [u, v] \in A.$$

Equivalently

$$(u - x_\alpha, Bx_\alpha + v) \geqslant 0 \qquad \forall [u, v] \in A_\alpha. \tag{1.16}$$

The operator B being coercive and monotone, it follows that the set of all x_α satisfying (1.16) is bounded in X. Since the space X is reflexive we may pass to an infinite subsequence $x_n = x_{\alpha_n}$ and assume that

$$x_n \rightharpoonup x, \quad Bx_n \rightharpoonup y.$$

It follows that

$$\limsup_{n \to \infty} (x_n, Bx_n) \leqslant (u - x, v) + (u, y), \qquad \forall [u, v] \in A. \tag{1.17}$$

Without any loss of generality we may assume that A is maximal in the class of all monotone operators $\tilde{A} \subset X \times X^*$ such that $D(\tilde{A}) \subset K$. (By Zorn's lemma each monotone subset $A \subset X \times X^*$ with its domain in K admits a maximal extension in this class). We assert that

$$\limsup_{n\to\infty} (x_n - x, Bx_n) \leqslant 0. \tag{1.18}$$

Indeed, suppose that this is not the case. Then from inequality (1.17) it follows that

$$(u - x, v) + (u, y) \geqslant (x, y), \qquad \forall [u, v] \in A.$$

As $x \in K$ and A is maximal monotone with respect to $K \times X^*$, this implies that $[x, -y] \in A$.

In particular, setting $u = x$ in (1.17) we obtain (1.18) which contradicts the hypothesis.

For a given u in $D(A)$, let $u_t = tx + (1 - t)u$, $0 \leqslant t \leqslant 1$. By the monotonicity of B, we have

$$(x_n - u_t, Bx_n) \geqslant (x_n - u_t, Bu_t).$$

Therefore

$$(1 - t)(x_n - u, Bx_n) + t(x_n - x, Bx_n) \geqslant (1 - t)(x_n - u, Bu_t) + \\ + t(x_n - x, Bu_t).$$

Next, by (1.18)

$$\limsup_{n\to\infty} (x_n - u, Bx_n) \geqslant (x - u, Bu_t), \quad 0 \leqslant t \leqslant 1$$

and hence

$$\limsup_{n\to\infty} (x_n - u, Bx_n) \geqslant (x - u, Bx),$$

because B is hemicontinuous. We now apply inequality (1.17) and obtain

$$(u - x, Bx + v) \geqslant 0 \quad \text{for all } v \in Au,$$

as claimed.

Remark. From the proof it is clear that condition $D(B) = X$ is unnecessary if we require the demicontinuity of B instead of the hemicontinuity. In fact if $D(B) = K$ and B is demicontinuous then operators B_α defined above are obviously continuous on X_α so the existence of x satisfying (1.9) follows by the same argument.

We shall use Theorem 1.1 to prove the following characterization of maximal monotone operators in reflexive Banach spaces.

THEOREM 1.2. *Let X and X^* be reflexive and strictly convex. Let $F : X \to X^*$ be the duality mapping of X. Let A be any monotone subset of $X \times X^*$. Then A is maximal monotone in $X \times X^*$, if and only if, for any $\lambda > 0$ (equivalently for some $\lambda > 0$), $R(A + \lambda F)$ is all of X^*.*

Proof. "*If*" *part.* Assume that $R(A + \lambda F) = X^*$ for some $\lambda > 0$. We have to show that A is maximal monotone. Suppose that this is not the case and that there exists $[x_0, y_0] \in X \times X^*$ such that $[x_0, y_0] \bar{\in} A$ and

$$(x - x_0, y - y_0) \geqslant 0, \qquad \forall\, [x, y] \in A. \tag{1.19}$$

By hypothesis there exists $[x_1, y_1] \in A$ such that

$$\lambda F x_1 + y_1 = \lambda F x_0 + y_0$$

Substituting $[x_1, y_1]$ in place of $[x, y]$ in (1.19) yields

$$(x_1 - x_0, F(x_1) - F(x_0)) \leqslant 0.$$

Since X and X^* are strictly convex, the above inequality implies that $x_1 = x_0$, so that we have $[x_0, y_0] = [x_1, y_1] \in A$, which is a contradiction.

"*Only if*" *part.* The space X^* being strictly convex, the duality mapping F is single-valued and demicontinuous from X to X^* (see Proposition 1.3 in Chapter I). Obviously F is also monotone, bounded and coercive on X. Let y_0 be an arbitrary element of X^* and λ a positive number. We shall apply Theorem 1.1 where $K = X$ and $B(x) = \lambda F(x) - y_0$. There exists $x_0 \in X$ such that

$$(u - x_0, \lambda F(x_0) - y_0 + v) \geqslant 0 \qquad \forall\, [u, v] \in A.$$

Hence $x_0 \in D(A)$ and

$$y_0 - \lambda F x_0 \in A x_0$$

because A is maximal monotone. This concludes the proof.

The following perturbation result will be sharpened to Theorem 1.7 in the next section.

COROLLARY 1.1. *Let X be reflexive, and let B be a monotone, hemicontinuous and bounded operator from X to X^*. Let A be a maximal monotone subset of $X \times X^*$. Then $A + B$ is maximal monotone.*

Proof. In X and X^* we choose any strictly convex equivalent dual norms (see Theorem 1.1 in Chapter I).

By Theorem 1.2, to prove that $A + B$ is maximal monotone, we must show that for any $y_0^* \in X^*$ equation

$$F(x) + Ax + Bx \ni y_0^* \tag{1.20}$$

has at least a solution $x_0 \in D(A)$. Since the operator $x \to F(x) + B(x) - y_0^*$ is hemicontinuous, monotone, coercive and bounded from X to X^*, we can apply Theorem 1.1 to deduce the existence of $x_0 \in X$ such that

$$(u - x_0, F(x_0) + B(x_0) - y_0^* + v) \geq 0$$

for all $[u, v] \in A$. Since A is maximal monotone, this implies that x_0 is a solution of Eq. (1.20).

In particular, it follows from Corollary 1.1 that any monotone hemicontinuous and bounded operator from X to X^* is maximal monotone in $X \times X^*$. Subsequently we shall prove that this assumption of boundedness is unnecessary.

THEOREM 1.3. *Let X be reflexive and let B be monotone, everywhere defined and hemicontinuous from X to X^*. Then B is maximal monotone. If in addition B is coercive, then $R(B) = X^*$.*

Proof. Suppose that B is not maximal monotone. Then we may find $x_0 \in X$ and y_0^* in X^* such that $y_0^* \neq Bx_0$, and

$$(Bx - y_0^*, x - x_0) \geq 0 \quad \text{for every } x \in X = D(B).$$

For each $t \in [0, 1]$, we set $x_t = tx_0 + (1 - t)u$ where u is arbitrary in X. Then

$$(Bx_t - y_0^*, x_0 - u) \leq 0, \qquad \forall\, t \in [0, 1].$$

In particular, it follows from the hemicontinuity of B that

$$(Bx_0 - y_0^*, x_0 - u) \leq 0.$$

Thus the arbitrariness of $u \in X$ implies that $y_0^* = Bx_0$ which contradicts the assumption.

Suppose now that B is coercive, and let x_0^* be any fixed element of X^*. Invoking the renorming theorem, we may assume that X and X^* are strictly convex. It follows from Theorem 1.2 that for every $\lambda > 0$, equation

$$\lambda F x_\lambda + B x_\lambda = x_0^* \tag{1.21}$$

has a solution $x_\lambda \in X$. Once again using the monotonicity of B, we see, after multiplying Eq. (1.21) by x_λ, that

$$\lambda \|x_\lambda\| \leq \|B0\| + \|x_0^*\|.$$

Since B is coercive, this implies that $\{x_\lambda\}$ is bounded as $\lambda \to 0$. If we pass to an infinite subsequence, we may assume that x_λ converges weakly to x_0 in X and Bx_λ converges strongly to x_0^* in X^* as $\lambda \to 0$. Thus by the monotonicity of B we find that

$$(x_0^* - Bx, x_0 - x) \geq 0, \qquad \forall\, x \in X.$$

Since B is maximal monotone, it follows that $Bx_0 = x_0^*$. Thus, we have shown that the range of B is all of X^*.

We conclude this section with an extension theorem for monotone operators.

THEOREM 1.4. *Let X be a reflexive Banach space and let A be a monotone subset of $X \times X^*$. Let K be a nonempty closed convex subset of X containing $D(A)$. Then there exists a maximal monotone subset $\tilde{A} \subset X \times X^*$ such that $A \subset \tilde{A}$ and $D(\tilde{A}) \subset K$.*

Proof. We may assume without loss of generality that X and X^* are strictly convex. In virtue of Zorn's lemma, A admits a monotone extension $\tilde{A}$ which is maximal in the class of all monotone subsets of $X \times X^*$ having their domains contained in K. The proof will be complete if we show that $\tilde{A}$ is maximal monotone in $X \times X^*$, i.e. $R(F + \tilde{A}) = X^*$. Let $y^* \in X^*$ be arbitrarily given. We apply Theorem 1.1 where $B = F$ and A is the operator $u \to \tilde{A}u - y^*$. Consequently, there is an element $x \in K$ such that

$$(u - x, Fx + v - y^*) \geqslant 0 \quad \text{for all } [u, v] \in \tilde{A}.$$

Since $\tilde{A}$ has no monotone extension on $K \times X^*$, we infer that $x \in D(\tilde{A})$ and $y^* - Fx \in \tilde{A}x$.

1.2. A general perturbation theorem

Let X and X^* be reflexive and strictly convex and let A be a maximal monotone set in $X \times X^*$. In virtue of Corollary 1.1 equation

$$F(x_\lambda - x) + \lambda A x_\lambda \ni 0 \tag{1.22}$$

has a solution $x_\lambda \in D(A)$ for every $x \in X$ and $\lambda > 0$. Since $(x - y, Fx - Fy) \geqslant (\|x\| - \|y\|)^2$ by definition of the duality mapping F, and X is strictly convex, it follows that Eq. (1.22) has a unique solution x_λ. We define

$$x_\lambda = J_\lambda x, \tag{1.23}$$

$$A_\lambda x = -\lambda^{-1} F(x_\lambda - x), \tag{1.24}$$

for every $x \in X$ and $\lambda > 0$.

For later use we have collected some properties of J_λ and A_λ in the following proposition.

PROPOSITION 1.1. *Let X and X^* be strictly convex and reflexive. Then*

(i) *A_λ is a single-valued monotone operator of all of X into X^*.*

(ii) *A_λ and J_λ are bounded. Moreover, A_λ is demicontinuous from X to X^*.*

(iii) *for every $x \in D(A)$, $\|A_\lambda x\| \leqslant |Ax|$ and, for every $x \in \overline{\text{conv } D(A)}$, $\lim_{\lambda \to 0} J_\lambda x = x$.*

(iv) *If $\lambda_n \to 0$, $x_n \rightharpoonup x$, $A_{\lambda_n} x_n \rightharpoonup y^*$ and*

$$\limsup_{n, m \to \infty} (x_n - x_m, A_{\lambda_n} x_n - A_{\lambda_m} x_m) \leqslant 0,$$

then $[x, y] \in A$ and

$$\lim_{n, m \to \infty} (x_n - x_m, A_{\lambda_n} x_n - A_{\lambda_m} x_m) = 0.$$

In (iii) we have denoted by $\overline{\text{conv } D(A)}$ the closure of the convex hull of $D(A)$ and

$$|Ax| = \inf \{\|y\|;\ y \in Ax\}.$$

For the proof the following lemma is necessary.

LEMMA 1.3. *Let X be a reflexive Banach space and let A be a maximal monotone subset in $X \times X^*$. If $[u_n, v_n] \in A$, $u_n \rightharpoonup u$, $v_n \rightharpoonup v$ and either*

$$\limsup_{n, m \to \infty} (u_n - u_m, v_n - v_m) \leqslant 0 \tag{1.25}$$

or

$$\limsup_{n \to \infty} (u_n - u, v_n - v) \leqslant 0 \tag{1.26}$$

then $[u, v] \in A$ and $(u_n, v_n) \to (u, v)$.

Proof. First we assume that condition (1.25) holds. Since the operator A is monotone, we find that

$$\lim_{n, m \to \infty} (u_n - u_m, v_n - v_m) = 0. \tag{1.27}$$

Let $\{n_i\}$ be a subsequence of $\{n\}$ such that $(u_{n_i}, v_{n_i}) \to \mu$. Then from Eq. (1.27) we obtain

$$0 = \lim_{n_i \to \infty} [\lim_{n_k \to \infty} (u_{n_i} - u_{n_k}, v_{n_i} - v_{n_k})] = 2\mu - 2(u, v).$$

Hence $\mu = (u, v) = \lim_{n \to \infty} (u_n, v_n)$. Since A is monotone this implies that

$$(x - u, y - v) \geqslant 0 \text{ for every } [x, y] \in A.$$

By the maximal monotonicity of A, $[u, v] \in A$. Proof of the lemma with condition (1.26) is quite similar.

Proof of Proposition 1.1. (i) By definition we have

$$(x - y, A_\lambda x - A_\lambda y) = (J_\lambda x - J_\lambda y, A_\lambda x - A_\lambda y) +$$
$$+ ((x - J_\lambda x) - (y - J_\lambda y), A_\lambda x - A_\lambda y).$$

Since $A_\lambda x \in AJ_\lambda x$, this yields

$$(x - y, A_\lambda x - A_\lambda y) \geqslant 0$$

because A and F are monotone.

(ii) Let $[u, v] \in A$. Multiplying Eq. (1.22) by $J_\lambda x - u$ yields

$$(J_\lambda x - u, F(J_\lambda x - x)) \leqslant \lambda(u - J_\lambda x, v).$$

Hence

$$\|J_\lambda x - x\|^2 \leqslant \|x - u\| \, \|J_\lambda x - x\| + \lambda\|x - u\| \, \|v\| + \lambda\|v\| \, \|x - J_\lambda x\|$$

which implies that J_λ maps bounded sets into bounded sets. From Eq. (1.24) it is obvious that A_λ also has this property. Now we shall prove that A_λ is demicontinuous. Let $x_n \to x_0$ in X. Let $u_n = J_\lambda x_n$ and $v_n = A_\lambda x_n$. Since

$$F(u_n - x_n) + \lambda v_n = 0$$

it follows that

$$((u_n - x_n) - (u_m - x_m), F(u_n - x_n) - F(u_m - x_m)) +$$
$$+ \lambda(u_n - u_m, v_n - v_m) = (x_m - x_n, F(u_n - x_n) - F(u_m - x_m)).$$

Therefore,

$$\lim_{m, n \to \infty} (u_n - u_m, v_n - v_m) = 0$$

and

$$\lim_{m, n \to \infty} ((u_n - x_n) - (u_m - x_m), F(u_n - x_n) - F(u_m - x_m)) = 0.$$

Let $\{n_k\}$ be a subsequence of $\{n\}$ such that $u_{n_k} \rightharpoonup u$, $v_{n_k} \rightharpoonup v$ and $F(u_{n_k} - x_{n_k}) \rightharpoonup w$. Then $[u, v] \in A$ and $F(u - x_0) + \lambda v = 0$ by Lemma 1.3. In other words, $u = J_\lambda x_0$, $v = A_\lambda x_0$ and since the limits are unique, $J_\lambda x_n \rightharpoonup J_\lambda x_0$ and $A_\lambda x_n \rightharpoonup A_\lambda x_0$.

(iii) Let $[x, x^*] \in A$. Again by the monotonicity of A it follows that

$$0 \leqslant (x - J_\lambda x, x^* - A_\lambda x) \leqslant \|x^*\| \, \|x - x_\lambda\| - \lambda^{-1}\|x - x_\lambda\|^2.$$

and hence

$$\lambda\|A_\lambda x\| = \|x - x_\lambda\| \leqslant \lambda\|x^*\|.$$

Since x^* is arbitrary in Ax, $\|A_\lambda x\| \leqslant |Ax|$. Let $[u, u^*] \in A$, then

$$(J_\lambda x - u, A_\lambda x - u^*) \geqslant 0$$

and hence

$$\|J_\lambda x - x\|^2 \leqslant \lambda(J_\lambda x - u, u^*) + (u - x, F(J_\lambda x - x)). \tag{1.28}$$

Let $\{\lambda_n\}$ be a subsequence of $\{\lambda\}$ converging to 0 such that $F(J_{\lambda_n} x - x) \rightharpoonup y$. It follows from relation (1.28) that

$$\overline{\lim_{n\to\infty}} \|J_{\lambda_n} x - x\|^2 \leqslant (u - x, y), \qquad \forall\, u \in D(A) \tag{1.29}$$

and therefore also for any $u \in \overline{\operatorname{conv} D(A)}$. In particular relation (1.29) holds for $u = x$ which implies the desired result.

COROLLARY 1.2. *Let X be a reflexive Banach space and let A be a maximal monotone set in $X \times X^*$. Then both $\overline{D(A)}$ and $\overline{R(A)}$ are convex.*

Proof. Again using Asplund's renorming theorem we may assume that X and X^* are strictly convex. Then Proposition 1.1 implies that $\lim_{\lambda\to 0} J_\lambda x = x$ for every $x \in \overline{\operatorname{conv} D(A)}$. Since J_λ maps X onto $D(A)$ for any $\lambda > 0$, we conclude that $\overline{D(A)} = \overline{\operatorname{conv} D(A)}$ ($\overline{D(A)}$ denotes the closure of $D(A)$ in the strong topology of X).

To complete the proof it suffices to observe that $R(A) = D(A^{-1})$ and that A^{-1} is also maximal monotone in $X^* \times X$.

THEOREM 1.5. *Let A be a monotone set in $X \times X^*$. If x_0 is an interior point of $D(A)$, then A is locally bounded at x_0.*

Let us first prove the following lemma.

LEMMA 1.4. *Let $\{x_n\} \subset X$ and $\{x_n^*\} \subset X^*$ be such that $x_n \to 0$ in X and $\|x_n^*\| \to \infty$ as $n \to \infty$. Then for every closed sphere $B(0, r) = \{x \in X; \|x\| \leqslant r\}$ there exist $y_0 \in B(0, r)$, $\{x_{n_k}\} \subset \{x_n\}$ and $\{x_{n_k}^*\} \subset \{x_n^*\}$ such that*

$$\lim_{n_k\to\infty} (x_{n_k} - y_0, x_{n_k}^*) = -\infty.$$

Proof. Suppose that the lemma is false. Then for every $u \in B(0, r)$ there exists $C_u > -\infty$ such that

$$(x_n - u, x_n^*) \geqslant C_u, \qquad \text{for all } n \in N.$$

We set $E_k = \{u \in B(0, r); (x_n - u, x_n^*) \geqslant -k$ for all $n \in N\}$. Since $B(0, r) = \cup E_k$, it follows by the Baire-Hausdorff theorem (see K. Yosida [1], pp. 11) that there are $\varepsilon > 0$, $k_0 \in N$ and $y \in B(0, r)$ such that $B(y, \varepsilon) = \{x; \|y - x\| \leqslant \varepsilon\} \subset E_{k_0}$. Setting $C_{-y} = C$ we get

$$(2x_n + y - u, x_n^*) \geqslant C - k_0, \qquad \forall n \in N, \ u \in B(y, \varepsilon).$$

Let n_0 be such that $\|x_n\| \leqslant \varepsilon/4$ for $n \geqslant n_0$. This clearly implies that $u = 2x_n + y - v \in B(y, \varepsilon)$ if $n \geqslant n_0$ and $\|v\| \leqslant \varepsilon/2$. In other words, we have

$$(v, x_n^*) \geqslant C - k_0, \text{ for } n \geqslant n_0 \text{ and } \|v\| \leqslant \varepsilon/2$$

contrary to $\|x_n^*\| \to +\infty$.

Proof of Theorem 1.4. Clearly, we may assume without loss of generality that $x_0 = 0$. This can be achieved by shifting $D(A)$. Assuming that A is not locally bounded at x_0, we shall derive a contradiction. Thus there exist $\{x_n\} \subset D(A)$ and $x_n^* \in Ax_n$ such that $x_n \to 0$ and $\|x_n^*\| \to \infty$ as $n \to \infty$. Let r be positive and sufficiently small such that $B(0, r) \subset D(A)$. By virtue of the preceding lemma, there are subsequences $\{x_{n_k}\} \subset \{x_n\}$, $\{x_{n_k}^*\} \subset \{x_n^*\}$ and $x \in B(0, r) \subset D(A)$ such that

$$(x_{n_k} - x, x_{n_k}^*) \to -\infty \quad \text{as} \quad n_k \to \infty.$$

By the monotonicity of A, we have

$$(x_{n_k} - x, Ax) \to -\infty \quad \text{as} \quad n_k \to \infty.$$

This contradicts the fact that $\{\|x_{n_k}\|\}$ is bounded, which concludes the proof.

THEOREM 1.6. *Let X be reflexive and let A be a maximal monotone subset of $X \times X^*$. Then $R(A) = X^*$, if and only if, A^{-1} is locally bounded at every point x^* in $R(A) = D(A^{-1})$.*

Proof. If A is maximal monotone, then as mentioned before, A^{-1} is a maximal monotone set in $X^* \times X$. If $R(A) = D(A^{-1}) = X^*$, then in virtue of theorem 1.5, A^{-1} is locally bounded at every point $x_0^* \in X^* = R(A)$. Conversely, assume that A^{-1} is locally bounded at every point $x^* \in D(A^{-1})$. The proof will be complete if we show that the set $R(A)$ is open and closed in space X^*. Let $x_0^* \in \overline{R(A)} = \overline{D(A^{-1})}$ and let $\{x_n^*\} \subset R(A)$ be such that $x_n^* \to x_0^*$ as $n \to \infty$. Since A^{-1} is locally bounded at x_0^*, $\|y_n\| \leqslant M$, for some $M > 0$, where $y_n \in A^{-1}x_n^*$. Since X is reflexive we may assume that $y_n \rightharpoonup y_0$. By the monotonicity of A, we have

$$(x_n^* - x^*, y_n - y) \geqslant 0, \quad \text{for all } [x^*, y] \in A^{-1}$$

and hence $(x_0^* - x^*, y_0 - y) \geqslant 0$ for all $[x^*, y] \in A^{-1}$, which implies that $[x_0^*, y_0] \in A^{-1}$ because A^{-1} is maximal. Consequently, $x_0^* \in R(A)$. Thus we have shown that $R(A)$ is a closed subset of X^*. Next, we shall prove that $R(A)$ is open in X^*. Let $[x_0, y_0] \in A$ and $\rho > 0$ be such that A^{-1} is bounded on $\{y \in R(A);\ \|y - y_0\| \leqslant \rho\}$. In virtue of Corollary 1.1 the equation

$$Ax_\varepsilon + \varepsilon F(x_\varepsilon - x_0) \in y$$

has a solution $x_\varepsilon \in D(A)$ for every $\varepsilon > 0$. (Just as in the proofs of preceding theorems, no loss of generality results in assuming that X and X^* are strictly convex). We have

$$(x_\varepsilon - x_0, y_\varepsilon - y_0) \geqslant 0, \quad y_\varepsilon = y - \varepsilon F(x_\varepsilon - x_0).$$

because A is monotone. Let $y \in X^*$ be such that $\|y - y_0\| \leqslant \rho/2$. Then the above inequality implies that

$$\varepsilon \|x_\varepsilon - x_0\| \leqslant \|y - y_0\| \leqslant \rho/2, \ \varepsilon > 0.$$

Hence

$$\|y_\varepsilon - y_0\| \leqslant \rho.$$

Since A^{-1} is bounded on $\{y \in X^*;\ \|y - y_0\| \leqslant \rho\}$, we deduce that $\{x_\varepsilon\}$ is bounded in X. Thus we may assume that $x_\varepsilon \rightharpoonup x$ as $\varepsilon \to 0$. Since y_ε converges (strongly) to y as $\varepsilon \to 0$, this implies that $[x, y] \in A$ in virtue of the maximal monotonicity of A. Thus we have proved that $\{y \in X^*;\ \|y - y_0\| \leqslant \rho/2\} \subset R(A)$ which concludes proof.

Now we shall prove the following remarkable theorem.

THEOREM 1.7. *Let X be reflexive, and let A and B be maximal monotone sets in $X \times X^*$. Suppose the following condition is satisfied:*

$$(\text{int } D(A)) \cap D(B) \neq \emptyset. \tag{1.30}$$

Then $A + B$ is maximal monotone in $X \times X^*$.

Proof. As in the preceding cases we shall assume that X and X^* are strictly convex. Without loss of generality we may also assume that $0 \in (\text{int } D(A)) \cap D(B)$, $0 \in A0$ and $0 \in B0$. This can be achieved by shifting the domains and ranges of A and B. Let x^* be any element of X^* and consider equation

$$F(x_\lambda) + Ax_\lambda + B_\lambda x_\lambda \ni x^*. \tag{1.31}$$

Since B_λ is demicontinuous and bounded (see Proposition 1.1), it follows by Corollary 1.1 and Theorem 1.2 that (1.31) has a solution $x_\lambda \in D(A)$ for any $\lambda > 0$.

Since A and B_λ are monotone, we see that by multiplying Eq. (1.31) by x_λ

$$\|x_\lambda\| \leqslant \|x^*\|, \quad \forall \lambda > 0.$$

Moreover, since $0 \in \operatorname{int} D(A)$, it follows from Theorem 1.5 that A is locally bounded at 0. Hence there exist constants $\rho > 0$ and $M > 0$ such that

$$\|x^*\| \leqslant M, \forall x^* \in \cup \{Ax; \|x\| \leqslant \rho\}.$$

For $\lambda > 0$, define $y_\lambda = \rho F^{-1}(x_\lambda^*)/2\|x_\lambda^*\|$ where $x_\lambda^* = x^* - B_\lambda x_\lambda - F(x_\lambda) \in Ax_\lambda$. We note that F^{-1} is defined on all of X^* and it is just the duality mapping of X^*. Thus F^{-1} is single-valued and demicontinuous from X^* onto X. As $\|y_\lambda\| \leqslant \rho/2$, $y_\lambda \in D(A)$ and $\|y_\lambda^*\| \leqslant M$ for any $y_\lambda^* \in Ay_\lambda$. Again using the monotonicity of A we obtain

$$0 \leqslant (x_\lambda - y_\lambda, x_\lambda^* - y_\lambda^*) = (x_\lambda, x_\lambda^*) + (y_\lambda, y_\lambda^*) - (y_\lambda, x_\lambda^*) - \\ - (x_\lambda, y_\lambda^*), \quad \forall y_\lambda^* \in Ay_\lambda \tag{1.32}$$

which implies that

$$\rho/2\,\|x_\lambda^*\| = (y_\lambda, x_\lambda^*) \leqslant \|x^*\|^2 + M(\rho/2 + \|x^*\|).$$

Thus we have shown that $\{x_\lambda\}$, $\{B_\lambda x_\lambda\}$ and $\{x_\lambda^*\} \subset \{Ax_\lambda\}$ are bounded subsets of X and X^*, respectively. As X is reflexive we may assume that $x_\lambda \rightharpoonup x_0$, $F(x_\lambda) \rightharpoonup x_0^*$, $B_\lambda x_\lambda \rightharpoonup x_1^*$ and $x_\lambda^* \rightharpoonup x_2^*$ as $\lambda \to 0$. Since A is monotone, equality

$$F(x_\lambda) + B_\lambda x_\lambda + x_\lambda^* = x^*, \; x_\lambda^* \in Ax_\lambda$$

implies that

$$0 \geqslant (x_\lambda - x_\mu, F(x_\lambda) + x_\lambda^* - F(x_\mu) - x_\mu^*) + (x_\lambda - x_\mu, B_\lambda x_\lambda - B_\mu x_\mu). \tag{1.33}$$

Therefore

$$\limsup_{\lambda, \mu \to 0} (x_\lambda - x_\mu, B_\lambda x_\lambda - B_\mu x_\mu) \leqslant 0. \tag{1.34}$$

In virtue of Proposition 1.1, this implies that $[x_0, x_1^*] \in B$ and

$$\lim_{\lambda, \mu \to 0} (x_\lambda - x_\mu, B_\lambda x_\lambda - B_\mu x_\mu) = 0.$$

Consequently,

$$\limsup_{\lambda, \mu \to 0} (x_\lambda - x_\mu, F(x_\lambda) + x_\lambda^* - F(x_\mu) - x_\mu^*) \leqslant 0. \tag{1.35}$$

A being maximal monotone, also $F + A$ has this property. Then (1.35) and Lemma 1.3 imply that $[x_0, x_0^* + x_2^*] \in A + F$. Passing to limit through λ in Eq. (1.31), we obtain $F(x_0) + x_2^* + x_1^* = x^*$, where $x_2^* \in A\,x_0$ and $x_1^* \in Bx_0$, i.e. $x^* \in R(F + A + B)$.

Since $R(F + A + B)$ is all of X^*, it follows by Theorem 1.2 that $A + B$ is maximal monotone in $X \times X^*$.

Theorem 1.7 contains as a very special case the following sharpened form of Corollary 1.1.

COROLLARY 1.3. *Let X be reflexive and let B be a monotone and hemicontinuous operator from X to X^*. Let A be a maximal monotone subset of $X \times X^*$. Then $A + B$ is maximal monotone in $X \times X^*$. Moreover, if $A + B$ is coercive then $R(A + B) = X^*$.*

Proof. Since, in virtue of Theorem 1.3, B is maximal monotone, we may apply Theorem 1.7 to conclude that $A + B$ is maximal monotone. If $A + B$ is coercive, then $(A + B)^{-1}$ is locally bounded at every point of $D((A + B)^{-1})$. From Theorem 1.6 this implies that $R(A + B)$ is all of X^*.

1.3. A nonlinear elliptic boundary problem

The theory of nonlinear monotone operators allows for a detailed treatment of existence problems for solutions of nonlinear partial differential equations (particularly the elliptic and parabolic type). We shall now demonstrate this in an example. This example has been chosen mainly to illustrate how the theory can be used in some important cases, and it does not pretend to give the most general problems to which the theory is applicable. Further examples of these problems can be found in J. L. Lions [1].

Ω is an open and bounded set in an euclidean space R^n, where the boundary Γ of Ω is assumed to be a C^∞ manifold.

We shall consider a nonlinear differential operator of the form

$$Au = \sum_{|\alpha| \leqslant m} (-1)^\alpha D^\alpha A_\alpha(x, u \ldots D^m u), \tag{1.36}$$

where $A_\alpha(x, \xi)$ are real functions defined on $\Omega \times R^N$ and satisfy the following conditions:

(I) A_α are measurable in x and continuous in ξ. There exist $p > 1$, $g \in L^{p'}(\Omega)$ $(1/p + 1/p' = 1)$ and a positive constant C such that

$$|A_\alpha(x, \xi)| \leqslant C(|\xi|^{p-1} + g(x)), \quad \text{a.e. } x \in \Omega, \tag{1.37}$$

where $\xi = (\xi_\alpha; |\alpha| \leqslant m)$.

(II) For any $(\xi, \eta) \in R^N \times R^N$ and for almost every $x \in \Omega$ the following inequality holds

$$\sum_{|\alpha| \leqslant m} (A_\alpha(x, \xi) - A_\alpha(x, \eta))\ (\xi_\alpha - \eta_\alpha) \geqslant 0. \tag{1.38}$$

Let a: $W^{m,p}(\Omega) \times W^{m,p}(\Omega) \to R^1$ be defined by

$$a(u, v) = \sum_{|\alpha| \leqslant m} \int_\Omega A_\alpha(x, u \,..\, D^m u) D^\alpha v \,\mathrm{d}x.$$

It follows from (I) that a is well defined on $W^{m,p}(\Omega) \times W^{m,p}(\Omega)$ and

$$|a(u, v)| \leqslant \varphi(\|u\|_{m,p}) \, \|v\|_{m,p}, \qquad \forall u, \;\; v \in W^{m,p}(\Omega), \tag{1.39}$$

where φ is some scalar positive function on R^1_+ and $\|\cdot\|_{m,p}$ denotes the norm of Sobolev space $W^{m,p}(\Omega)$.

Let V be a closed linear subspace of $W^{m,p}(\Omega)$ such that

$$W_0^{m,p}(\Omega) \subset V \subset W^{m,p}(\Omega)$$

and let V' be its dual.

Definition 1.5. Let $f \in V'$. The function $u \in V$ is called a variational solution of equation $Au = f$ if

$$a(u, v) = (v, f) \quad \text{for any } v \in V. \tag{1.40}$$

Here (,) denotes the pairing between V and V'.

If $V = W_0^{m,p}(\Omega)$ then any $u \in V$ satisfying Eq. (1.40) is a variational solution of the Dirichlet problem

$$\sum_{|\alpha| \leqslant m} (-1)^\alpha D^\alpha A_\alpha(x, u \,\ldots\, D^m u) = f, \; f \in W^{-m,p'}(\Omega)$$

$$\frac{\partial^j u}{\partial n^j} = 0, \quad j = 0, 1, \ldots, m-1. \tag{1.41}$$

That every strict solution u of the Dirichlet problem (1.40) is also a variational solution follows by integrating by parts in equation $(Au, v) = (f, v)$. If $V = W^{m,p}(\Omega)$ then $u \in V$ may be interpreted as the generalized solution for a Neumann-type problem associated with differential operator A.

From condition (1.39) it follows that for every $u \in V$, $v \to a(u, v)$ is a linear continuous functional on V. Then $a(u, v)$ may be written as

$$a(u, v) = (\tilde{A}u, v), \qquad \forall u, v \in V \tag{1.42}$$

where $\tilde{A}$: $V \to V'$.

Proposition 1.2. *Let A satisfy conditions* (I) *and* (II). *Then the operator $\tilde{A}$ defined as above is monotone and demicontinuous from V to V'.*

Proof. $\tilde{A}$ is clearly on all of V and monotone in $V \times V'$. Let $\{u_n\} \subset V$ be such that $u_n \to u$ in V. We shall prove that $\tilde{A}u_n \rightharpoonup \tilde{A}u$ in V'. Since $\tilde{A}$ is bounded on V and V is reflexive, it suffices to prove this for an arbitrary subsequence of $\{u_n\}$. Since $\{u_n\}$ is strongly convergent in V, it also converges in $W^{m,p}(\Omega)$. Consequently, there exists $\{u_{n_k}\} \subset \{u_n\}$ such that

$$D^\alpha u_{n_k}(x) \to D^\alpha u(c) \text{ a.e. } x \in \Omega.$$

By condition I, this implies that

$$\lim_{n_k \to \infty} A_\alpha(x, u_{n_k} \ldots D^m u_{n_k}) = A_\alpha(x, u, \ldots D^m u), \text{ a.e. } x \in \Omega.$$

Next, from the dominated convergence theorem of Lebesgue, we infer that

$$\lim_{n_k \to \infty} (v, \tilde{A} u_{n_k}) = \sum_{|\alpha| \leqslant m} \int_\Omega A_\alpha(x, u, \ldots, D^m u) D^\alpha v \, dx$$

which completes the proof.

COROLLARY 1.4. *Let A be the nonlinear differential operator as above satisfying conditions* (I), (II) *and*

$$\lim_{\|u\|_{m,p} \to \infty} (a(u, u))/\|u\|_{m,p} = \infty. \tag{1.43}$$

Then for every $f \in V'$, the variational problem (1.40) *has at least one solution $u \in V$.*

Proof. There is nothing left to do, except to combine Proposition 1.2 with Corollary 1.3.

§ 2. Subdifferential mappings

2.1. Lower semicontinuous convex functions

Let X be a real Banach space with dual X^*. A *proper convex function on X* is a function φ from X to $]-\infty, +\infty]$, not identically $+\infty$ such that

$$\varphi((1-\lambda)x + \lambda y) \leqslant (1-\lambda)\varphi(x) + \lambda\varphi(y)$$

whenever $x \in X$, $y \in X$ and $0 \leqslant \lambda \leqslant 1$.

Function $\varphi : X \to]-\infty, +\infty]$ is said to be *lower-semicontinuous* on X if

$$\liminf_{y \to x} \varphi(y) \geqslant \varphi(x), \text{ for any } x \in X.$$

A convex function φ is lower-semicontinuous, if and only if, it is lower-semicontinuous with respect to the weak topology on X. This is immediate from the fact that φ is lower-semicontinuous with respect to some topology, if and only if, all the level sets of the form $\{x \in X; \varphi(x) \leqslant \lambda\}$; $\lambda \in R^1$ are closed with respect to that topology. If φ is convex, these level sets are convex subsets of X, and, as is well-known, a convex subset of a locally convex space is closed iff it is weakly closed.

In this section we shall collect for later use some elementary properties of lower-semicontinuous proper convex functions defined on a real Banach space.

PROPOSITION 2.1. *Let φ be a proper lower-semicontinuous convex function on X. Then φ is bounded from below by an affine function*, i.e., *there exist $x^* \in X^*$ and $\mu \in R^1$ such that*

$$\varphi(x) \geqslant (x, x^*) + \mu \quad \textit{for every } x \in X. \tag{2.1}$$

Proof. Denote by $K(\varphi)$ the epigraf of φ:

$$K(\varphi) = \{[x, \lambda]) \in X \times R^1; \ \varphi(x) \leqslant \lambda\}.$$

Let $x_0 \in X$ and $r \in R^1$ be such that $\varphi(x_0) > r$. Since φ is lower-semicontinuous on X, there exists an open neighbourhood $V(x_0)$ of x_0 such that $\varphi(x) > r$ for all x in $V(x_0)$. It is immediately clear that $K(\varphi)$ is a closed convex subset of the product space $X \times R^1$. In virtue of a well-known separation theorem for convex sets, there exists a closed hyperplan $H \subset X \times R^1$ which separates $K(\varphi)$ and $V(x_0) \times]-\infty, r[$. Suppose that H has the following form

$$H = \{[x, \lambda] \in X \times R^1; (x, x_0^*) + \lambda = \alpha\}$$

where $x_0^* \in X^*$ and $\alpha \in R^1$. Since $\{x_0\} \times]-\infty, r[\subset \{[x, \lambda] \in X \times R^1, (x, x_0^*) + \lambda < \alpha\}$, it follows that

$$(x, x_0^*) + \lambda > \alpha, \qquad \forall [x, \lambda] \in K(\varphi)$$

which implies (2.1) with suitably chosen x^* and μ.

The subset $D(\varphi) \subset X$ defined by

$$D(\varphi) = \{x \in X; \ \varphi(x) < +\infty\} \tag{2.2}$$

is called the *effective domain* of φ.

PROPOSITION 2.2. *Let $\varphi : X \to]-\infty, +\infty]$ be proper, convex and lower-semicontinuous. Then φ is continuous on the interior of $D(\varphi)$.*

Proof. Let $x_0 \in \text{int } D(\varphi)$. We have to prove that φ is continuous at x_0. Since φ is lower-semicontinuous, it suffices to show that for some $\lambda > 0$, $\{x \in X; \varphi(x_0 + x) \leqslant \varphi(x_0) + \lambda\}$ is a neighbourhood of the origin. We set $C = \{x \in X; \varphi(x_0 + x) \leqslant \varphi(x_0) + \lambda\} \cap \{x \in X; \varphi(x_0 - x) \leqslant \varphi(x_0) + \lambda\}$. Obviously C is a closed and balanced convex subset of X. Moreover, C absorbs every bounded set of X because the function $t \to \varphi(x_0 + tx)$ is convex and finite in a neighbourhood of the origin for any $x \in X$. (We recall that this implies the continuity.) Thus, we infer that C is a neighbourhood of the origin in X. Obviously, also $\{x \in X; \varphi(x_0 + x) \leqslant \varphi(x_0) + \lambda\}$ has this property. This completes the proof.

The function φ^* on X^* defined by

$$\varphi^*(x^*) = \sup\{(x, x^*) - \varphi(x); x \in X\} \tag{2.3}$$

is called the *conjugate* of φ.

PROPOSITION 2.3. *Let φ be a lower-semicontinuous proper convex function on X. Then the conjugate φ^* is convex, lower-semicontinuous and proper on the dual space X^*.*

Proof. It is immediately obvious that φ^* is convex and $\varphi^*(x^*) > -\infty$ for all $x^* \in X^*$.
Moreover, it follows from Proposition 1.1 that $\varphi^* \not\equiv \infty$. The lower-semicontinuity of φ^* is immediately obvious because it is the supremum of a family of affine functions on X^*.

PROPOSITION 2.4. *Let X be reflexive and let φ be a lower-semicontinuous proper convex function defined on X. Suppose that*

$$\lim_{\|x\| \to \infty} \varphi(x) = +\infty. \tag{2.4}$$

Then there exists $x_0 \in X$ such that

$$\varphi(x_0) = \inf\{\varphi(x); x \in X\}.$$

Proof. Denote $d = \inf\{\varphi(x); x \in X\}$ and choose a sequence $\{x_n\} \subset D(\varphi)$ such that $\lim_{n\to\infty} \varphi(x_n) = d$. Condition (2.4) clearly implies that $\{x_n\}$ is bounded in X. The space X being reflexive, a subsequence $\{x_{n_k}\}$ can be extracted from $\{x_n\}$ such that $x_{n_k} \rightharpoonup x_0$ as $n_k \to \infty$. Let $\varepsilon > 0$. For sufficiently large n_k we have $d \leqslant \varphi(x_{n_k}) \leqslant d + \varepsilon$. Because the set $\{x \in X; \varphi(x) \leqslant d + \varepsilon\}$ is closed and convex, it is also weakly closed. Hence $\varphi(x_0) \leqslant d + \varepsilon$. The arbitrariness of ε, then implies that $\varphi(x_0) = d$.

2.2. Subdifferentials of convex functions

Given a proper convex function φ on X and a point $x \in X$, we denote by $\partial\varphi(x)$ the set of all $x^* \in X^*$ such that

$$\varphi(x) \leqslant \varphi(y) + (x - y, x^*), \qquad \text{for every } y \in X. \tag{2.5}$$

Such elements x^* are called *subgradients* of φ at x, and $\partial\varphi(x)$ is called the *subdifferential* of φ at x.

It is immediately obvious that $\partial\varphi$ is a monotone operator (in general multivalued) from X to X^*. From (2.5) it is also obvious that $\partial\varphi(x)$ is always a closed convex set (possibly empty). We note also that $\varphi(x) = \min\{\varphi(y); y \in X\}$ iff $0 \in \partial\varphi(x)$.

Now we shall illustrate the above definition in some typical cases. Other examples will be considered in the next section.

Example 1. Let φ be Gâteaux differentiable at x. Then $\partial\varphi(x)$ consists of a single element, namely the Gâteaux differential of φ at x. Indeed the convexity of φ implies that

$$\varphi(x + \lambda(y - x)) \leqslant \varphi(x) + \lambda(\varphi(y) - \varphi(x))$$

for every $x \in X$ and $0 \leqslant \lambda \leqslant 1$. Hence

$$\frac{\varphi(x + \lambda(y - x)) - \varphi(x)}{\lambda} \leqslant \varphi(y) - \varphi(x), \qquad 0 < \lambda < 1,$$

which implies that $x \in D(\partial\varphi)$ and $\nabla_x\varphi(x) \in \partial\varphi(x)$. Now, let w be arbitrary in $\partial\varphi(x)$. From the definition, it follows that

$$\varphi(x) - \varphi(y) \leqslant (x - y, w), \qquad \forall y \in X$$

or equivalently

$$\frac{\varphi(x + \lambda y) - \varphi(x)}{\lambda} \geqslant (y, w), \qquad \forall\lambda > 0.$$

Therefore

$$(y, \nabla_x\varphi(x) - w) \geqslant 0 \quad \text{for every } y \in X.$$

which proves that $w = \nabla_x\varphi(x)$.

Example 2. Let $\varphi(x) = \frac{1}{2}\|x\|^2$. In this case we shall prove that the subdifferential $\partial\varphi$ coincides with the duality mapping $F: X \to X^*$. In fact, if $w \in F(x)$ then we have

$$(x - y, w) = \|x\|^2 - (y, w) \geqslant \frac{1}{2}(\|x\|^2 - \|y\|^2), \quad \forall\ y \in X.$$

Hence $w \in \partial\varphi(x)$. Conversely, let $w \in \partial\varphi(x)$. Then we have

$$\|x\|^2 \leqslant \|y\|^2 + 2(x - y, w), \qquad \forall\ y \in X$$

and hence

$$\|x\|^2 \leqslant \|x + ty\|^2 - 2t(y, w)$$

for every $y \in X$ and real t. Arguing as in the proof of Proposition 1.1 (Chapter I), one finally obtains $\|w\| = \|x\|$ and $(w, x) = \|x\|^2$, i.e., $w \in F(x)$.

Example 3. Let K be a closed convex subset of X and let I_K be the *indicator function* of K, i.e.

$$I_K(x) = \begin{cases} 0 & \text{if } x \in K \\ +\infty & \text{otherwise.} \end{cases}$$

Clearly I_K is convex and lower-semicontinuous on X. By definition, we have

$$\partial I_K(x) = \{x^* \in X^*;\ \ (x - u, x^*) \geqslant 0 \text{ for every } u \in K\}. \tag{2.6}$$

This shows that $D(\partial I_K) = D(I_K) = \mathrm{K}$ and $\partial I_K(x) = \{0\}$ for each $x \in \operatorname{int} K$. Moreover, if x lies on the boundary of K, then $\partial I_K(x)$ coincides with the cone of normals to K at point x.

We note that the subdifferential $\partial\varphi^*$ of the conjugate function φ^* of φ coincides with $(\partial\varphi)^{-1}$. In fact, if $x \in (\partial\varphi)^{-1}(x^*)$ then $x^* \in \partial\varphi(x)$, so that

$$\varphi(x) - \varphi(u) \leqslant (x - u, x^*), \qquad \text{for every } u \in X.$$

This implies that $\varphi^*(x^*) = (x, x^*) - \varphi(x)$. Hence

$$\varphi^*(x^*) - \varphi^*(y^*) \leqslant (x^* - y^*, x) \text{ for any } y \in X^*, \text{ i.e., } \ x \in \partial\varphi^*(x^*).$$

As mentioned, the subdifferential operator $\partial\varphi$ is a monotone subset of $X \times X^*$. In this context a more precise result can be proved.

THEOREM 2.1. *Let X be a real Banach space. If φ is a lower-semicontinuous proper convex function on X, then $\partial\varphi$ is a maximal monotone operator from X to X^*.*

Proof. To reduce the length of the exposition, we shall prove this theorem in the reflexive case and refer the reader to R.T. Rockafellar [6] for the proof of the general case. Then no loss of generality results in assuming X and X^* to be strictly convex. From Theorem 1.2 the maximal monotonicity of $\partial\varphi$ is equivalent to $R(F + \partial\varphi) = X^*$. Let x_0^* be arbitrary but fixed in X^*. We must show that equation

$$Fx + \partial\varphi(x) \ni x_0^* \tag{2.7}$$

has at least a solution $x_0 \in D(\partial\varphi)$. Let $f : X \to]-\infty, +\infty]$ be the proper convex function defined by

$$f(x) = \|x\|^2/2 + \varphi(x) - (x, x_0^*).$$

By hypothesis, f is lower-semicontinuous on X and from Proposition 2.1 it follows that

$$\lim_{\|x\| \to \infty} f(x) = +\infty.$$

Thus, by Proposition 2.4 the *infimum* of f on X is attained, i.e., there exists $x_0 \in D(f)$ such that

$$f(x_0) \leqslant f(x), \qquad \text{for every } x \in X.$$

We write this inequality in the form

$$\varphi(x_0) - \varphi(x) \leqslant (x_0 - x, x_0^*) + (x - x_0, F(x)), \qquad \forall\ x \in X. \tag{2.8}$$

In inequality (2.8) we set $x = tx_0 + (1 - t)u$ where $0 < t < 1$ and u is any element of X. Since φ is convex, it follows that

$$\varphi(x_0) - \varphi(u) \leqslant (x_0 - u, x_0^*) + (u - x_0, F(tx_0 + (1 - t)u)).$$

Passing to the limit $t \to 1$, we obtain

$$\varphi(x_0) - \varphi(u) \leqslant (x_0 - u, x_0^*) + (u - x_0, F(x_0))$$

because F is demicontinuous. Then from (2.5) we conclude that $x_0^* - F(x_0) \in \in \partial\varphi(x_0)$. This completes the proof.

COROLLARY 2.1. *Let φ be a lower-semicontinuous proper convex function on X. Then $D(\partial\varphi)$ is a dense subset of $D(\varphi)$.*

Proof. Let x be any element of $D(\varphi)$ and $x_\lambda = J_\lambda x$ be the solution of the equation (see Section 1.2)

$$F(x_\lambda - x) + \lambda\partial\varphi(x_\lambda) \ni 0. \tag{2.9}$$

By definition of $\partial\varphi$, we see by multiplying Eq. (2.9) by $x_\lambda - x$ that

$$\|x_\lambda - x\|^2 = -\lambda(x_\lambda - x, \partial\varphi(x_\lambda)) \leqslant \lambda(\varphi(x) - \varphi(x_\lambda)).$$

Since φ is bounded from below by an affine function (Proposition 1.1), this inequality implies that $\lim_{\lambda \to 0} x_\lambda = x$. As $x_\lambda \in D(\partial\varphi)$ and x is arbitrary in $D(\varphi)$ we conclude that $\overline{D(\partial\varphi)} = \overline{D(\varphi)}$ as claimed.

PROPOSITION 2.5. *Let φ be a lower-semicontinuous proper convex function on X. Then* $\operatorname{int} D(\varphi) = \operatorname{int} D(\partial\varphi)$.

Proof. Let $x_0 \in \operatorname{int} D(\varphi)$ and V be an open convex neighbourhood of x_0 which is contained in $D(\varphi)$. By Proposition 2.2, φ is continuous on V. Hence $\{(x, \lambda) \in X \times R^1; x \in V, \varphi(x) < \lambda\}$ is an open convex subset of $X \times R^1$. Thus there exists a closed hyperplane which separates this set from point $\{x_0, \varphi(x_0)\}$. From this fact one easily follows that $x_0 \in D(\partial\varphi)$. Thus the proof is complete.

The following proposition is closely related to Theorem 1.5.

PROPOSITION 2.6. *Let X be reflexive and $A = \partial\varphi$ where $\varphi: X \to]-\infty, +\infty]$ is a proper convex and lower-semicontinuous function on X. Then the following conditions are equivalent*

$$\lim_{\substack{\|x\| \to \infty \\ x \in D(\varphi)}} \varphi(x)/\|x\| = +\infty. \tag{2.10}$$

$$R(A) = X^* \text{ and } A^{-1} \text{ is bounded.} \tag{2.11}$$

Proof. Condition (2.10) $\Rightarrow$ (2.11). Let $x_0 \in D(A)$. By definition we have

$$\varphi(x) - \varphi(x_0) \leqslant (x - x_0, y), \qquad \forall [x, y] \in A.$$

Then condition (2.10) implies that A is coercive, i.e.,

$$\lim_{\substack{\|x\| \to \infty \\ [x, y] \in D(A)}} (x - x_0, y)/\|x\| = +\infty.$$

Using Theorem 1.6 we then conclude that $R(A) = X^*$. Obviously, in this case A^{-1} is bounded on X^* (i.e. maps bounded sets in bounded sets).
Condition (2.11) $\Rightarrow$ (2.10). Since φ is bounded from below by an affine function, no loss of generality results in assuming $\varphi \geqslant 0$ on X. Let $r > 0$. For every $z \in X^*$, $\|z\| \leqslant r$ there exist $v \in D(A)$ and $M > 0$ such that

$$z \in Av, \ \|v\| \leqslant M.$$

Next, by

$$\varphi(u) - \varphi(v) \geqslant (u - v, z), \qquad \forall u \in D(\varphi)$$

it follows that $(u, z) \leqslant \varphi(u) + Mr$ for any $u \in D(\varphi)$. Consequently

$$\varphi(u)/\|u\| \geqslant r - Mr/\|u\|$$

and hence

$$\liminf_{\substack{\|u\| \to \infty \\ u \in D(\varphi)}} \varphi(u)/\|u\| = \infty.$$

Remark. In particular, condition (2.10) implies that the conjugate function φ^* of φ is everywhere finite and continuous on X^*. To see this it suffices to recall that $(\partial\varphi)^{-1} = \partial\varphi^*$ and to apply Proposition 2.6.

For every $\lambda > 0$ define

$$\varphi_\lambda(x) = \inf_{u\in X} \{\|x - u\|^2/2\lambda + \varphi(u)\}, \qquad x \in X.$$

Function φ_λ is obviously convex and everywhere finite on X. Let $A = \partial\varphi$ and $J_\lambda: X \to X$, $A_\lambda: X \to X^*$ be the operators defined by Eqs. (1.23) and (1.24), respectively. We have already seen (Proposition 1.1) that A_λ is demicontinuous and in a certain sense it can be regarded as an approximation of A. It turns out that if A is a *subdifferential*, then A_λ remains in the same class.

THEOREM 2.2. *Let X and X^* be reflexive and strictly convex and let φ be a proper convex lower-semicontinuous function on X. Let $A = \partial\varphi$. Then function φ_λ is Gâteaux differentiable on X and $A_\lambda = \partial\varphi_\lambda$. In addition,*

$$\varphi_\lambda(x) = \|x - J_\lambda x\|^2/2\lambda + \varphi(J_\lambda x), \qquad \textit{for every } \lambda > 0 \textit{ and } x \in X. \tag{2.12}$$

$$\lim_{\lambda\to 0} \varphi_\lambda(x) = \varphi(x), \quad \textit{for every } x \in X. \tag{2.13}$$

$$\varphi(J_\lambda x) \leqslant \varphi_\lambda(x) \leqslant \varphi(x), \qquad \textit{for every } \lambda > 0 \textit{ and } x \in X. \tag{2.14}$$

Proof. We observe that the subdifferential of function $u \to \|x - u\|^2/2\lambda + \varphi(u)$ is just the operator $u \to \lambda^{-1}F(u-x) + \partial\varphi(u)$. This implies that every solution u_0 of equation

$$\lambda^{-1}F(u - x) + \partial\varphi(u) \ni 0$$

is a minimum point of $u \to \|x - u\|^2/2\lambda + \varphi(u)$, i.e., $\varphi_\lambda(x) = \|x - u_0\|^2/2\lambda + \varphi(u_0)$. Since $u_0 = J_\lambda x$ we obtain the equality (2.12). Inequality (2.14) is immediate in view of the definition and (2.12). To prove (2.13) we shall consider two cases. If $x \in D(\varphi)$, then $\lim_{\lambda\to 0} J_\lambda x = x$ in virtue of Proposition 1.1 and of Corollary 2.1. Then inequality (2.14) and the lower-semicontinuity of φ imply that $\varphi_\lambda(x) \to \varphi(x)$ as $\lambda \to 0$. Now suppose that $x \bar\in D(\varphi)$. We must show that $\varphi_\lambda(x) \to +\infty$ as $\lambda \to 0$. Suppose that this is not the case and that for example

$$\varphi_{\lambda_n}(x) \leqslant C < \infty \quad \text{where} \quad \lambda_n \to 0.$$

Again using Eq. (2.12), it would then follow under the present circumstances that $J_{\lambda_n}x \to x$ and $\varphi(J_{\lambda_n}x) \leqslant C$. Then the lower-semicontinuity of φ would imply that $\varphi(x) \leqslant C \leqslant \infty$, which is a contradiction. To conclude the proof it remains to be shown that φ_λ is Gâteaux differentiable at every point $x \in X$ and $\partial\varphi_\lambda(x) = A_\lambda x$. By an elementary calculation involving Eq. (2.12) one obtains

$$\varphi_\lambda(y) - \varphi_\lambda(x) - (A_\lambda x, y - x) \leqslant (A_\lambda y - A_\lambda x, y - x)$$

for all $\lambda > 0$ and x, y in X. Hence

$$0 \leqslant \varphi_\lambda(y) - \varphi_\lambda(x) - (A_\lambda x, y - x) \leqslant (A_\lambda y - A_\lambda x, y - x). \tag{2.15}$$

In inequality (2.15) we set $y = x + tu$, where $t>0$, and divide by t. We obtain

$$\lim_{t\to 0} (\varphi_\lambda(x + tu) - \varphi_\lambda(x))/t = (A_\lambda x, u), \qquad \forall\, u \in X$$

because A_λ is demicontinuous in virtue of Proposition 1.1. Hence φ_λ is Gâteaux differentiable at any $x \in X$ and $\partial\varphi_\lambda(x) = A_\lambda x$.

COROLLARY 2.2. *In Theorem 2.2 assume that $X = H$ is a real Hilbert space. Then the function φ_λ is Fréchet differentiable on H and its Fréchet differential $\partial\varphi_\lambda = A_\lambda$ is Lipschitz on H.*

Proof. Denote by I the identity operator in H. Then $F = I$ and J_λ, respectively A_λ, can be expressed as

$$J_\lambda = (I + \lambda A)^{-1}$$

respectively

$$A_\lambda = \lambda^{-1}(I - J_\lambda).$$

By the monotonicity of A we obtain

$$\|J_\lambda x - J_\lambda y\| \leqslant \|x - y\|$$

for any $\lambda > 0$ and x, y in X. This clearly implies that A_λ is Lipschitz continuous on H with Lipschitz constant $2/\lambda$ (in fact $1/\lambda$). Then inequality (2.15) implies that

$$|\varphi_\lambda(y) - \varphi_\lambda(x) - (A_\lambda x, y - x)| \leqslant \|x - y\|^2/2\lambda$$

for all $\lambda > 0$ and $x, y \in X$. This shows that φ_λ is Fréchet differentiable on H.

DEFINITION 2.1. A (multivalued) operator $A: X \to X^*$ is said to be *cyclically monotone* if

$$(x_0 - x_1, x_0^*) + \dots + (x_{n-1} - x_n, x_{n-1}^*) + (x_n - x_0, x_n^*) \geqslant 0 \tag{2.16}$$

whenever $x_i^* \in Ax_i$, $i = 0, \dots, n$.

A is said to be *maximal cyclically monotone* if it is cyclically monotone and has no cyclically monotone extensions in $X \times X^*$.

Obviously, every cyclically monotone operator is also monotone. If φ is a proper convex function on X, then a simple calculation involving the definition of $\partial\varphi$ shows that operator $\partial\varphi$ is cyclically monotone. Moreover, it follows from Theorem 2.1 that if φ is in addition lower-semicontinuous on X then its subdifferential $\partial\varphi$ is cyclically maximal monotone. Surprisingly, it turns out that condition (2.16) is both necessary and sufficient in order that an operator A be the subdifferential of some proper convex function.

THEOREM 2.3. *Let X be a real Banach space and let $A \subset X \times X^*$. In order that there exists a lower-semicontinuous proper convex function φ on X such that $A = \partial\varphi$, it is necessary and sufficient that A be a maximal cyclically monotone operator. Moreover, in this case A uniquely determines φ up to an additive constant.*

Proof. The necessity of the condition was proved in the above remarks. To prove the sufficiency, we suppose, therefore, that A is maximal cyclically monotone in $X \times X^*$. Without any loss of generality we may suppose that $D(A) \neq \emptyset$ and fix some $[x_0, x_0^*]$ in A. For every $x \in X$, define

$$\varphi(x) = \sup \{(x - x_n, x_n^*) + \ldots + (x_1 - x_0, x_0^*)\},$$

where $x_i^* \in Ax_i$ for $i = 1, \ldots, n$ and the supremum is taken over all possible finite sets of pairs $[x_i, x_i^*] \in A$. We shall prove that $A = \partial\varphi$. Clearly, $\varphi(x) > -\infty$, for all $x \in X$. Note also that φ is convex and lower-semicontinuous on X. Furthermore, $\varphi(x_0) = 0$ because A is cyclically montone. Hence $\varphi \not\equiv +\infty$. Let $[\tilde{x}, \tilde{x}^*] \in A$. To prove that $[\tilde{x}, \tilde{x}^*] \in \partial\varphi$ it suffices to show that for every $\lambda < \varphi(\tilde{x})$, we have

$$\varphi(x) \geqslant \lambda + (x - \tilde{x}, \tilde{x}^*), \qquad \forall\, x \in X. \tag{2.17}$$

Let $\lambda < \varphi(\tilde{x})$. Then, by the definition of φ there exist $[x_i, x_i^*] \in A, i = 1, \ldots, m$, such that

$$\lambda < (\tilde{x} - x_m, x_m^*) + \ldots + (x_1 - x_0, x_0^*).$$

Let $x_{m+1} = \tilde{x}$ and $x_{m+1}^* = \tilde{x}^*$. Then, again from the definition of φ we have

$$\varphi(x) \geqslant (x - x_{m+1}, x_{m+1}^*) + (x_{m+1} - x_m, x_m^*) + \ldots + (x_1 - x_0, x_0^*)$$

for all $x \in X$. This implies (2.17).

By the arbitrarinesss of $[\tilde{x}, \tilde{x}^*] \in A$ we conclude that $A \subset \partial\varphi$. Since A is maximal in the class of cyclically monotone sets of $X \times X^*$, it follows that $A = \partial\varphi$ as claimed. It remains to be proven that φ is uniquely determined up to an additive constant. Let ψ be another lower-semicontinuous proper convex function on X such that

$$\partial\varphi(x) = \partial\psi(x), \qquad \forall\, x \in X. \tag{2.18}$$

To complete the proof it suffices to demonstrate that $\varphi = \psi + \text{constant}$. For simplicity we shall restrict ourselves to the case X being reflexive. Then in view of the renorming theorem we may assume that X and X^* are strictly convex. We introduce functions φ_λ and ψ_λ as defined above. By Theorem 2.2, $\partial\varphi_\lambda = (\partial\varphi)_\lambda$ and $\partial\psi_\lambda = (\partial\psi)_\lambda$ for all $\lambda > 0$. Then Eq. (2.18) implies that

$$\partial\varphi_\lambda(x) = \partial\psi_\lambda(x) \quad \text{for all } x \in X.$$

Since φ_λ and ψ_λ are Gâteaux differentiable, this equality implies that $\varphi_\lambda \psi_\lambda + \text{constant}$. Then again applying Theorem 2.2 (relation (2.13)) we obtain Eq. (2.18) as claimed.

2.3. Some examples of cyclically monotone operators

Example 1. *Maximal monotone graphs in* R^2.

Every monotone graph in R^2 (in other words, every maximal monotone operator from real axis R^1 to itself) is cyclically monotone. Indeed, let β be a monotone subset of $R^1 \times R^1$ and let $[x_i, y_i] \in R^1 \times R^1$, $i = 0, 1, \ldots, n$ be a finite sequence of pairs such that $y_i \in \beta(x_i)$. We may assume that $x_0 \leqslant x_1 \leqslant \ldots \leqslant x_n$. Then the monotonicity of β implies that $y_0 \leqslant y_1 \leqslant \ldots \leqslant y_n$ and condition (2.16) is trivially satisfied. Thus in virtue of Theorem 2.3 we may conclude that: *if β is a maximal monotone subset* of R^2, *then $\beta = \partial j$ for some proper convex lower-semicontinuous function j: $R^1 \to]-\infty, +\infty]$.*

Example 2. *Self-adjoint operators in Hilbert spaces*

Let H be a real Hilbert space whose norm and inner product are denoted by $(,)$ and $|\cdot|$, respectively. Let A be a single-valued, linear and densely defined maximal monotone operator in H.

PROPOSITION 2.7. *A is cyclically monotone iff it is self-adjoint. Moreover, in this case, $A = \partial\varphi$ where*

$$\varphi(x) = \begin{cases} \dfrac{1}{2}|A^{1/2}x|^2 & \text{if } x \in D(A^{1/2}) \\ +\infty & \text{otherwise} \end{cases} \tag{2.19}$$

Proof. First suppose that A is self-adjoint. Then, φ defined by Eq. (2.19), $(A^{1/2})$ denotes the square root of operator A, is convex and lower-semicontinuous on H (because $A^{1/2}$ is closed). Let $x \in D(A)$. We have

$$\frac{1}{2}|A^{1/2}x| - \frac{1}{2}|A^{1/2}u|^2 \leqslant (Ax, x-u), \qquad \forall\, u \in D(A^{1/2})$$

because $(Ax, u) = (A^{1/2}x, A^{1/2}u)$ for all x in $D(A)$ and $u \in D(A^{1/2})$. Therefore, $A \subset \partial\varphi$. Since A is maximal we conclude that $A = \partial\varphi$.

Next, assume that A is cyclically monotone. According to Theorem 2.3 there exists $\psi: H \to]-\infty, +\infty]$ convex and lower-semicontinuous, such that $A = \partial\psi$. Since $A0 = 0$, we may choose the function ψ such that $\psi(0) = 0$. By Theorem 2.2 and Corollary 2.2, ψ_λ is Fréchet differentiable on H and $A_\lambda = \partial\psi_\lambda$ for all $\lambda > 0$. This implies that $\frac{\mathrm{d}}{\mathrm{d}t}\psi_\lambda(tx) = t(A_\lambda x, x)$. By integrating this equality over [0,1] we obtain $\psi_\lambda(x) = \frac{1}{2}(A_\lambda x, x)$. Hence $\partial\psi_\lambda = \frac{1}{2}(A_\lambda + A_\lambda^*)$ (A_λ^* is the adjoint operator of A_λ). Thus we have proved that $A_\lambda = A_\lambda^*$. Since $A_\lambda = \lambda^{-1}(I - (I + \lambda A)^{-1})$ the above equality shows that $A = A^*$ as claimed.

Example 3. *Convex integrands.*

Let $g: R^1 \to]-\infty, \infty]$ be a lower-semicontinuous proper convex function on $R^1 =]-\infty, +\infty[$ and let $\beta = \partial g$. Let $\varphi: L^2(\Omega) \to]-\infty, +\infty]$ be defined by

$$\varphi(u) = \begin{cases} \int_\Omega g(u(x))\,\mathrm{d}x & \text{if } g(u) \in L^1(\Omega) \\ +\infty & \text{otherwise} \end{cases}$$

where Ω is a bounded domain in R^n.

PROPOSITION 2.8. *The function φ is lower-semicontinuous and convex on $L^2(\Omega)$. Moreover, $w \in \partial\varphi(u)$ iff $w(x) \in \beta(u(x))$ a.e. on Ω and*

$$\overline{D(\varphi)} = \{u \in L^2(\Omega);\ u(x) \in \overline{D(g)},\ \text{a.e. on } \Omega\}. \tag{2.20}$$

Proof. As the function g is lower-semicontinuous, $g(u(x))$ is measurable in x for every measurable function u on Ω. Moreover, since g is bounded from below by an affine function, it follows that $\varphi(u) > -\infty$ for any $u \in L^2(\Omega)$. Also $\varphi \not\equiv +\infty$. The convexity of φ, is immediately obvious from the convexity of g. To prove that φ is lower-semicontinuous on $L^2(\Omega)$ it suffices to show that for every λ,

$$C = \left\{u \in L^2(\Omega);\ \int_\Omega g(u(x))\,\mathrm{d}x \leqslant \lambda\right\}$$

is a closed subset of $L^2(\Omega)$. Let $\{u_n\}$ be such that $u_n \to u$ in $L^2(\Omega)$ as $n \to \infty$.

There exists $\{u_{n_k}\} \subset \{u_n\}$ such that

$$\lim_{n_k \to \infty} u_{n_k}(x) = u(x), \quad \text{a.e.} \quad x \in \Omega.$$

Let α and β in R^1 be such that $g(u) \geqslant \alpha u + \beta$ for every $u \in R^1$. Then $g(u_{n_k}(x)) - \alpha u_{n_k}(x) - \beta \geqslant 0$ a.e. on Ω and $\liminf_{n_k \to \infty} (g(u_{n_k}(x)) - \alpha u_{n_k}(x) - \beta) \geqslant$ $\geqslant g(u(x)) - \alpha u(x) - \beta$ a.e. $x \in \Omega$. Then, Fatou's lemma implies that

$$\int_\Omega g(u)\,(x))\,\mathrm{d}x \leqslant \liminf_{n_k \to \infty} \int_\Omega g(u_{n_k}(x))\,\mathrm{d}x.$$

Therefore $u \in C$ as claimed. Let $[u, w] \in \partial\varphi \subset L^2(\Omega) \times L^2(\Omega)$. By the definition of $\partial\varphi$, we have

$$\int_\Omega (g(u(x)) - g(v(x)))\,\mathrm{d}x \leqslant \int_\Omega w(x)\,(u(x) - v(x))\,\mathrm{d}x, \quad \forall v \in L^2(\Omega).$$

Let E be any measurable subset of Ω and let

$$\tilde{u}(x) = \begin{cases} v(x) & \text{if } x \in E \\ u(x) & \text{if } x \in E^* \end{cases}$$

where E^* denotes the complement of E in Ω. In the preceding inequality we take $v = \tilde{u}$. Thus

$$\int_E (g(u(x)) - g(v(x)) - w(x)\,(u(x) - v(x)))\,\mathrm{d}x \leqslant 0, \quad \forall v \in L^2(\Omega).$$

Inasmuch as E was any measurable subset of Ω, we may conclude that

$$g(u(x)) - g(v(x)) \leqslant w(x)\,(u(x) - v(x)), \quad \text{a.e.} \quad x \in \Omega$$

for every $v \in L^2(\Omega)$. Hence $[u(x), w(x)] \in \partial g$ a.e. $x \in \Omega$. Conversely, let $[u, w] \in L^2(\Omega) \times L^2(\Omega)$ be such that

$$g(u(x)) - g(v) \leqslant w(x)\,(u(x) - v), \text{ a.e. } x \in \Omega, \quad \forall v \in R^1.$$

From this it is immediately clear that $[u, w] \in \partial\varphi$. To conclude the proof we have to prove (2.20). By the definition of φ, $D(\varphi) \subset \{u \in L^2(\Omega), u(x) \in D(g)$ a.e. $x \in \Omega\}$ so that it suffices to prove that any $u \in L^2(\Omega)$ such that $u(x) \in D(g)$ a.e. on Ω, lies in $\overline{D(\varphi)}$. For such a function u, define

$$u_\lambda = (1 + \lambda\beta)^{-1}\,u, \qquad \lambda > 0$$

where $\beta = \partial g$. From Proposition 1.1 it follows that

$$\lim_{\lambda \to 0} u_\lambda(x) = u(x) \qquad \text{a.e. } x \in \Omega.$$

Since β is monotone, $u \to (1+\lambda\beta)^{-1}u$ is nonexpansive on R^1. Then by the dominated convergence theorem of Lebesgue we infer that $u_\lambda \to u$ in $L^2(\Omega)$ as $\lambda \to 0$. Moreover, $u_\lambda \in D(\partial\varphi)$ because $\beta_\lambda(u) = \lambda^{-1}(u - u_\lambda) \in \beta(u_\lambda)$. So we have shown that $u \in \overline{D(\partial\varphi)} = \overline{D(\varphi)}$ which concludes the proof.

Example 4. Let Ω be a bounded and open domain of R^n with smooth boundary. Let j be a lower-semicontinuous proper convex function from R^1 to $]-\infty, +\infty]$ and let $\beta = \partial j$. For every $u \in L^2(\Omega)$ define

$$\varphi(u) = \begin{cases} \dfrac{1}{2}\displaystyle\int_\Omega |\text{grad } u|^2 \, dx + \int_\Gamma j(u) d\sigma & \text{if } u \in H^1(\Omega) \text{ and } j(u) \in L^1(\Gamma) \\ +\infty & \text{otherwise.} \end{cases}$$

Here $H^1(\Omega)$ is a usual Sobolev space defined on Ω and $L^1(\Gamma)$ denotes the space of all integrable functions on Γ.

It is easy to check if φ is proper and convex from $L^2(\Omega)$ to $]-\infty, +\infty]$. Moreover, by Fatou's lemma, it follows in much the same manner as in the proof of Proposition 2.8 that φ is lower-semicontinuous on $L^2(\Omega)$. The next proposition gives a precise description of the subdifferential $\partial\varphi$.

PROPOSITION 2.9. *We have,* $\partial\varphi = -\Delta$ *with*

$$D(\partial\varphi) = \{u \in H^2(\Omega);\ -\partial u/\partial n \in \beta(u) \text{ a.e. on } \Gamma\}$$

where $\partial/\partial n$ *is the outward normal derivative. Moreover, there are some constants* C_1 *and* C_2 *such that*

$$\|u\|_{H^2(\Omega)} \leqslant C_1\|u - \Delta u\|_{L^2(\Omega)} + C_2, \quad \forall\, u \in D(\partial\varphi). \tag{2.21}$$

Proof. Let $A: L^2(\Omega) \to L^2(\Omega)$ be the nonlinear operator defined by

$$Au = -\Delta u$$

where $D(A) = \{u \in H^2(\Omega);\ -\partial u/\partial n \in \beta(u) \text{ a.e. on } \Gamma\}$. Note that in the definition of $D(A)$, $-\partial u/\partial n \in \beta(u)$ a.e. on Γ makes sense because $u \in H^2(\Omega)$ implies that $u|_\Gamma \in H^{3/2}(\Gamma)$ and $\partial u/\partial n \in H^{1/2}(\Gamma) \subset L^2(\Gamma)$ (see Theorem 2.5 in Ch. I). Then Green's formula (see J. L. Lions and E. Magenes [1], Ch. II) yields

$$\int_\Omega -\Delta u(v - u)\, dx = \int_\Omega \text{grad } u\, (\text{grad } v - \text{grad } u)\, dx -$$

$$- \int_\Gamma \frac{\partial u}{\partial n}(v - u)\, d\sigma \leqslant \varphi(v) - \varphi(u).$$

for every $u \in D(A)$ and $v \in L^2(\Omega)$. This shows that $A \subset \partial\varphi$. To prove the converse inclusion relation it suffices to demonstrate that A is maximal monotone, i.e. $R(I+A) = L^2(\Omega)$. To this end we consider the approximate equations

$$\begin{aligned} -\Delta u_\lambda + u_\lambda &= f \qquad \text{on } \Omega \\ -\partial u_\lambda/\partial n &= \beta_\lambda(u_\lambda) \qquad \text{on } \Gamma \end{aligned} \tag{2.22}$$

where $\beta_\lambda = \lambda^{-1}(1-(1+\lambda\beta)^{-1})$ and f is given in $L^2(\Omega)$. First we shall prove that Eq. (2.22) has a unique solution $u_\lambda \in H^2(\Omega)$ for every $\lambda > 0$. Consider the operator $u \in L^2(\Gamma) \to Lu \in L^2(\Gamma)$ where $Lu = v|_\Gamma$ and v is the solution of the linear equation

$$-\Delta v + v = f \text{ on } \Omega, \quad v + \lambda\partial v/\partial n = (1+\lambda\beta)^{-1}u \text{ on } \Gamma.$$

From the linear theory of elliptic boundary problems it is well known (see e.g. J. L. Lions and E. Magenes [1]) that this problem has a unique solution $v \in H^2(\Omega)$. (We note that $(1+\lambda\beta)^{-1}u \in L^2(\Gamma)$ because $(1+\lambda\beta)^{-1}$ is nonexpansive on R^1.) Thus L is well defined on all of $L^2(\Gamma)$. Moreover, using Green's formula it follows that

$$\|Lu - L\bar{u}\|_{L^2(\Gamma)} \leqslant \rho\|u-\bar{u}\|_{L^2(\Gamma)}, \qquad \forall\, u, \bar{u} \in L^2(\Gamma),\ \rho < 1.$$

Then by the fixed-point contraction theorem we conclude that there exists $u \in L^2(\Gamma)$ such that $Lu = u$. Then the corresponding $v \in H^2(\Omega)$ from the definition of L will satisfy Eq. (2.22).

For the time being, let us assume that there exists a constant $C > 0$ independent with respect to λ such that

$$\|u_\lambda\|_{H^2(\Omega)} \leqslant C(\|f\|_{L^2(\Omega)} + 1), \qquad \forall\, \lambda > 0. \tag{2.23}$$

Inasmuch as the mapping $u \to \{u|_\Gamma, \partial u/\partial n\}$ is continuous from $H^2(\Omega)$ to $H^{3/2}(\Gamma) \times H^{1/2}(\Gamma)$, we may now conclude that $\{u_\lambda\}$ and $\{\partial u_\lambda/\partial n\}$ are bounded in $L^2(\Gamma)$. Since the injection of $H^2(\Omega)$ into $L^2(\Omega)$ is compact (see Theorem 2.3 in Chapter I) we may assume that as $\lambda \to 0$, we have

$$\begin{aligned} &u_\lambda \rightharpoonup u \qquad \text{in } H^2(\Omega) \\ &u_\lambda \to u \qquad \text{in } L^2(\Omega) \\ &\partial u_\lambda/\partial n \rightharpoonup \partial u/\partial n \quad \text{in } L^2(\Gamma). \end{aligned} \tag{2.24}$$

Moreover,

$$(1+\lambda\beta)^{-1}u_\lambda \to u \quad \text{in} \quad L^2(\Gamma) \text{ as } \lambda \to 0 \tag{2.25}$$

because $\beta_\lambda(u_\lambda)$ are uniformly bounded in $L^2(\Gamma)$.

Passing to limit in Eq. (2.22) we obtain

$$-\Delta u + u = f \quad \text{on} \quad \Omega.$$

As for the boundary conditions we proceed as follows. Denote by $\tilde{\beta} \subset L^2(\Gamma) \times L^2(\Gamma)$ the operator defined by

$$\tilde{\beta}(u) = \{w \in L^2(\Gamma);\ w \in \beta(u), \text{ a.e. on } \Gamma\},\ u \in L^2(\Gamma)$$

It is easy to see that $\tilde{\beta}$ is maximal monotone in $L^2(\Gamma) \times L^2(\Gamma)$ and $(\tilde{\beta}_\lambda(u))(x) = \beta_\lambda(u(x))$ a.e. $x \in \Gamma$. Moreover, $\tilde{\beta}_\lambda(u) \in \tilde{\beta}(I + \lambda\tilde{\beta})^{-1}u)$ and (2.24) imply that

$$\tilde{\beta}_\lambda(u_\lambda) \rightharpoonup \partial u/\partial n \text{ in } L^2(\Gamma) \text{ as } \lambda \to 0.$$

Next, by (2.25)

$$(I + \lambda\tilde{\beta})^{-1}u_\lambda \to u \quad \text{in } L^2(\Gamma) \text{ as } \lambda \to 0.$$

Thus the maximal monotonicity of $\tilde{\beta}$ implies using a standard argument, that $\partial u/\partial n \in \tilde{\beta}(u)$ as claimed. Hence $f \in R(I + A)$. To complete the proof we must demonstrate estimate (2.23). Multiplying Eq. (2.22) by $u_\lambda - u_0$ where $u_0 \in D(\beta)$ and applying Green's formula, we get

$$\|u_\lambda\|_{H^1(\Omega)} \leqslant C_1(\|f\|_{L^2(\Omega)} + 1), \qquad \forall \lambda > 0,$$

because $|\beta_\lambda(u_0)| \leqslant |\beta(u_0)|$ (see Proposition 1.1). Interior estimates in $H^2(\Omega)$ are obtained as follows. Let $\Omega' \subset \Omega$ be such that $\bar{\Omega}' \subset \Omega$ and let $\rho \in \mathscr{D}(\Omega)$ be such that $\rho = 1$ on $\bar{\Omega}'$. Let $v_\lambda = \rho u_\lambda$. Then we can write

$$v_\lambda - \Delta v_\lambda = \rho f + g_\lambda,$$

where g_λ is a linear combination of $\partial u_\lambda/\partial x_i$. Since $v_\lambda \in H^2(R^n)$, Parseval's formula implies that

$$\|v_\lambda\|_{H^2(R^n)} \leqslant C_2(\|v_\lambda\|_{L^2(\Omega)} + \|\rho f\|_{L^2(\Omega)} + \|u_\lambda\|_{H^1(\Omega)}).$$

Therefore,

$$\|u_\lambda\|_{H^2(\Omega')} \leqslant C_3(\|f\|_{L^2(\Omega)} + 1), \qquad \text{for all } \lambda > 0. \tag{2.26}$$

Next we obtain estimates near the boundary. Let $x_0 \in \Gamma$ and let V be a neighbourhood of x_0. Let $y = Tx$ be a local chart such that $Tx_0 = 0$, $T(V \cap \Omega) = \Omega_R$ and $T(V \cap \Gamma) = \Gamma_R$ where

$$\Omega_R = \{y = (y'; y_n), \qquad |y'| < R, 0 < y_n < R\},$$

and $\Gamma_R = \{y = (y', y_n);\ |y'| < R, y_n = 0\}$. We may assume that $\dfrac{\partial T_n}{\partial x_i} = -\cos(n, x_i)$ on Γ for $i = 1, \dots n$.

Substituting $\tilde{u}_\lambda(y) = u_\lambda(T^{-1}y)$ and $\tilde{f}(y) = f(T^{-1}y)$ in problem (2.22) one obtains

$$\tilde{u}_\lambda - \sum_{i,j=1}^{n} \frac{\partial}{\partial y_i}\left(a_{ij}(y)\frac{\partial \tilde{u}_\lambda}{\partial y_j}\right) = \tilde{f} \text{ on } \Omega_R$$

$$-\frac{\partial \tilde{u}_\lambda}{\partial n} = \beta_\lambda(\tilde{u}_\lambda) \quad on \quad \Gamma_R, \tag{2.27}$$

where

$$a_{ij}(y) = \sum_{k=1}^{n} \frac{\partial T_i}{\partial x_k}(x)\frac{\partial T_j}{\partial x_k}(x)$$

and $\partial\tilde{u}/\partial n = -\sum_{i=1}^{n}\frac{\partial \tilde{u}}{\partial y_i} a_{in}$, $T = (T_1, \ldots, T_n)$.

For simplicity we shall write u instead of $\tilde{u}$. Let $\eta \in C_0^\infty(\Omega_R)$ be such that $\eta = 1$ on a neighbourhood of $\Omega_{R/2}$. Multiplying Eq. (2.27) by $-\sum_{i=1}\frac{\partial}{\partial y_i}\left(\eta^2\frac{\partial u}{\partial y_i}\right)$ and integrating by parts we get

$$\sum_{i=1}^{n}\int_{\Gamma_R}\frac{\mathrm{d}}{\mathrm{d}u_\lambda}\beta_\lambda(u_\lambda)\eta^2\left|\frac{\partial u_\lambda}{\partial y_i}\right|^2\mathrm{d}\sigma + \sum_{i=1}^{n-1}\sum_{k,l=1}^{n}\int_{\Omega_R}\frac{\partial}{\partial y_i}\left(a_{kl}\frac{\partial u_\lambda}{\partial y_k}\right)\cdot$$

$$\frac{\partial}{\partial y_l}\left(\eta^2\frac{\partial u_\lambda}{\partial y_i}\right)\mathrm{d}y = -\sum_{i=1}^{n-1}\int_{\Omega_R}\tilde{f}\frac{\partial}{\partial y_i}\left(\eta^2\frac{\partial u_\lambda}{\partial y_i}\right)\mathrm{d}y.$$

Since β_λ is a monotone function on R^1 and $\sum a_{ij}\,\xi_i\,\xi_j \geqslant \mu|\xi|^2$ for all $\xi \in R^n$ and some $\mu > 0$, from the preceding inequality after some rearrangements one obtains

$$\sum_{i=1}^{n-1}\sum_{k=1}^{n}\left\|\eta\frac{\partial^2 u_\lambda}{\partial y_i \partial y_k}\right\|^2_{L^2(\Omega_R)} \leqslant C(\|f\|_{L^2(\Omega_R)} +$$

$$+ \|u_\lambda\|_{H^1(\Omega_R)})\left(\|u_\lambda\|_{H^1(\Omega_R)} + \sum_{i=1}^{n-1}\sum_{k=1}^{n}\left\|\eta\frac{\partial^2 u_\lambda}{\partial y_i\,\partial y_k}\right\|_{L^2(\Omega_R)}\right).$$

Hence

$$\sum_{i=1}^{n-1}\sum_{k=1}^{n}\left\|\eta\frac{\partial^2 u_\lambda}{\partial y_i \partial y_k}\right\|_{L^2(\Omega_R)} \leqslant C\,\|f\|_{L^2(\Omega_R)} + \|u_\lambda\|_{H^1(\Omega_R)}).$$

Using Eq. (2.27), a similar estimate is obtained for $\partial^2 u/\partial x_n^2$. Then going back to $u(x)$ we see that

$$\|u_\lambda\|_{H^2(\Omega)} \leqslant C(\|f\|_{L^2(\Omega)} + \|u_\lambda\|_{H^1(\Omega)}).$$

Taking a partition of unity subordinate to a finite system of $\{V\}$ and using the interior estimates (2.26), one finally obtains estimate (2.23). Thus the proof of Proposition 2.9 is complete.

Example 5. Let $H_0^1(\Omega)$ and $H^{-1}(\Omega)$ be usual Sobolev spaces on Ω. As mentioned in Chapter I, $H^{-1}(\Omega)$ is the dual of $H_0^1(\Omega)$. Moreover the Laplace operator $-\Delta$ is the *canonical isomorphism* from $H_0^1(\Omega)$ onto $H^{-1}(\Omega)$. $H^{-1}(\Omega)$ is a Hilbert space with an inner product

$$((u, v)) = (J^{-1}u, v), \quad \forall\ u, v \in H^{-1}(\Omega)$$

where $J = -\Delta$ and (,) is the usual pairing between $H_0^1(\Omega)$ and $H^{-1}(\Omega)$.

Let j be a lower-semicontinuous proper convex function from R^1 into $]-\infty, +\infty]$ and let $\partial j = \beta$. We shall assume that

$$\lim_{|r|\to\infty} j(r)/|r| = \infty. \tag{2.28}$$

Let $\varphi : H^{-1}(\Omega) \to]-\infty, +\infty]$ be defined by

$$\varphi(u) = \begin{cases} \int_\Omega j(u(x))\,\mathrm{d}x & \text{if } u \in L^1(\Omega) \text{ and } j(u) \in L^1(\Omega) \\ +\infty & \text{otherwise} \end{cases}$$

PROPOSITION 2.10. *The function* φ *is convex and lower-semicontinuous on* $H^{-1}(\Omega)$. *Moreover,* $f \in \partial\varphi(u)$, *if and only if,* $f \in J\beta(u)$.

Proof. First we shall prove that φ is lower-semicontinuous on $H^{-1}(\Omega)$. Let $\{u_n\} \subset H^{-1}(\Omega) \cap L^1(\Omega)$ be such that $u_n \to u$ in $H^{-1}(\Omega)$ and $\int_\Omega j(u_n(x))\,\mathrm{d}x \leqslant \lambda$.

We must prove that $\int_\Omega j(u(x))\,\mathrm{d}x \leqslant \lambda$. We have already seen in the proof of Proposition 2.8 that the function $u \to \int_\Omega j(u(x))\,\mathrm{d}x$ is lower-semicontinuous on $L^1(\Omega)$. Since every convex and lower-semicontinuous function on a Banach space is also weakly lower-semicontinuous, it suffices to show that the sequence $\{u_n\}$ is weakly compact in $L^1(\Omega)$. By Dunford-Pettis criterion for weak compactness in L^1-spaces (see e.g. N. Dunford and J. Schwartz [1], p. 294) it suffices to prove that the integrals $\int |u_n|\,\mathrm{d}x$ are uniformly absolutely continuous, i.e. there exists for every $\varepsilon > 0$ some δ such that $\int_T |u_n|\,\mathrm{d}x < \varepsilon$

whenever meas. $T < \delta$. Let $M > 2\lambda/\varepsilon$ and let R be such that $j(r)/|r| \geqslant M$ for $|r| > R$. Here we have used condition (2.28). If $\delta < \varepsilon/2R$ then we have

$$\int_T |u_n(x)|\,dx \leqslant \int_{\{x\in T;\,|u_n(x)|\geqslant R\}} |u_n(x)|\,dx + \int_{\{x\in T;\,|u_n(x)|<R\}} |u_n(x)|\,dx \leqslant$$

$$\leqslant M^{-1}\int_\Omega j(u_n(x))\,dx + R\delta < \varepsilon$$

as desired.

Now let $A : H^{-1}(\Omega) \to H^{-1}(\Omega)$ be the operator defined by

$$Au = \{Jv;\ v \in H_0^1(\Omega) \text{ and } v(x) \in \beta(u(x)), \quad \text{a.e. } x \in \Omega\}$$

where $D(A)$ is the set of all $u \in H^{-1}(\Omega) \cap L^1(\Omega)$ with the property that there is some $v \in H_0^1(\Omega)$ such that $v(x) \in \beta(u(x))$ a.e. on Ω. We must prove that $A = \partial\varphi$.

We need the following lemma.

Lemma 2.1. *Let $f \in H^{-1}(\Omega) \cap L^1(\Omega)$ and let $v \in H_0^1(\Omega)$. Let $g \in L^1(\Omega)$ and let h be measurable on Ω such that*

$$f(x)\,v(x) \geqslant h(x) \geqslant g(x), \quad \text{a.e. } x \in \Omega. \tag{2.29}$$

Then $h \in L^1(\Omega)$ and $(f, v) \geqslant \int_\Omega h(x)\,dx$.

Proof of Lemma 2.1. Define,

$$v_n = \begin{cases} n & \text{if } v \geqslant n \\ v & \text{if } |v| \leqslant n \\ -n & \text{if } v \leqslant -n. \end{cases}$$

Let $h_n = h\dfrac{v_n}{v}$ and let $g_n = g\dfrac{v_n}{v}$. Multiplying condition (2.29) by $\dfrac{v_n}{v}$ we obtain

$$fv_n \geqslant h_n \geqslant g_n.$$

Hence $0 \leqslant h_n - g_n \leqslant fv_n - g_n$. Clearly, $h_n - g_n \to h - g$, a.e. on Ω as $n \to \infty$ and

$$\int_\Omega (h_n - g_n)\,dx \leqslant \int_\Omega fv_n\,dx - \int_\Omega g_n\,dx = (f, v_n) - \int_\Omega g_n\,dx. \tag{2.30}$$

(We recall that (,) denotes the pairing between H_0^1 and H^{-1}.) On the other hand, $v_n \to v$ in $H_0^1(\Omega)$ and $g_n \to g$ in $L^1(\Omega)$. Again using Fatou's lemma, from condition (2.30) we conclude that $h - g \in L^1(\Omega)$ and

$$\int_\Omega (h - g)\, dx \leqslant (f, v) - \int_\Omega g\, dx$$

which concludes the proof of Lemma 2.1.

Now we continue the proof of Proposition 2.10 by showing that $A \subset \partial\varphi$. Let $f \in Au$. This implies that $f = Jv$ where $v \in H_0^1(\Omega)$ and $v(x) \in \beta(u(x))$ a.e. $x \in \Omega$. Let $w \in H^{-1}(\Omega) \cap L^1(\Omega)$ be such that $j(w) \in L^1(\Omega)$ (i.e. $w \in D(\varphi)$). Since $\beta = \partial j$ we have

$$j(w) - j(u) \geqslant v(w - u), \quad \text{a.e. on } \Omega.$$

We apply Lemma 2.1. with $f = u - w$, $h = j(u) - j(w)$ and $g = -c_1|u| - c_2 - j(w)$ where c_1 and c_2 are some constants such that $j(r) \geqslant -c_1|r| - c_2$ for all $r \in R^1$ (see Proposition 2.1). Hence $j(u) \in L^1(\Omega)$ and

$$\int_\Omega j(w)\, dx - \int_\Omega j(u)\, dx \geqslant (v, w - u) = (J^{-1}f, w - u)$$

which shows that $f \in \partial\varphi(u)$. The arbitrariness of f implies that $A \subset \partial\varphi$. Next we shall prove that A is maximal monotone in $H^{-1}(\Omega)$. Let f_0 be given in $H^{-1}(\Omega)$. We must show that there are $u \in H^{-1}(\Omega) \cap L^1(\Omega)$ and $v \in H_0^1(\Omega)$ such that

$$u + Jv = f_0, \ v(x) \in \beta(u(x)), \text{ a.e. } x \in \Omega$$

or equivalently

$$\gamma(v) + Jv = f_0 \tag{2.31}$$

where $\gamma = \beta^{-1}$. Consider the approximate equation

$$\gamma_\lambda(v_\lambda) + Jv_\lambda = f_0 \tag{2.32}$$

where $\gamma_\lambda = \lambda^{-1}(1 - (1 + \lambda\gamma)^{-1})$. We have already seen that γ_λ is Lipschitz on $X = R^1$. Recalling that $J = -\Delta$ it follows using standard arguments that Eq. (2.32) has a unique solution $v_\lambda \in H_0^1(\Omega)$ for every $\lambda > 0$. We multiply Eq. (2.32) by v_λ and integrate over Ω. We obtain

$$\int_\Omega |\text{grad } v_\lambda|^2\, dx + \int_\Omega \gamma_\lambda(v_\lambda)v_\lambda\, dx = (f_0, v_\lambda). \tag{2.33}$$

Without any loss of generality we may assume that $0 \in \gamma(0)$. (We note that condition (2.28) implies that $D(\gamma) = R^1$ in virtue of Proposition 2.6.) Then the last inequality implies that v_λ is bounded in $H_0^1(\Omega)$ as $\lambda \to 0$. Thus a subsequence (denoted again v_λ) can be extracted from $\{v_\lambda\}$ such that $v_\lambda \to v$ in $H_0^1(\Omega)$, and $v_\lambda \to v$ in $L^2(\Omega)$. Extracting further subsequences if necessary we may also assume that

$$v_\lambda(x) \to v(x), \quad \text{a.e. } x \in \Omega$$

$$(1 + \lambda\gamma)^{-1} v_\lambda(x) \to v(x), \quad \text{a.e. } x \in \Omega \tag{2.34}$$

because $\lim_{\lambda \to 0} (1 + \lambda\gamma)^{-1} v = v$, for every $v \in \overline{D(\gamma)} = R^1$.

We set $f_\lambda = \gamma_\lambda(v_\lambda)$ and $w_\lambda = (1 + \lambda\gamma)^{-1} v_\lambda$. We note that $f_\lambda(x) \in \gamma(w_\lambda(x))$ a.e. $x \in \Omega$ and $f_\lambda \in L^2(\Omega)$. In particular this implies that $f_\lambda v_\lambda \in L^1(\Omega)$ and by (2.33)

$$\int_\Omega f_\lambda v_\lambda \, dx \leqslant C, \quad \forall \lambda > 0 \tag{2.35}$$

where C is some positive constant. On the other hand, for some $v_0 \in D(j)$ we have

$$j(f_\lambda(x)) \leqslant j(v_0) + (f_\lambda(x) - v_0)v, \qquad \forall v \in \beta(f_\lambda(x)).$$

Next by (2.34) and (2.35),

$$\int_\Omega j(f_\lambda(x)) \, dx \leqslant C.$$

Using once again the Dunford-Pettis theorem, we find that $\{f_\lambda\}$ is weakly compact in $L^1(\Omega)$. Consequently we may assume that

$$f_\lambda \rightharpoonup f \qquad \text{in } L^1(\Omega) \text{ as } \lambda \to 0.$$

Passing to limit through $\{\lambda\}$ in Eq. (2.32) we obtain $f + Jv = f_0$. To conclude the proof it remains to be shown that $f(x) \in \gamma(v(x))$ a.e. $x \in \Omega$. It suffices to prove that for every N, $f(x) \in \gamma(v(x))$ a.e. $x \in \Omega_N = \{x \in \Omega; |v(x)| \leqslant N\}$. According to Egorov's lemma for every $\varepsilon > 0$ there is $E \subset \Omega_N$ such that meas. $E < \varepsilon$, $v_\lambda(x) \to v(x)$ uniformly on Ω_N/E. (We note that $v_\lambda \to v$ a.e. on Ω in virtue of (2.34).) Thus no loss of generality results in assuming that $v_\lambda \to v$ uniformly on Ω and $v \in L^\infty(\Omega)$. Let $g : R^1 \to]-\infty, +\infty]$ be such that $\partial g = \gamma$. Then

$$\int_\Omega f_\lambda(x)(w_\lambda(x) - \tilde{v}(x)) \, dx \geqslant \int_\Omega g(w_\lambda(x)) \, dx - \int_\Omega g(\tilde{v}(x)) \, dx$$

for every $\tilde{v} \in L^\infty(\Omega)$. Again using Fatou's lemma we get

$$\int_\Omega f(x)(v(x) - \tilde{v}(x)) \, dx \geqslant \int_\Omega g(v(x)) \, dx - \int_\Omega g(\tilde{v}(x)) \, dx$$

because f_λ converges weakly to f in $L^1(\Omega)$. Then by using a standard argument we obtain the pointwise inequality

$$f(x)(v(x) - \tilde{v}) \geqslant g(v(x)) - g(\tilde{v}), \quad \text{a.e. } x \in \Omega$$

for all $\tilde{v} \in R^1$. Hence $f(x) \in \partial g(x) = \gamma(v(x))$ a.e. $x \in \Omega$ and the proof of Proposition 2.10 is complete.

§3. Dissipative sets in Banach spaces

3.1. Basic properties of dissipative sets

As in the preceding paragraphs, X will denote a real Banach space, $\| \ \|$ its norm, X^* the dual space and $(,)$ the pairing between X and X^*. Denote also by $F : X \to X^*$ the duality mapping of X.

DEFINITION 3.1. A subset A of $X \times X$ (equivalently a multivalued operator from X to itself) is called *dissipative* if for any $[x_i, y_i] \in A$, $i = 1, 2$ there exists $f \in F(x_1 - x_2)$ such that

$$(y_1 - y_2, f) \leqslant 0. \tag{3.1}$$

A dissipative set A is said to be *maximal dissipative* if it is not properly contained in any dissipative subset of $X \times X$.

A dissipative set A is called *m-dissipative* if

$$R(1 - A) = X. \tag{3.2}$$

When there is no danger of confusion we shall simply write "1" instead of "I", the identity operator in X.

A subset A of $X \times X$ is called *accretive* (*maximal accretive* and *m-accretive*, respectively) if $-A$ is dissipative (*maximal dissipative* and *m-dissipative*, respectively). In particular, if X is a Hilbert space, then the notion of accretive set coincides with that of monotone set.

We shall begin with the following simple lemma which will be frequently used in the following.

LEMMA 3.1. *Let* $x, y \in X$. *Then*

$$\|x\| \leqslant \|x - \lambda y\|, \qquad \forall \lambda > 0 \tag{3.3}$$

if and only if there exists $f \in F(x)$ *such that* $(y, f) \leqslant 0$.

Proof. *"If" part.* Let x and y in X be such that $(y,f) \leqslant 0$ for some $f \in F(x)$. Using the definition of F we obtain

$$\|x\|^2 = (x,f) \leqslant (x - \lambda y, f) \leqslant \|x - \lambda y\| \, \|f\|, \qquad \forall \lambda > 0,$$

from which (3.3) follows.

"Only if" part. Suppose that (3.3) holds. For every $\lambda > 0$ let f_λ be an arbitrary element in $F(x - \lambda y)$. Without any loss of generality we may assume that $x \neq 0$. Then $f_\lambda \neq 0$ for all $\lambda > 0$. Let $g_\lambda = f_\lambda / \|f_\lambda\|$. Since the unit sphere of the dual space X^* is compact in the weak-star topology of X^*, we may assume that

$$\lim_{\lambda \to 0} (u, g_\lambda) = (u, g), \qquad \forall u \in X. \tag{3.4}$$

where g is some element of X^*.

Next by

$$\|x\| \leqslant \|x - \lambda y\| = (x - \lambda y, g_\lambda) \leqslant \|x\| - \lambda(y, g_\lambda), \qquad \forall \lambda > 0, \tag{3.5}$$

it follows that $(y, g_\lambda) \leqslant 0$ for all $\lambda > 0$, and using (3.4) we see that $(y, g) \leqslant 0$. We note that (3.5) also implies that $\|x\| \leqslant (x, g)$. Hence $(x, g) = \|x\|$. This shows that $g\|x\| \in F(x)$ and the desired inequality follows with $f = g\|x\|$. Using Lemma 3.1 we see that the dissipative sets can be characterized by a simple metric condition.

PROPOSITION 3.1. *A subset A of $X \times X$ is dissipative, if and only if, for every $\lambda > 0$ and any $[x_i, y_i] \in A$, $i = 1, 2$ the following inequality holds*

$$\|x_1 - x_2\| \leqslant \|x_1 - \lambda y_1 - (x_2 - \lambda y_2)\|. \tag{3.6}$$

Proof. We simply apply Lemma 3.1 where $x = x_1 - x_2$ and $y = y_1 - y_2$. In particular, Proposition 3.1 shows that if A is dissipative, then for every $\lambda > 0$, the operator $(1 - \lambda A)^{-1}$ is well defined and *nonexpansive* on $R(1 - \lambda A)$, i.e.,

$$\|(1 - \lambda A)^{-1}x - (1 - \lambda A)^{-1}y\| \leqslant \|x - y\| \text{ for any } x, y \in R(1 - \lambda A).$$

Define,

$$J_\lambda x = (1 - \lambda A)^{-1}x, \qquad x \in R(1 - \lambda A). \tag{3.7}$$

$$A_\lambda x = \lambda^{-1}(J_\lambda x - x), \qquad x \in R(1 - \lambda A). \tag{3.8}$$

If A is accretive in $X \times X$ we also denote by J_λ and A_λ the operators $(1 + \lambda A)^{-1}$ and $\lambda^{-1}(1 - J_\lambda)$, respectively. We observe that if $X = X^* = H$ is a Hilbert space, these operators are just J_λ and A_λ defined in Section 1.2.

Let C be a closed subset of X. We shall use the following notation

$$|C| = \inf \{\|x\|;\ x \in C\}. \tag{3.9}$$

We collect some elementary properties of J_λ and A_λ in the following proposition.

PROPOSITION 3.2. *Let A be a dissipative subset of $X \times X$ and let $\lambda > 0$. Then*

(a) $\|J_\lambda x - J_\lambda y\| \leqslant \|x - y\|$, *for all* x, y *in* $R(1 - \lambda A)$.

(b) A_λ *is dissipative in* $X \times X$ *and Lipschitz on* $R(1 - \lambda A)$ (*with Lipschitz constant* $2/\lambda$).

(c) $A_\lambda x \in AJ_\lambda x$, *for every* $x \in R(1 - \lambda A)$.

(d) $\|A_\lambda x\| \leqslant |Ax|$, *for every* $x \in D(A) \cap R(1 - \lambda A)$.

(e) $\lim_{\lambda \to 0} J_\lambda x = x$, *for any* $x \in \overline{D(A)} \bigcap_{\lambda > 0} R(1 - \lambda A)$.

Proof. The properties (*a*), (*b*) are immediately clear.

(c) Let $x \in R(1 - \lambda A)$. By (3.7) and (3.8) we see that $A_\lambda x \in \lambda^{-1}(J_\lambda x - (1 - \lambda A)J_\lambda x) = AJ_\lambda x$.

(d) Let $x \in D(A) \cap R(1 - \lambda A)$. Then $A_\lambda x = \lambda^{-1}(J_\lambda x - J_\lambda(1 - \lambda Ax))$. Since J_λ is nonexpansive on $R(1 - \lambda A)$, this implies that $\|A_\lambda x\| \leqslant \|y\|$ for any $y \in X$, i.e. $\|A_\lambda x\| \leqslant |Ax|$.

(e) For every $x \in D(A) \cap R(1 - \lambda A)$ we have

$$\|J_\lambda x - x\| = \lambda \|A_\lambda x\| \leqslant \lambda |Ax|$$

and hence $\lim_{\lambda \to 0} J_\lambda x = x$. Obviously, this extends to all

$$x \in \overline{D(A)} \bigcap_{\lambda > 0} R(1 - \lambda A).$$

PROPOSITION 3.3. *A dissipative set $A \subset X \times X$ is m-dissipative, if and only if, for every (equivalently for some) $\lambda > 0$*

$$R(1 - \lambda A) = X. \tag{3.10}$$

Proof. Suppose that A is m-dissipative. Let $y \in X$ and $\lambda > 0$. The equation

$$(1 - \lambda A)x \ni y \tag{3.11}$$

may be written as

$$x = (1 - A)^{-1}(y/\lambda + (1 - 1/\lambda)x) \tag{3.12}$$

because $R(1 - A) = X$. Moreover, the operator $(1 - A)^{-1}$ is nonexpansive on X. Thus the operator defined by the right-hand side of Eq. (3.12) is Lipschitzian on X with Lipschitz constant $|1 - 1/\lambda|$. By applying the standard fixed-point contraction theorem we conclude that Eq. (3.12) has a unique solution $x \in X$. Since y is arbitrary we may conclude that $R(1 - \lambda A) = X$ for every $\lambda > 0$. Conversely, suppose that A is dissipative and that for some $\lambda > 0$, $R(1 - \lambda A)$ is all of X. We must prove that $R(1 - A) = X$. Let y be arbitrarily given in X and let $\mu > 0$. The equation

$$\mu x - Ax \ni y \tag{3.13}$$

can be written in the following equivalent form

$$x = (1 - \lambda A)^{-1}((1 - \lambda\mu)x + \lambda y).$$

Since $(1 - \lambda A)^{-1} = J_\lambda$ is nonexpansive on X, we see that the above equation has a unique solution x whenever $0 < \mu < 2/\lambda$. In particular, $\left(1 - \frac{3}{4}\lambda A\right)^{-1}$ is nonexpansive on X.

Now taking $\left(1 - \frac{3}{4}\lambda A\right)^{-1}$ instead of $(1-\lambda A)^{-1}$ we find that $R(\mu - A) = X$ for $0 < \mu < 8/3\lambda$. Continuing in this way we finally obtain $R(1 - A) = X$.

THEOREM 3.1. *Any m-dissipative set in $X \times X$ is maximal dissipative. If $X = X^* = H$ is a Hilbert space then the notions of maximal dissipative sets and m-dissipative sets coincide.*

Proof. Let A be an m-dissipative subset of $X \times X$. We must prove that A is maximal dissipative. Suppose that this is not the case. Then there exists $[x_0, y_0] \in X \times X$ such that $[x_0, y_0] \bar{\in} A$ and

$$(y_0 - y, f) \leqslant 0 \qquad \forall\ [x, y] \in A, \tag{3.14}$$

with some $f \in F(x_0 - x)$. Since $R(1 - A) = X$, we may choose $[x_1, y_1] \in A$ such that $x_1 - y_1 = x_0 - y_0$. Then condition (3.14) implies that $[x_0, y_0] = [x_1, y_1] \in A$ which is a contradiction.

Now suppose that $X = H$ is a Hilbert space identified with its own dual. If A is maximal dissipative, then $-A$ is maximal monotone in $H \times H$ so that Theorem 1.2 implies $R(1 - A) = H$. This completes the proof.

Using Zorn's lemma it is obvious that every dissipative set $A \subset X \times X$ has a maximal dissipative extension in $X \times X$. If X is actually a Hilbert space, we obtain a more precise result, namely the following.

THEOREM 3.2. *Let C be a closed convex subset of a real Hilbert space H. Let A be a dissipative subset of $H \times H$ such that $D(A) \subset C$. Then there exists a maximal dissipative set $\tilde{A} \subset H \times H$ such that $D(A) \subseteq D(\tilde{A}) \subset C$ and*

$$Ax \subset \tilde{A}x \quad \textit{for every } x \in D(A). \tag{3.15}$$

Proof. This is an immediate consequence of Theorem 1.4 where $X = X^* = H$. Let C be a closed subset of H and let A be a dissipative subset of $H \times H$ such that $D(A) \subset C$. We denote the set of all maximal dissipative sets $\tilde{A}$ such that $D(\tilde{A}) \subset C$ and $Ax \subset \tilde{A}x$ for $x \in D(A)$ by $E(A; C)$. Then, Theorem 3.2 simply says that

$$E(A; C) \neq \varnothing \tag{3.16}$$

Definition 3.2. A subset A of $X \times X$ is said to be *demiclosed* if it is strongly-weakly closed in $X \times X$. In other words, $x_n \to x$, $y_n \rightharpoonup y$ and $[x_n, y_n] \in A$ imply that $[x, y] \in A$. (We recall that $\rightharpoonup$ denotes the weak convergence in X.) A set A is closed if $[x_n, y_n] \in A$, $x_n \to x$ and $y_n \to y$ imply $[x, y] \in A$.

Proposition 3.4. *Let A be an m-dissipative subset of $X \times X$. Then*

(a) *A is closed*

(b) *If $\{x_\lambda\} \subset X$ is such that $x_\lambda \to x$ and $A_\lambda x_\lambda \to y$ as $\lambda \to 0$ then $[x, y] \in A$.*

Proof. (a) Let $[x_n, y_n] \in A$ be such that $x_n \to x_0$ and $y_n \to y_0$ as $n \to \infty$. Since A is dissipative, Proposition 3.1 implies that

$$\|x_n - x\| \leqslant \|x_n - \lambda y_n - (x - \lambda y)\|, \qquad \forall \lambda > 0, \quad \forall [x, y] \in A.$$

Thus letting $n \to \infty$ yields

$$\|x_0 - x\| \leqslant \|x_0 - \lambda y_0 - (x - \lambda y)\|, \qquad \forall \lambda > 0, \quad \forall [x, y] \in A,$$

or equivalently

$$(y_0 - x, f) \leqslant 0$$

for every $[x, y] \in A$ and some $f \in F(x_0 - x)$ (see Lemma 3.1). Thus $A_1 = [x_0, y_0] \cup A$ is dissipative. Since A is maximal dissipative and $A_1 \supset A$ we conclude that $[x_0, y_0] \in A$ which is the desired conclusion.

(b) Since $\{A_\lambda x_\lambda\}$ is bounded as $\lambda \to 0$, by Eq. (3.8) it follows that

$$\lim_{\lambda \to 0} \|J_\lambda x_\lambda - x\| = 0.$$

But $A_\lambda x \in AJ_\lambda x_\lambda$ so that we can use (a) to obtain $[x, y] \in A$. The proof of Proposition 3.4 is complete.

When the space X is "sufficiently smooth" these results can be considerably improved, namely as follows

Proposition 3.5. *Let A be an m-dissipative subset of $X \times X$ and let X^* be uniformly convex. Then*

(i) *A is demiclosed.*

(ii) *Let $\{x_\lambda\} \subset X$ be such that $x_\lambda \to x_0$ and $\{A_\lambda x_\lambda\}$ is bounded as $\lambda \to 0$. Then $x_0 \in D(A)$. Moreover, if for some sequence $\lambda_n \to 0$, $A_{\lambda_n} x_{\lambda_n} \rightharpoonup y_0$ then $y_0 \in Ax_0$.*

(iii) *For any $x \in X$, $\|A_\lambda x\|$ is monotone nonincreasing in λ, $0 < \lambda < \infty$ and*

$$\lim_{\lambda \to 0} \|A_\lambda x\| = |Ax|. \tag{3.17}$$

(iv) *For every $x \in D(A)$, Ax is a closed convex subset of X.*

Proof. (i) Let $[x_n, y_n] \in A$ be such that $x_n \to x_0$ and $y_n \rightharpoonup y_0$ as $n \to \infty$. Then

$$(y_n - y, F(x_n - x)) \leqslant 0, \qquad \forall [x, y] \in A.$$

Since X^* is uniformly convex, the duality mapping $F: X \to X^*$ is single-valued and uniformly continuous on the bounded subsets of X (see Proposition 1.5, Chapter I). Thus, letting $n \to \infty$ in the preceding inequality, we obtain

$$(y_0 - y, F(x_0 - x)) \leqslant 0, \quad \forall [x, y] \in A$$

which implies $[x_0, y_0] \in A$ by the maximality of A.

(ii) Let $u_\lambda = J_\lambda x_\lambda$ and $v_\lambda = A_\lambda x_\lambda$. Since by Milman's theorem (see Theorem 1.2 in Chapter I), X is reflexive, there exists $\{\lambda_n\} \subset \{\lambda\}$ such that $A_{\lambda_n} x_{\lambda_n} \to y_0$ as $n \to \infty$. But $A_{\lambda_n} x_{\lambda_n} \in AJ_{\lambda_n} x_{\lambda_n}$ and $J_{\lambda_n} x_{\lambda_n} \to x_0$ because $\{A_\lambda x_\lambda\}$ is bounded. By virtue of (i) we have $[x_0, y_0] \in A$.

(iii) We set $u(\lambda) = J_\lambda x$ and $v(\lambda) = A_\lambda x$. Clearly $[u(\lambda), v(\lambda)] \in A$ and $x = u(\lambda) - \lambda v(\lambda)$ for $\lambda > 0$. We have

$$(v(\lambda) - v(\mu), F(u(\lambda) - u(\mu))) = (v(\mu) - v(\lambda), F(\lambda v(\lambda) - \mu v(\mu)))$$

for $\lambda, \mu > 0$. Multiplying this equality by $\lambda > 0$, we find after some rearrangements that

$$\begin{aligned} \|\lambda v(\lambda) - \mu v(\mu)\|^2 &= (\lambda - \mu)(v(\lambda), F(\lambda v(\lambda) - \mu v(\mu)) \leqslant \\ &\leqslant |\lambda - \mu| \; \|v(\lambda)\| \; \|\lambda v(\lambda) - \mu v(\mu)\|. \end{aligned} \tag{3.18}$$

Hence

$$\|\lambda v(\lambda) - \mu v(\mu)\| \leqslant |\lambda - \mu| \; \|v(\lambda)\| \tag{3.19}$$

which implies that $\|v(\lambda)\| \leqslant \|v(\mu)\|$ for $\lambda > \mu$ as desired. Let $r = \lim_{\lambda \to 0} \|A_\lambda x\|$ where $x \in D(A)$ is given. Since $\|A_\lambda x\| \leqslant |Ax|$, it follows that $r \leqslant |Ax|$. Let $\{\lambda_n\}$ be a subsequence of $\{\lambda\}$ such that $A_{\lambda_n} x \rightharpoonup y$ and $\lambda_n \to 0$ as $n \to \infty$. Clearly, $\|y\| \leqslant r$. On the other hand it follows by (ii) that $y \in Ax$. Hence $|Ax| \leqslant r$. Thus we have proved that $r = |Ax|$.

(iv) Since A is maximal dissipative for any $x_0 \in D(A)$, Ax_0 is given by

$$Ax_0 = \{y_0 \in X; \; (y - y_0, F(x - x_0)) \leqslant 0, \quad \forall [x, y] \in A\} \tag{3.20}$$

and this implies the desired result. The proof is complete.

Remark. To prove (i) and (iv) it suffices only to suppose that A is maximal dissipative in $X \times X$.

With each subset A of $X \times X$ we associate a subset A^0, as follows

$$A^0 x = \{y \in Ax; \; \|y\| = |Ax|\}. \tag{3.21}$$

In the event that A is maximal dissipative, Ax is closed and convex for any $x \in D(A)$. Recall that if C is a closed convex subset of a reflexive Banach space X, then for every $x \in X$ there exists at least one nearest point to x in C. If X happens to be strictly convex, then each x has one and only one nearest point in C. Thus, if X is reflexive (in particular if X^* is uniformly convex), $D(A^0) = D(A)$. If X is also strictly convex, then A^0 is single-valued. The operator A^0 is called the *minimal section* of A.

PROPOSITION 3.6. *Let X^* be uniformly convex and let $A \subset X \times X$ be m-dissipative. Then*

(a) $\lim_{\lambda \to 0} F(A_\lambda x) = F(A^0 x)$, *for each* $x \in D(A)$.

If X is also uniformly convex, then

(b) $\lim_{\lambda \to 0} A_\lambda x = A^0 x$, *for each* $x \in D(A)$.

(c) $\overline{D(A)}$ *is a convex subset of* X.

Proof. (a) First we shall prove that the operator $x \to F(A^0 x)$ is single-valued. If C is a closed convex subset of X, then y is an element of minimal norm in C, if and only if,

$$\|y\| \leqslant \|(1 - \alpha)y + \alpha z\|, \quad \text{for all } z \in C \text{ and } 0 \leqslant \alpha \leqslant 1. \tag{3.22}$$

Hence

$$A^0 x = \{y \in Ax;\ \|y\| \leqslant \|(1 - \alpha)y + \alpha z\|, \quad \forall z \in Ax;\ 0 \leqslant \alpha \leqslant 1\} \tag{3.23}$$

and invoking Lemma 3.1 we find that

$$A^0 x = \{y \in Ax;\ (z, F(y)) \geqslant |Ax|^2, \quad \forall\, z \in Ax\}. \tag{3.24}$$

Let y_1, y_2 be arbitrary but fixed in $A^0 x$. By Eq. (3.24) it follows that

$$\|y_1\|^2 \leqslant (y_2, F(y_1)) \leqslant \|y_1\|\ \|y_2\|$$

which implies that $(y_2, F(y_1)) = \|y_2\|^2$ and $\|F(y_1)\| = \|y_2\|$. Therefore, $F(y_1) = F(y_2)$ as desired. Next, we shall prove that $F(A_\lambda x) \to F(A^0 x)$ as $\lambda \to 0$. A being dissipative, we have

$$(A_\lambda x - u, F(J_\lambda x - x)) \leqslant 0, \quad \forall\, u \in Ax.$$

Hence

$$(u, F(A_\lambda x)) \geqslant \|A_\lambda x\|^2, \quad \forall\, u \in Ax.$$

Since X^* is reflexive, we may assume that $F(A_\lambda x) \rightharpoonup y$ in X^* as $\lambda \to 0$, so Proposition 3.5 (iii) yields

$$(u, y) \geqslant |Ax|^2, \quad \forall\, u \in Ax,$$

which implies $\|y\| = |Ax|$ and $(A^0x, y) = |Ax|^2$ because $\|A_\lambda x\| = \|F(A_\lambda x)\| \leqslant |Ax|$ for all $\lambda > 0$. This shows that $y = F(A^0x)$. Moreover, since X^* is uniformly convex, Proposition 1.4 (Chapter I) implies that

$$\lim_{\lambda \to 0} F(A_\lambda x) = F(A^0x)$$

as claimed.

Now suppose that X is also uniformly convex. Then $F^{-1} : X^* \to X$, the duality mapping of X^*, is continuous in virtue of Proposition 1.5 (Chapter I). Moreover, in this case A^0x is a singleton set. Thus we deduce (b) as an immediate consequence of (a).

(c) Let $x_1, x_2 \in D(A)$ and $0 \leqslant \alpha \leqslant 1$. Set

$$x_\alpha = \alpha x_1 + (1 - \alpha)x_2.$$

Then, as is easily verified,

$$\|J_\lambda(x_\alpha) - x_1\| \leqslant \|x_\alpha - x_1\| + \lambda|Ax_1|, \quad \forall\, \lambda > 0$$

$$\|J_\lambda(x_\alpha) - x_2\| \leqslant \|x_\alpha - x_2\| + \lambda|Ax_2|, \quad \forall\, \lambda > 0.$$

Inasmuch as space X is uniformly convex the preceding inequalities imply by a standard device that

$$\|J_\lambda(x_\alpha) - x_\alpha\| \leqslant \delta(\lambda), \quad \forall\, \lambda > 0,$$

where $\delta(\lambda) > 0$ for $\lambda > 0$ and $\lim_{\lambda \to 0} \delta(\lambda) = 0$. Since $R(J_\lambda) \subset D(A)$, this implies that $x_\alpha \in \overline{D(A)}$, i.e. $\overline{D(A)}$ is convex.

DEFINITION 3.3. Let X^* be uniformly convex and let $A \subset X \times X$ be m-dissipative. A subset A' of A is called *a principal section* of A if $D(A') = D(A)$ and has the following property: if $x_0 \in \overline{D(A)}$, $y_0 \in X$ and

$$(y_0 - A'x, F(x_0 - x)) \leqslant 0, \quad \forall\, x \in D(A) \tag{3.25}$$

then $[x_0, y_0] \in A$.

In other words, $A' \subset A$ is a principal section of A if $D(A') = D(A)$ and if any dissipative extension $\tilde{A}$ of A with $D(\tilde{A}) \subset \overline{D(A)}$ coincides with A'.

The characterisation of the *principal sections* of an m-dissipative set remains an open problem. Here are some partial results.

THEOREM 3.3. *Let X^* be uniformly convex and let A be an m-dissipative subset of $X \times X$. Let $A' \subset A$ be such that $D(A) = D(A')$, A' is single-valued and*

$$\|A'x\| \leqslant \omega(|Ax|), \quad \forall\, x \in D(A), \tag{3.26}$$

where $\omega : [0, \infty[\to R^1$ *is positive and locally bounded on* $[0, \infty[$. *If* $[x_0, y_0] \in \in D(A) \times X$ *is such that*

$$(y_0 - A'x,\ F(x_0 - x)) \leqslant 0, \quad \forall\, x \in D(A), \tag{3.27}$$

then $[x_0, y_0] \in A$.

Proof. Clearly, we may assume without loss of generality that $y_0 = 0$. This can be achieved by shifting the range of A. Substituting $x = J_\lambda x_0$ in (3.27), we obtain

$$(A'J_\lambda x_0,\ F(A_\lambda x_0)) \leqslant 0, \quad \forall\, \lambda > 0.$$

Since $|AJ_\lambda x_0| \leqslant \|A_\lambda x_0\| \leqslant |Ax_0|$, (3.26) implies that $\{\|A'J_\lambda x_0\|\}$ is bounded as $\lambda \to 0$. Let $\{A'J_{\lambda_n} x_0\} \subset \{A'J_\lambda x_0\}$ be such that $A'J_{\lambda_n} x_0 \rightharpoonup y$ as $\lambda_n \to 0$. Since A is demiclosed (Proposition 3.5), this implies $y \in Ax_0$. From Proposition 3.6 (a) we deduce that

$$(y,\ F(A^0 x_0) \leqslant 0.$$

Then it follows from (3.24) that $A^0 x_0 = 0$, i.e., $0 \in Ax_0$. This completes the proof.

COROLLARY 3.1. *Let* X^* *be uniformly convex. Let* A *and* B *be two m-dissipative sets in* $X \times X$ *such that* $A^0 x \cap B^0 x \neq \emptyset$ *for each* $x \in D(A) = D(B)$. *Then* $A = B$. *In particular,* $A^0 = B^0$ *implies that* $A = B$.
Proof. Suppose that there exists $[x_0, y_0] \in A$ such that $[x_0, y_0] \in B$ and argue from this step to a contradiction. Let $A' : X \to X$ be a single-valued operator such that $D(A') = D(A) = D(B)$ and $A'x \in A^0 x \cap B^0 x$ for every $x \in D(A)$. Since A is dissipative, we have

$$(y_0 - A'x,\ F(x_0 - x)) \leqslant 0, \quad \forall\, x \in D(A').$$

On the other hand

$$\|A'x\| = \|B^0 x\| = |Bx|.$$

Thus the previous theorem implies $[x_0, y_0] \in B$, which concludes the proof.

THEOREM 3.4. *Let* X^* *be uniformly convex and let* $A \subset X \times X$ *be m-dissipative. Then* A^0 *is a principal section of* A.

Proof. We shall give here a simple proof of this theorem under the assumption that $X = X^* = H$ is a Hilbert space. Proof of the general case will be given in the next chapter (Section 1.4). Let $[x_0, y_0] \in \overline{D(A)} \times H$ be such that

$$(y_0 - A^0 x,\ x_0 - x) \leqslant 0, \quad \forall\, x \in D(A).$$

Taking $x = J_\lambda x_0$, where $\lambda > 0$, yields

$$(A^0 J_\lambda x_0, A_\lambda x_0) \leqslant \|y_0\| \; \|A_\lambda x_0\|.$$

Next, by (3.24)

$$\|A_\lambda x_0\|^2 \leqslant (A^0 J_\lambda x_0, A_\lambda x_0) \leqslant \|y_0\| \; \|A_\lambda x_0\|, \qquad \lambda > 0,$$

which shows that $\|A_\lambda x_0\| \leqslant \|y_0\|$ for all $\lambda > 0$. Then, from Proposition 3.5 (ii) we deduce that $x_0 \in D(A)$. Thus we can apply Theorem 3.3 to obtain $[x_0, y_0] \in A$.

COROLLARY 3.2. *Let H be a real Hilbert space and let A be an m-dissipative (equivalently maximal dissipative) subset of $H \times H$. Then*

$$E(A^0; \overline{D(A)}) = A. \tag{3.28}$$

Proof. We have already shown that $\overline{D(A)}$ is convex. Moreover, in virtue of Theorem 3.2, $E(A^0; \overline{D(A)}) \neq \emptyset$. Let B be arbitrary in $E(A^0; \overline{D(A)})$ and let $[x_0, y_0] \in B$. Since $A^0 \subset B$ is dissipative, we have

$$(y_0 - A^0 x, x_0 - x) \leqslant 0, \quad \forall\, x \in D(A).$$

But $x_0 \in \overline{D(A)}$ and A^0 is a principal section of A in virtue of Theorem 3.4. Thus the previous inequality implies $[x_0, y_0] \in A$. Therefore, $A = B$ and the arbitrariness of B in $E(A^0; \overline{D(A)})$ implies Eq. (3.28).

3.2. Perturbations of dissipative sets

It is apparent that, given any two dissipative operators A and B from X to itself, the operator $A + B$ with the domain $D(A) \cap D(B)$ is again dissipative. If A and B are m-dissipative it does not necessarily follow, however, that $A + B$ is again m-dissipative. The problem we shall study in this section is that of determining the conditions under which $A + B$ is m-dissipative. For maximal monotone operators this problem was treated in Section 1.2 and it is of a fundamental importance in the existence theory for nonlinear partial differential equations.

LEMMA 3.2. *Let X^* be uniformly convex, and let A and B be m-dissipative subsets of $X \times X$ such that $D(A) \cap D(B) \neq \emptyset$. Then for every $y \in X$ and $\lambda > 0$ equation*

$$x_\lambda - Ax_\lambda - B_\lambda x_\lambda \ni y \tag{3.29}$$

has a unique solution $x_\lambda \in D(A)$. *Moreover* $\|x_\lambda\|$ *is bounded as* $\lambda \to 0$. *If* $\|B_\lambda x_\lambda\|$ *is bounded as* $\lambda \to 0$, *then equation*

$$x - Ax - Bx \ni y \tag{3.30}$$

has a unique solution $x \in D(A) \cap D(B)$ *and* $\lim\limits_{\lambda \to 0} x_\lambda = x$.

Proof. Equation (3.29) may be written in the following equivalent form

$$x_\lambda = \left(1 - \frac{\lambda}{\lambda + 1} A\right)^{-1} \left(\frac{\lambda}{\lambda + 1} y + \frac{1}{\lambda + 1} (1 - \lambda B)^{-1} x_\lambda\right). \tag{3.31}$$

Since $\left(1 - \dfrac{\lambda}{\lambda + 1} A\right)^{-1}$ and $(1 - \lambda B)^{-1}$ are nonexpansive on X, by Banach's fixed-point theorem it follows that Eq. (3.31) (and Eq. (3.29)) has a unique solution $x_\lambda \in D(A)$.

Next, assume that $\|B_\lambda x\|$ is bounded as $\lambda \to 0$. Let $x_1 \in D(A) \cap D(B)$ and let $y_\lambda^1 \in x_1 - Ax_1 - B_\lambda x_1$. Since A and B are dissipative, we get

$$\|x_\lambda - x_1\|^2 \leqslant (y - y_\lambda^1, F(x_\lambda - x_1)) \leqslant \|y - y_\lambda^1\| \, \|x_\lambda - x_1\|$$

and hence

$$\|x_\lambda\| \leqslant C, \quad \text{for all } \lambda > 0,$$

because $\|y_\lambda^1\|$ is bounded. (We recall that $\|B_\lambda x_1\| \leqslant |Bx_1|$). Now, let $\lambda, \mu > 0$. We have

$$\|x_\lambda - x_\mu\|^2 \leqslant (B_\lambda x_\lambda - B_\mu x_\mu, \; F(x_\lambda - x_\mu))$$

and hence

$$\|x_\lambda - x_\mu\|^2 \leqslant (B_\lambda x_\lambda - B_\mu x_\mu, \; F(x_\lambda - x_\mu) - F(J_\lambda x_\lambda - J_\mu x_\mu)) \tag{3.32}$$

because $B_\lambda x_\lambda \in BJ_\lambda x_\lambda$ and B is dissipative. (We have denoted $J_\lambda = (1 - \lambda B)^{-1}$).) Recalling that $\lambda B_\lambda x_\lambda = J_\lambda x_\lambda - x_\lambda$, the boundedness of $\|B_\lambda x_\lambda\|$ implies that

$$\|J_\lambda x_\lambda - x_\lambda\| \leqslant C_1 \lambda, \quad \text{for all } \lambda > 0.$$

Since F is uniformly continuous on every bounded subset of X (see Proposition 1.4, Chapter I) it follows from Eq. (3.32) that $\{x_\lambda\}$ is convergent as $\lambda \to 0$. Let

$$x = \lim_{\lambda \to 0} x_\lambda.$$

To conclude the proof we shall demonstrate that x is a solution of Eq. (3.30). Let $\{\lambda_n\}$ be a subsequence of $\{\lambda\}$ such that $\lambda_n \to 0$ as $n \to \infty$, $B_{\lambda_n}x_{\lambda_n} \rightharpoonup z$, and $w_{\lambda_n} \rightharpoonup w$, where $w_{\lambda_n} = y + B_{\lambda_n}x_{\lambda_n} - x_{\lambda_n} \in -Ax_{\lambda_n}$. From Proposition 3.5 we see the $x \in D(A) \cap D(B)$, $z \in Bx$ and $w \in -Ax$. Then, passing to limit through the sequence λ_n in the approximate equation (3.29) we obtain $x - Ax - Bx \ni y$ as claimed. The uniqueness of the solution is immediate from the dissipativeness of A and B.

THEOREM 3.5. *Let A and B be two m-dissipative subsets of $X \times X$ where X^* is uniformly convex. Suppose that the following two conditions hold:*

(a) $D(A) \subset D(B)$.

(b) *For each $r > 0$ there are constants $L < 1$ and M such that*

$$|Bx| \leqslant L|Ax| + M, \quad \forall\, x \in D(A),\ \|x\| \leqslant r. \tag{3.33}$$

Then $A + B$ is m-dissipative.

Proof. Since $A + B$ is dissipative, the theorem will be proved if we show that Eq. (3.30) has a solution $x \in D(A) \cap D(B)$ for each $y \in X$. To this end we consider the solution x_λ of the approximate equation (3.29). Since $x_\lambda \in D(A) \subset D(B)$ and $\|B_\lambda x\| \leqslant |Bx|$ for $x \in D(B)$ (by Proposition 3.2), condition (3.33) implies that

$$(1 - L)\,|Ax_\lambda| \leqslant M, \quad \forall\, \lambda > 0,$$

with a modified constant M. Then, again using (3.33) we obtain that $\|B_\lambda x_\lambda\| \leqslant \leqslant |Bx_\lambda|$ are bounded. Applying Lemma 3.2 we obtain $y \in R(1 - A - B)$ as claimed.

Remark. Condition (b) of Theorem 3.5 can be replaced by the following local condition (see T. Kato [2]):

(b′) For each $x_0 \in D(A)$ there is a neighbourhood V of x_0 and constants $L < 1$ and M such that

$$|Bx| \leqslant L|Ax| + M, \quad \text{for } x \in V \cap D(A).$$

Theorem 3.5, although a legitimate result from the theoretical point of view, appears to be of little value for application. A more convenient criterion for maximality is furnished by the following theorem.

THEOREM 3.6. *Let X be uniformly convex. Let A and B be m-dissipative sets in $X \times X$ such that*

(a) $D(A) \cap D(B) \neq \emptyset$,

(b) $(y, F(B_\lambda x)) \geqslant 0$, *for all* $[x, y] \in A$ *and* $\lambda > 0$.

Then $A + B$ is m-dissipative.

Proof. Let $x_\lambda \in D(A)$ be the solution of Eq. (3.29) where $y \in X$. Multiplying this equation by $F(B_\lambda x_\lambda)$ and using condition (b) one easily sees that $\{\|B_\lambda x_\lambda\|\}$ is bounded as $\lambda \to 0$. Thus we conclude the proof by applying Lema 3.2.

If X is a Hilbert space, this theorem can be sharpened as follows.

THEOREM 3.7. *Let φ be a proper lower-semicontinuous convex function from a real Hilbert space H into $]-\infty, +\infty]$, $\varphi \not\equiv +\infty$ and let $A \subset H \times H$ be maximal monotone. In addition, assume that there are $h \in H$ and a constant M such that*

$$\varphi((1 + \lambda A)^{-1}(x + \lambda h)) \leqslant \varphi(x) + M\lambda(1 + \varphi(x)) \tag{3.34}$$

for every $x \in H$ and $\lambda > 0$. Then $A + \partial\varphi$ is a maximal monotone subset of $H \times H$ and

$$\overline{D(A + \partial\varphi)} = \overline{D(A)} \cap \overline{D(\varphi)}. \tag{3.35}$$

Proof. First we shall show that $A + \partial\varphi$ is maximal monotone in $H \times H$, i.e., equation

$$x + Ax + \partial\varphi(x) \ni y \tag{3.36}$$

has a solution x for each $y \in H$. Consider the approximate equation

$$x_\lambda + A_\lambda x_\lambda + \partial\varphi(x_\lambda) \ni y, \quad \lambda > 0 \tag{3.37}$$

where $A_\lambda = \lambda^{-1}(1 - (1 + \lambda A)^{-1})$. According to Lemma 3.2 it suffices to show that $\{\|A_\lambda x_\lambda\|\}$ is bounded as $\lambda \to 0$. By the definition of the subdifferential $\partial\varphi$, we have

$$(A_\lambda x_\lambda + x_\lambda - y, x_\lambda - u) \leqslant \varphi(u) - \varphi(x_\lambda), \text{ for every } u \in H. \tag{3.38}$$

Define $u_\lambda = (1 + \lambda A)^{-1}(x_\lambda + \lambda h)$, $\lambda > 0$. Substituting u_λ for u in relation (3.38) and using condition (3.34), we obtain

$$(A_\lambda x_\lambda + x_\lambda - y, x_\lambda - (1 + \lambda A)^{-1}(x_\lambda + \lambda h)) \leqslant M\lambda(1 + \varphi(x_\lambda)).$$

Hence

$$(A_\lambda x_\lambda + x_\lambda - y, A_\lambda x_\lambda) \leqslant M(1 + \varphi(x_\lambda)) + \|A_\lambda x_\lambda + x_\lambda - y\| \|h\| \tag{3.39}$$

because $(1 + \lambda A)^{-1}$ is nonexpansive.

Let $u \in D(\varphi)$. Using again (3.38) we have

$$\varphi(x_\lambda) \leqslant \varphi(u) + \|A_\lambda x_\lambda + x_\lambda - y\| \|x_\lambda - u|. \tag{3.40}$$

Since $\{\|x_\lambda\|\}$ is bounded (by Lemma 3.2) from (3.39) and (3.40) it follows after some rearrangements that $\{\|A_\lambda x_\lambda\|\}$ is bounded, as desired.

To conclude the proof Eq. (3.35) must be verified. First we shall prove that $\overline{D(A)} \cap \overline{D(\varphi)} \subset \overline{D(A) \cap D(\varphi)}$. Let $u \in \overline{D(A)} \cap \overline{D(\varphi)}$ and $h \in H$ be given by condition (3.34). Then there exists $u_\varepsilon \in X$ such that $u_\varepsilon + \varepsilon h \in D(\varphi)$ and $u_\varepsilon \to u$ as $\varepsilon \to 0$. Define,

$$v_\varepsilon = (1 + \varepsilon A)^{-1}(u_\varepsilon + \varepsilon h).$$

But $v_\varepsilon \in D(\varphi)$ by condition (3.34). Furthermore,

$$\|v_\varepsilon - u\| \leqslant \|(1 + \varepsilon A)^{-1}(u_\varepsilon + \varepsilon h) - (1 + \varepsilon A)^{-1}u\| + \|u - (1 + \varepsilon A)^{-1} u\| \leqslant$$

$$\leqslant \|u_\varepsilon - u\| + \varepsilon\|h\| + \|u - (1 + \varepsilon A)^{-1} u\|.$$

Since $u \in \overline{D(A)}$, we deduce that $v_\varepsilon \to u$ as $\varepsilon \to 0$ as we wished to show. Now we shall prove that $D(A) \cap D(\varphi) \subset \overline{D(A) \cap D(\partial\varphi)}$. Let $u \in D(A) \cap D(\varphi)$ and let $u_\varepsilon \in D(A) \cap D(\partial\varphi)$ be the solution of the following equation

$$u_\varepsilon + \varepsilon(Au_\varepsilon + \partial\varphi(u_\varepsilon)) \ni u.$$

We have

$$\varphi(u) - \varphi(u_\varepsilon) \geqslant \left(\frac{u - u_\varepsilon}{\varepsilon} - Au_\varepsilon, u - u_\varepsilon\right) \geqslant \left\| \frac{u - u_\varepsilon}{\sqrt{\varepsilon}} \right\|^2 - (A^0u, u - u_\varepsilon).$$

Hence $\lim_{\varepsilon \to 0} u_\varepsilon = u$. This completes the proof of Theorem 3.7.

We conclude this section with the following perturbation result.

THEOREM 3.8. *Let X^* be uniformly convex and let C be a closed convex subset of X. Let A and B be two dissipative subsets of $X \times X$ which are assumed to satisfy the following conditions:*

(i) *A is a closed subset of $X \times X$ and $C \subset R(1 - \lambda A)$ for all $\lambda > 0$. Furthermore,*

$$(1 - \lambda A)^{-1}C \subset C, \quad \forall \lambda > 0. \tag{3.41}$$

(ii) *B is single-valued, linear, densely defined and m-dissipative from X to itself. Moreover,*

$$(1 - \lambda B)^{-1}C \subset C, \quad \forall \lambda > 0. \tag{3.42}$$

(iii) *$(1 - \lambda A)^{-1}(D(B) \cap C) \subset D(B)$ for all $\lambda > 0$ and there exists $M > 0$ such that*

$$\|B(1 - \lambda A)^{-1}x\| \leqslant \|Bx\| + M\lambda(1 + \|x\| + \|A_\lambda x\|), \tag{3.43}$$

for every $x \in D(B) \cap C$ and $\lambda > 0$.

Then

$$C \subset \overline{R(1 - \lambda A + B)}, \quad \forall \lambda > 0 \tag{3.44}$$

and

$$D(B) \cap C \subset R(1 - \lambda(\overline{A + B})), \quad 0 < \lambda \leqslant \omega, \tag{3.45}$$

for a sufficiently small $\omega > 0$. *Here* $\overline{A + B}$ *denotes the closure of* $A + B$ *in* $X \times X$.

In particular, using Theorem 3.8 we deduce the following criterion for m-dissipativeness

COROLLARY 3.3. *Let* X^* *be uniformly convex and let* A *and* B *be two m-dissipative subsets of* $X + X$ *satisfying the following conditions:*

(a) B *is linear and densely defined in* X.

(b) $(1 - \lambda A)^{-1}D(B) \subset D(B)$ *for all* $\lambda > 0$, *and there is a constant* M *such that*

$$\|B(1 - \lambda A)^{-1} x\| \leqslant \|Bx\| + M\lambda(\|x\| + \|A_\lambda x\|), \qquad \forall\, \lambda > 0. \tag{3.46}$$

Then, $\overline{A + B}$ *is m-dissipative and there exists* $\omega > 0$ *such that*

$$D(B) \subset R(1 - \lambda)\,(\overline{A + B})),\ 0 < \lambda < \omega. \tag{3.47}$$

Proof of Theorem 3.8. First we shall show that $\lambda y \in (\lambda - A - B)u$ has a solution u for each $y \in D(B) \cap C$ and for all sufficiently large λ. To this end we consider the approximate equations

$$\lambda u_n - Bu_n - A_n u_n = \lambda y,\ n = 1, 2, \ldots \tag{3.48}$$

where $A_n = n((1 - n^{-1}A)^{-1} - 1)$. Equation (3.48) may be written as

$$u_n = \lambda(\lambda + n - B)^{-1} y + n(\lambda + n - B)^{-1}(1 - n^{-1}A)^{-1} u_n. \tag{3.49}$$

By hypotheses (i) and (ii) the operator defined by the right-hand side of Eq. (3.49) maps the convex closed set $\overline{D(B) \cap C}$ into itself and is Lipschitz with constant $n/n + \lambda$. (We apply Proposition 3.2.) Therefore, Eq. (3.48) has a unique solution $u_n \in D(B) \cap C$ for every n. Let $u_0 \in D(A) \cap D(B) \cap C$. (It follows by hypothesis (iii) that $D(A) \cap D(B) \cap C \neq \emptyset$). Since A_n and B are dissipative, we see by multiplying (3.48) by $F(u_n - u_0)$ that

$$\lambda\|u_n - u_0\| \leqslant \lambda\|y\| + \|Bu_0\| + |Au_0| + \lambda\|u_0\|, \quad n = 1, 2, \ldots$$

because $\|A_n u_0\| \leqslant |Au_0|$. This shows that $\{\|u_n\|\}$ is bounded as $n \to \infty$. On the other hand, by a simple calculation involving inequality (3.43) and Eq. (3.48) we obtain

$$\|Bu_n\| \leqslant \lambda(n - \lambda)^{-1}\|By\| + n(n + \lambda)^{-1}\|Bu_n\| +$$

$$+ M(n + \lambda)^{-1}(\|u_n\| + \|A_n u_n\|)$$

and hence

$$\|Bu_n\| \leqslant \lambda(\lambda - M)^{-1}\|By\| + M(\lambda + 1)(\lambda - M)^{-1}\|u_n\| +$$

$$+ M\lambda(\lambda - M)^{-1}\|y\|$$

for every $\lambda > M$. By the same argument as that used in the proof of Lemma 3.2 we then conclude that $\lim_{n\to\infty} u_n = u$ exists. Since B is demiclosed, we see that

$$\lambda u - Bu - v = \lambda y \tag{3.50}$$

where $v = w - \lim_{n\to\infty} A_n u_n$. We must prove that $[u, v] \in D(A)$. Let $\tilde{A}$ be the smallest demiclosed extension of A. In other words, $\tilde{A}$ is the strong-weak closure of A in $X \times X$. Obviously, $D(\tilde{A}) \subset \overline{D(A)}$. Furthermore, using the continuity of the duality mapping F, one can easily see that $\tilde{A}$ is dissipative in $X \times X$. But $A_n u_n \in A(1 - n^{-1}A)^{-1} u_n$ and $\|u_n - (1 - n^{-1}A)^{-1}u_n\| = n^{-1} \|A_n u_n\|$. Hence $\lim_{n\to\infty} (1 - n^{-1}A)^{-1}u_n = u$ and the demiclosedness of $\tilde{A}$ then implies that $u \in D(\tilde{A})$ and $v \in \tilde{A}u$. Let $\varepsilon_n > 0$ be such that $\varepsilon_n \to 0$ as $n \to \infty$. Define,

$$\Lambda_n = \varepsilon_n^{-1}(S(\varepsilon_n)\, u - u) - Bu, \quad n = 1, 2, \ldots,$$

where $S(t)$ is the semigroup of linear contractions generated by B (see Corollary 3.1, of the Hille-Yosida theorem in Chapter I). The exponential formula (see Theorem 2.3, Chapter III),

$$S(t)\, x = \lim_{n\to\infty} \left(1 - \frac{t}{n} B\right)^{-n} x, \quad t \geqslant 0,\ x \in X$$

and condition (3.42), then imply that $S(\varepsilon_n)C \subset C$ for all n. By assumption, this yields the existence of a point $[x_n, y_n] \in A$ such that

$$S(\varepsilon_n)\, u = x_n - \varepsilon_n y_n - \lambda \varepsilon_n (y - x_n).$$

Hence

$$\Lambda_n = \varepsilon_n^{-1}(x_n - u) - Bu - y_n - \lambda(y - x_n). \tag{3.51}$$

Since $\tilde{A}$ is dissipative, we see by multiplying (3.51) by $F(x_n - u)$ that

$$(\lambda + \varepsilon_n^{-1})\, \|x_n - u\| \leqslant \|\Lambda_n\|.$$

Inasmuch, as $\lim_{n\to\infty} \Lambda_n = 0$, the previous estimate shows that $\lim_{n\to\infty} y_n = \lambda u - - Bu - \lambda y$. Since A is closed, we infer that $u \in D(A)$ and

$$\lambda u - Au - Bu \ni \lambda y.$$

Summarizing to this point, we have shown that $\overline{D(B) \cap C} \subset R(1 - \lambda(A + B))$ for all $0 < \lambda < \omega = M^{-1}$. We note that $\overline{D(B) \cap C} = C$. Indeed, for every $x \in C$ and n, $x_n = (1 - n^{-1}B)^{-1} x$ belongs to $D(B) \cap C$ and $\lim_{n\to\infty} x_n = x$ because $D(B)$ is dense in X. Let $\overline{A + B}$ be the closure of $A + B$ in $X \times X$. Obviously, $\overline{A + B}$ is a dissipative subset of $X \times X$. In particular, $(1 - \lambda\overline{A+B})^{-1}$ is nonexpansive on $D(\overline{A+B})$ for every $\lambda>0$. Since $\overline{D(B) \cap C}=C$, it follows from Eq. (3.45) that

$$C \subset R(1 - \lambda\overline{A + B}), \quad 0 < \lambda < \omega. \tag{3.52}$$

To conclude the proof, it remains to show that (3.52) extends to all $\lambda > 0$. Let $y \in C$ and $\mu \geqslant \omega$ be given. Equation

$$x - \mu\overline{A + B}x \ni y, \quad x \in C$$

may be written as

$$x = (1 - \lambda\overline{A + B})^{-1}\left(\frac{\lambda}{\mu} y + \left(1 - \frac{\lambda}{\mu}\right)x\right) \tag{3.53}$$

where $0 < \lambda < \omega$. Since $(1 - \lambda\overline{A + B})^{-1}$ is nonexpansive and maps C into itself we deduce that Eq. (3.53) has a unique solution $x \in D(\overline{A + B}) \cap C$. Hence $C \subset R(1 - \mu\overline{A + B})$ as we wished to show. This completes the proof of Theorem 3.8.

We conclude this section with an example. Let Ω be a bounded and open domain in R^n with smooth boundary Γ. Let β be a maximal monotone subset of $R^1 \times R^1$ such that $0 \in D(\beta)$. Let $\tilde{\beta} \subset L^p(\Omega) \times L^p(\Omega)$, $1 \leqslant p < \infty$ be the operator defined by

$$D(\tilde{\beta}) = \{u \in L^p(\Omega);\ \exists v \in L^p(\Omega), \text{ such that } v(x) \in \beta(u(x)) \text{ a.e. on } \Omega\}$$

$$\tilde{\beta}(u) = \{v \in L^p(\Omega);\ v(x) \in \beta(u(x)), \text{ a.e. on } \Omega\},\ u \in D(\tilde{\beta}).$$

It is easy to see that $\tilde{\beta}$ is m-accretive in $L^p(\Omega) \times L^p(\Omega)$. Furthermore, we have

$$((1 + \lambda\tilde{\beta})^{-1}u)(x) = (1 + \lambda\beta)^{-1}u(x), \quad \text{a.e. } x \in \Omega$$

$$(\tilde{\beta}_\lambda u)(x) = \beta_\lambda(u(x)), \quad \text{a.e. } x \in \Omega$$

for every $u \in L^p(\Omega)$ and $\lambda > 0$.

PROPOSITION 3.7. *Let $1 < p < \infty$ and let B: $L^p(\Omega) \to L^p(\Omega)$ be the operator defined by*

$$D(B) = W_0^{1,p}(\Omega) \cap W^{2,p}(\Omega) \cap D(\tilde{\beta}).$$

$$Bu = -\Delta u + \tilde{\beta}(u), \quad u \in D(B).$$

Then B is m-accretive.

Proof. We first notice that the duality mapping $F: L^p(\Omega) \to L^{p'}(\Omega)$, $1/p + 1/p' = 1$ is given by

$$F(u)(x) = u(x)\,|u(x)|^{p-2}/\|u\|^{p-2}_{L^p(\Omega)}, \quad \text{a.e. } x \in \Omega.$$

Then, as is easily verified, the operator $-\Delta$ with the domain $W_0^{1,p}(\Omega) \cap W^{2,p}(\Omega)$ is accretive in $L^p(\Omega)$. Moreover, by a well-known result from the existence theory of partial differential equations, $R(1 - \Delta) = L^p(\Omega)$ (see e.g. S. Agmon, A. Douglas and L. Nirenberg [1]). Thus $-\Delta$ is m-accretive. To conclude the proof we shall apply Theorem 3.6 where $X = L^p(\Omega)$, $A = \Delta$, $D(A) = W_0^{1,p}(\Omega) \cap W^{2,p}(\Omega)$ and $B = \tilde{\beta}$. (We recall that for $1 < p < \infty$, $L^p(\Omega)$ is a uniformly convex Banach space.) We have to prove that

$$\int_\Omega \Delta u(x)\, \beta_\lambda(u(x))\, |\beta_\lambda(u(x))|^{p-2}\, dx \leqslant 0 \tag{3.54}$$

for every $u \in W_0^{1,p}(\Omega) \cap W^{2,p}(\Omega)$ and $\lambda > 0$.

But $d/dr\, \beta_\lambda(r) \geqslant 0$ because β_λ is monotone and Lipschitzian on R^1. This implies (3.54) by using a standard argument.

Remark. In particular, Proposition 3.7 implies that for any $f \in L^p(\Omega)$ and every $\lambda > 0$, the nonlinear boundary value problem

$$\lambda u(x) - \Delta u(x) + \beta(u(x)) \ni f(x) \qquad \text{a.e. } x \in \Omega$$
$$u = 0, \qquad \text{a.e. on } \Gamma$$

has a unique solution $u \in W_0^{1,p}(\Omega) \cap W^{2,p}(\Omega)$, $1 < p < \infty$. The case $p = 1$ which is more delicate and requires additional assumptions on β, will be treated in the next chapter.

If $p = 2$ then the operator $-\Delta$ with the domain $H_0^1(\Omega) \cap H^2(\Omega)$ is the subdifferential of $\varphi: L^2(\Omega) \to]-\infty, +\infty]$ defined by

$$\varphi(u) = \begin{cases} \dfrac{1}{2}\displaystyle\int_\Omega |\text{grad } u|^2\, dx & \text{if } u \in H_0^1(\Omega) \\ +\infty & \text{otherwise.} \end{cases}$$

Let h be a real number such that $h \in \beta(0)$. Then $(1 + \lambda\tilde{\beta})^{-1}(u + \lambda h) \in H_0^1(\Omega)$ for all $u \in H_0^1(\Omega)$ and $\varphi((1 + \lambda\tilde{\beta})^{-1}(u + \lambda h)) \leqslant \varphi(u)$, for all $\lambda > 0$. Thus in this case we can also apply Theorem 3.7 where $H = L^2(\Omega)$, $A = \tilde{\beta}$ and φ defined as above, to deduce that $\tilde{\beta} - \Delta$ is maximal monotone in $L^2(\Omega) \times L^2(\Omega)$. Let $j: R^1 \to]-\infty, +\infty]$ be such that $\beta = \partial j$ and let $\psi: L^2(\Omega) \to -\infty, +\infty]$ be the convex function defined by

$$\psi(u) = \begin{cases} \dfrac{1}{2}\displaystyle\int_\Omega |\text{grad } u|^2\, dx + \int_\Omega j(u)\, dx, & \text{if } u \in H_0^1(\Omega),\ j(u) \in L^1(\Omega). \\ +\infty & \text{otherwise.} \end{cases} \tag{3.55}$$

Since $u \to \int_\Omega j(u)\,dx$ is lower-semicontinuous on $L^2(\Omega)$ (see Proposition 2.8) ψ also has this property. But, as is easily verified, $-\Delta + \tilde{\beta} \subset \partial\psi$. Hence $\partial\psi = -\Delta + \tilde{\beta}$ and

$$\overline{D(\partial\psi)} = \overline{D(\psi)} = \overline{D(\tilde{\beta})}.$$

To summarize, we have proved the following.

PROPOSITION 3.8. *Let* $\psi: L^2(\Omega) \to]-\infty, +\infty]$ *be defined by* (3.55). *Then* $[u, w] \in \partial\psi$ *iff* $u \in H_0^1(\Omega) \cap H^2(\Omega)$ *and* $w(x) \in -\Delta u(x) + \beta(u(x))$, a.e. *on* Ω. *Moreover,*

$$\overline{D(\psi)} = \{u \in L^2(\Omega);\, u(x) \in \overline{D(\beta)}, \quad \text{a.e. } x \in \Omega\}.$$

3.3. Riccati equations in Hilbert spaces

We shall begin this section with some preliminary results concerning *Hilbert-Schmidt* operators. Since this subject is frequently treated in books, we shall only state the definitions and the main results and refer the reader to the Bibliography for proofs.

Let us assume that H_1, H_2 are separable Hilbert spaces. A linear continuous operator P with domain in all of H_1, mapping H_1 into H_2 is said to be *Hilbert-Schmidt*, if for some complete orthonormal system $\{e_n\}$ in H_1,

$$\sum_{n=1}^{\infty} (Pe_n, Pe_n)_2 < \infty$$

where $(,)_i$ denotes the inner product of H_i, $i = 1,2$. We shall denote by $(H_1, H_2)_{\text{H.S.}}$ the space of Hilbert-Schmidt operators mapping H_1 into H_2.

The space $(H_1, H_2)_{\text{H.S.}}$ is Hilbert with the inner product defined by

$$[P, Q] = \sum_{1}^{\infty} (Pe_n, Qe_n)_2$$

and the norm

$$\|P\|^2_{(H_1, H_2)_{\text{H.S.}}} = \sum_{1}^{\infty} |Pe_n|_2^2$$

where $\{e_n\}$ is a complete orthonormal system in H_1. The definitions are easily seen to be independent of the particular sequence $\{e_n\}$ used. We also observe that

$$\|P\|_{L(H_1, H_2)} \leqslant \|P\|_{(H_1, H_2)_{\text{H.S.}}}, \qquad \text{for every } P \in (H_1, H_2)_{\text{H.S.}}$$

where $L(H_1, H_2)$ is the space of all linear continuous operators from H_1 into H_2. In the following lemma we list some properties of Hilbert-Schmidt operators (for proof see N. Dunford and J. Schwartz [2], Chapter XI).

LEMMA 3.3. *Let H_1, H_2, H_3 be real separable Hilbert spaces.*

(a) *If $P_1 \in (H_1, H_2)_{\text{H.S.}}$ and $P_2 \in L(H_2, H_3)$, then $P_2P_1 \in (H_1, H_3)_{\text{H.S.}}$ and $\|P_2P_1\|_{(H_1,H_3)_{\text{H.S.}}} \leqslant \|P_2\|_{L(H_2, H_3)} \|P_1\|_{(H_1, H_2)_{\text{H.S.}}}$.*

(b) *If $P_1 \in L(H_1, H_2)$ and $P_2 \in (H_2, H_3)_{\text{H.S.}}$, then $P_2P_1 \in (H_1, H_3)_{\text{H.S.}}$ and $\|P_2P_1\|_{(H_1, H_3)} \leqslant \|P_1\|_{L(H_1, H_2)} \|P_2\|_{(H_2, H_3)_{\text{H.S.}}}$.*

(c) *If $P \in (H_1, H_2)_{\text{H.S.}}$ then $P^* \in (H_2^*, H_1^*)_{\text{H.S.}}$ and $\|P^*\|_{(H_2^*, H_1^*)_{\text{H.S.}}} = \|P\|_{(H_1, H_2)_{\text{H.S.}}}$.*

(d) *If $P \in (H_1, H_2)_{\text{H.S.}}$ then P is compact.*

(e) *If $P_1 \in (H_1, H_1)_{\text{H.S.}}$, $P_2 \in L(H_1, H_1)$ and $P_1 \geqslant 0$ then $[P_2P_1, P_1] \geqslant 0$ and $[P_1P_2, P_1] \geqslant 0$.*

Let V and H be a pair of real and separable Hilbert spaces such that V is dense in H and the injection mapping of V into H is continuous. The scalar product in V and H will be denoted by $((,))$ and $(,)$ and the norms by $\| \quad \|$ and $| \quad |$, respectively. We identify H with its own dual and denote by V' the dual space of V with the norm $\| \quad \|_*$.

Let (v_1, v_2) be the usual pairing between v_1 in V and v_2 in V'; if $v_1, v_2 \in H$, this is the ordinary inner product in H. The following inclusion relation holds

$$V \subset H \subset V'$$

We shall assume that

The injection mapping of V into H is completely continuous. (3.56)

Let J be the *canonical isomorphism* from V onto V', i.e.,

$$(Ju, v) = ((u, v)), \text{ for } u, v \in V. \tag{3.57}$$

Then $J^{-1} \in L(V', V) \cap L(H, V)$. Moreover, assumption (3.56) implies that J^{-1} is self-adjoint and compact on H.

Then, we can choose an orthonormal basis $\{e_n\}$ in H such that

$$J^{-1}e_n = \rho_n^{-2}e_n, \quad n = 1,2, \ldots \tag{3.58}$$

We now introduce the following spaces: $(H, H)_{\text{H.S.}}$, $(H, V)_{\text{H.S.}}$, $(H, V')_{\text{H.S.}}$ and $(V', H)_{\text{H.S.}}$. Clearly, the following inclusion relations hold

$$(V', H)_{\text{H.S.}} \subset (H, H)_{\text{H.S.}} \subset (V, H)_{\text{H.S.}}$$

$$(H, V)_{\text{H.S.}} \subset (H, H)_{\text{H.S.}} \subset (H, V')_{\text{H.S.}}$$

and each space to the left of an inclusion sign is dense in the space to its right and the corresponding inclusion map is continuous. Moreover, $(V', H)_{\text{H.S.}}$ and $(V, H)_{\text{H.S.}}$, respectively, $(H, V)_{\text{H.S.}}$ and $(H, V')_{\text{H.S.}}$, are dual pairs of spaces with respect to the pairing P_1, P_2 defined by

$$[P_1, P_2] = \sum_{n=1}^{\infty} (P_1 e_n, P_2 e_n) \tag{3.59}$$

where $\{e_n\}$ is given by Eq. (3.58).

For simplicity, we set

$$\mathscr{V} = (V', H)_{\text{H.S.}} \cap (H, V)_{\text{H.S.}}, \quad \mathscr{H} = (H, H)_{\text{H.S.}}.$$

Thus the dual space $\mathscr{V}'$ of $\mathscr{V}$ can be identified with $(V, H)_{\text{H.S.}} + (H, V')_{\text{H.S.}}$ and we have

$$\mathscr{V} \subset \mathscr{H} \subset \mathscr{V}'$$

where each injection mapping is continuous and densely defined. In what follows we shall denote by [,] the natural pairing between $\mathscr{V}$ and $\mathscr{V}'$, respectively, the inner product in $\mathscr{H}$. The norms in $\mathscr{V}$, $\mathscr{H}$ and $\mathscr{V}'$ will be denoted by $\| \ \|_{\mathscr{V}}$, $\| \ \|_{\mathscr{H}}$ and $\| \ \|_{\mathscr{V}'}$, respectively.

Besides the spaces V and H we are given a linear operator $A \in L(V, V')$ which is assumed to satisfy

$$(Au, u) \geqslant \omega \|u\|^2 \text{ for every } u \in V, \text{ where } \omega > 0. \tag{3.60}$$

Here we are interested in the following nonlinear equation in the space $L(H, H)$

$$A^*P + PA + P^2 = F.$$

This equation arises in the optimal control theory for elliptic partial equations with quadratic cost criterion (see e.g. J. L. Lions [2], Chapter I) and in other problems of practical interest. We shall prove the following existence result

THEOREM 3.9. *Let $F \in (H, H)_{\text{H.S.}}$ be given satisfying*

$$F = F^*, \ F \geqslant 0 \tag{3.61}$$

Under the above hypotheses, there exists a unique $P \in (H, H)_{\text{H.S.}}$ such that

$$A^*P + PA + P^2 = F \tag{3.62}$$

$$P = P^*, \ P \geqslant 0 \tag{3.63}$$

$$P \in (V', H)_{\text{H.S.}} \cap (H, V)_{\text{H.S.}} \tag{3.64}$$

Proof. We shall divide the proof of this theorem into several steps.

LEMMA 3.4. *There is $\alpha > 0$ such that*

$$[P A, P] \geqslant \alpha \|P\|^2_{(V', H)_{\text{H.S.}}}, \quad \forall P \in (V', H)_{\text{H.S.}}. \tag{3.65}$$

$$[A^*P, P] \geqslant \alpha \|P\|^2_{(H, V)_{\text{H.S.}}}, \quad \forall P \in (H, V)_{\text{H.S.}}. \tag{3.66}$$

Proof. Let $P \in (V', H)_{\text{H.S.}}$ Then from Lemma 3.3 part (a), $PA \in (V, H)_{\text{H.S.}}$ so that the inner product $[PA, P]$ makes sense. Let $\{e_n\}$ be the orthonormal basis in H defined by (3.58). Then the operator $P \in (V', H)_{\text{H.S.}} \subset (H, H)_{\text{H.S.}}$ can be represented as

$$P = \sum_{i, j=1}^{\infty} \alpha_{ij} e_i \otimes e_j, \quad \alpha_{ij} = (P e_i, e_j)$$

where the infinite series converges strongly in $L(H, H)$.

Define the operator P_n by

$$P_n = \sum_{i, j=1}^{n} \alpha_{ij} e_i \otimes e_j.$$

We note that

$$P_n A = \sum_{i, j=1}^{n} \alpha_{ij} A^* e_i \otimes e_j$$

so that

$$[P_n A, P_n] = \sum_{i, j, k=1}^{n} \alpha_{ij} \alpha_{kl} (A^* e_i, e_k)(e_j, e_l) = \sum_{j=1}^{n} \left(A^* \sum_{i=1}^{n} \alpha_{ij} e_i, \sum_{k=1}^{n} \alpha_{kj} e_k \right)$$

and hence

$$[P_n A, P_n] \geqslant \omega \sum_{j=1}^{n} \left\| \sum_{i=1}^{n} \alpha_{ij} e_i \right\|^2.$$

We also note that

$$\| P_n \|_{(V', H)_{\text{H.S.}}} = \sum_{i=1}^{n} \rho_i^2 |P_n e_i|^2 = \sum_{i=1}^{n} \sum_{j=1}^{n} \rho_i^2 |\alpha_{ij}|^2 = \sum_{j=1}^{n} \left\| \sum_{i=1}^{n} \alpha_{ij} e_i \right\|^2$$

because $\{\rho_i e_i\}$ is an orthonormal basis in V' and $((e_i, e_j)) = (Je_i, e_j) = \rho_i^2 \delta_{ij}$. (We have denoted by ((,)) the inner product in V.) Hence

$$[P_n A, P] \geqslant \omega \|P_n\|^2_{(V', H)_{\text{H.S.}}}.$$

Since $P_n \to P$ in $(V', H)_{\text{H.S.}}$ as $n \to \infty$ we can pass to limit to obtain the desired inequality (3.65). Proof (3.66) is similar.

Let $\mathscr{L} \subset \mathscr{H} \times \mathscr{H}$ be the operator defined by

$$\mathscr{L}(P) = A^*P + PA, \quad P \in D(\mathscr{L})$$

where

$$D(\mathscr{L}) = \{P \in \mathscr{V};\ \mathscr{L}(P) \in \mathscr{H}\}.$$

We note that Lemma 3.4 implies that $[\mathscr{L}P, P] \geqslant \alpha \|P\|^2_{\mathscr{V}}$, for all $P \in D(\mathscr{L})$. Define

$$C = \{P \in \mathscr{H};\ P = P^*,\ P \geqslant 0\}$$

and denote $\mathscr{I}$ as the identity operator in $\mathscr{H}$.

Lemma 3.5. *The operator $\mathscr{L}$ is maximal monotone in $\mathscr{H} \times \mathscr{H}$. Moreover*

$$(\mathscr{I} + \lambda \mathscr{L})^{-1} C \subset C \text{ for every } \lambda > 0.$$

Proof. It follows from Lemma 3.3 and Lemma 3.4 that the linear operator $P \to A^*P + PA$ is continuous, monotone and coercive from $\mathscr{V}$ to $\mathscr{V}'$. In particular, this implies (see e.g. Corollary 1.1) that for every $F \in \mathscr{H}$ equation

$$P + \lambda A^*P + \lambda PA = F, \quad \lambda > 0 \tag{3.67}$$

has a solution $P \in D(\mathscr{L})$. Suppose now, that $F \in C$. It is necessary to prove that $P \in C$. Let P^* be the adjoint of P. Since F is self-adjoint, (3.67) implies that

$$P^* + \lambda P^*A + \lambda A^*P = F.$$

Inasmuch as (3.67) has a unique solution, we infer that $P = P^*$. Since P is compact and self-adjoint in H there exists an orthonormal system of eigenvectors $\{\tilde{e}_n\}$ of P, i.e.

$$P\tilde{e}_n = \gamma_n \tilde{e}_n, \quad n = 1, \ldots$$

which is complete in the closure of the range space of P. If we use Eq. (3.67) we then have

$$(P\tilde{e}_n, \tilde{e}_n) + 2\lambda(P\tilde{e}_n, A\tilde{e}_n) = (F\tilde{e}_n, \tilde{e}_n)$$

so that

$$\gamma_n = (1 + 2\lambda(A\tilde{e}_n, \tilde{e}_n))^{-1}F(\tilde{e}_n, \tilde{e}_n) \geqslant 0.$$

Hence $P \geqslant 0$ and the proof of Lemma 3.5 is complete.

Define

$$\mathscr{B}(P) = P^2, \quad \forall\, P \in D(\mathscr{B}) = \mathrm{C}.$$

From Lemma 3.3 it follows that $\mathscr{B}$ is continuous. Moreover, we have

$$\|P^2\|_{\mathscr{H}} \leqslant \|P\|^2_{\mathscr{H}}, \quad \text{for every } P \in \mathscr{H}. \tag{3.68}$$

On the other hand, we have

$$[P_1^2 - P_2^2, P_1 - P_2] = [P_1(P_1 - P_2), P_1 - P_2] +$$

$$+ [(P_1 - P_2)P_2, P_1 - P_2]$$

so that Lemma 3.3 (part (e)) implies that $\mathscr{B}$ is monotone in $\mathscr{H} \times \mathscr{H}$.

LEMMA 3.6. *$C \subset R(\mathscr{I} + \lambda\mathscr{B})$ for every $\lambda > 0$.*

Proof. Let $F \in \mathscr{H}$ be such that $F = F^*$ and $F \geqslant 0$. We must prove that the equation

$$P + \lambda P^2 = F \tag{3.69}$$

has a solution $P \in C$, i.e., $P \in \mathscr{H}$, $P = P^*$ and $P \geqslant 0$. Let $\{f_j\}$ be the orthonormal system of eigenvectors of operator F and let $\{\alpha_j\}$ be the corresponding system of eigenvalues, i.e.,

$$F(f_j) = \alpha_j f_j, \quad j = 1,2, \ldots \tag{3.70}$$

For every j the equation

$$\gamma + \lambda\gamma^2 = \alpha_j \tag{3.71}$$

has a unique positive solution γ_j (because $\alpha_j \geqslant 0$). Define

$$Px = \sum_{j=1}^{\infty} \gamma_j(x, f_j)f_j, \quad \forall\, x \in H. \tag{3.72}$$

It is easy to see that the operator P defined by (3.72) is Hilbert-Schmidt in H and that it belongs to C. Moreover, (3.70) and (3.71) clearly imply that P satisfies (3.69). The proof of Lemma 3.6 is complete.

We can now turn to the proof of Theorem 3.9. Consider the approximated equations

$$\mathscr{L}(P_\lambda) + \mathscr{B}_\lambda(P_\lambda) = F, \; F \in C, \; \lambda > 0, \tag{3.73}$$

where $\mathscr{B}_\lambda = \lambda^{-1}(\mathscr{I} - (\mathscr{I} + \lambda\mathscr{B}))^{-1}$. In virtue of Lemma 3.5, this equation can be written under the following equivalent form

$$P_\lambda = \frac{1}{\lambda}\left(\mathscr{I} + \frac{1}{\lambda}\mathscr{L}\right)^{-1} F + \left(\mathscr{I} + \frac{1}{\lambda}\mathscr{L}\right)^{-1} (\mathscr{I} + \lambda\mathscr{B})^{-1} P_\lambda. \tag{3.74}$$

By Lemma 3.5 and Lemma 3.6 the operator defined by the right-hand side of Eq. (3.74) maps the closed convex cone C into itself. Moreover, this operator is Lipschitz with constant $1/1 + \alpha$ because

$$[\mathscr{L}P, P] \geqslant \alpha\|P\|^2_{\mathscr{V}} \geqslant \alpha\|P\|^2_{\mathscr{H}}. \tag{3.75}$$

Here we have used Lemma 3.4. We may therefore apply Banach's fixed-point theorem to conclude that Eq. 3.74 (equivalently (3.73)) has a unique solution $P_\lambda \in D(\mathscr{L}) \cap C$.

Now, multiplying Eq. (3.73) by P_λ and remembering (3.75), and the monotonicity of $\mathscr{B}$, we get

$$\alpha\|P_\lambda\|_{\mathscr{H}} \leqslant \|F\|_{\mathscr{H}}.$$

It follows, in particular, that $\|(\mathscr{I}+\lambda\mathscr{B})^{-1}P_\lambda\|_{\mathscr{H}}$ are bounded, such that relation (3.68) implies that

$$\|\mathscr{B}_\lambda(P_\lambda)\|_{\mathscr{H}} \leqslant M \text{ for all } \lambda > 0$$

because $\mathscr{B}_\lambda(P_\lambda) = \mathscr{B}(\mathscr{I} + \lambda\mathscr{B})^{-1}(P_\lambda)$. Thus by repeating the argument used in the proof of Theorem 3.8 (see also Lemma 3.2) one finds that P_λ converges in $\mathscr{H}$ as $\lambda \to 0$ to a solution P of Eq. (3.72). The uniqueness of P follows immediately by (3.75) and the monotonicity of $\mathscr{B}$.

Remark. Although Theorem 3.9 is rather general, it is not very convenient for application since it requires operator F to be compact which is not the case in many problems arising in the control theory.

To illustrate Theorem 3.9 we shall consider the following simple example: $H = L^2(\Omega)$, $V = H^1_0(\Omega)$, $V' = H^{-1}(\Omega)$ and $A = -\Delta$. It is well known that $(L^2(\Omega), L^2(\Omega))_{\text{H.S.}} = L^2(\Omega_x \times \Omega_\xi)$. Thus any $P \in (L^2(\Omega), L^2(\Omega))_{\text{H.S.}}$ can be represented as

$$(Pu)(x) = \int_\Omega P(x, \xi)\, u(\xi) \mathrm{d}\xi, \text{ a.e. on } \Omega, \; u \equiv L^2(\Omega)$$

where the kernel $P(x, \xi)$ belongs to $L^2(\Omega_x \times \Omega_\xi)$. Moreover, as is easily seen, we have

$$(H^{-1}(\Omega), L^2(\Omega))_{\text{H.S.}} \cap \{(L^2(\Omega), H_0^1(\Omega))_{\text{H.S.}}; P = P^*\} = H_0^1(\Omega_x \times \Omega_\xi).$$

Therefore, Theorem 3.9 yields the following result:

Given $F(x, \xi) \in L^2(\Omega_x \times \Omega_\xi)$ *such that*

$$F(x, \xi) = F(\xi, x), F(x, \xi) \geqslant 0, \quad \text{a.e. } on\ \Omega_x \times \Omega_\xi$$

there exists a unique $P(x, \xi) \in L^2(\Omega_x \times \Omega_\xi)$ *satisfying*

$$-\Delta_x P(x, \xi) - \Delta_\xi P(x, \xi) + \int_\Omega P(x, \zeta)P(\zeta, \xi)\mathrm{d}\zeta = F(x, \xi) \quad on\ \Omega_x \times \Omega_\xi$$

$$P(x, \xi) = 0 \qquad on\ \Gamma_x \times \Gamma_\xi.$$

$$P(x, \xi) = P(\xi, x); P(x, \xi) \geqslant 0 \qquad on\ \Omega_x \times \Omega_\xi.$$

Bibliographical notes

§ 1. The theory of nonlinear monotone operators has been the subject of many studies made in the last few years. The pioneer work of M.M. Vainberg [1], E.M. Zarantonello [1] and R.I. Kachurovski [1] was followed by the outstanding paper by G. J. Minty [1] which demonstrated clearly that monotonicity was a strong tool in the existence theory of nonlinear functional equations. In a long series of works, F.E. Browder has contributed remarkably to the rapid subsequent development of the monotonicity theory and stimulated much further work and interest in the subject.

Theorem 1.1 was first proved in the finite-dimensional case by H. Debrunner and Florr [1] and later extended to the infinite-dimensional case by G.J. Minty [2], [3] and H. Brézis [2]. Theorems 1.2 and 1.3 were originally given by G.J. Minty [1] in the case of Hilbert space and after various improvements they were extended by F. Browder [1], [2] to the general case.

Theorems 1.4 and 1.5 are due to R. T. Rockafellar [3], [4] (extending the previous results of F.E. Browder [3], [6]). The simple proofs that we have given here are due to P.M Fitzpatrick [1] and H. Brézis, M. Crandall and A. Pazy [1].

In this paragraph we left aside the maximal monotone operators defined in nonreflexive Banach spaces (see I.P. Gossez [1]), pseudo-monotone operators (see H. Brézis [1]) as well as the applications to the topological degree for noncompact mappings (see e.g. F. Browder [7], and S. Fucik, J. Nečas, J. Souček and V. Souček [1]).

For details and other significant results of the theory we refer the reader to the 1966 lecture notes of F.E.Browder [3], the surveys of R.I. Kachurovski [2] and F.E.Browder [7] as well as to the books by Gh. Marinescu [1] and D. Pascali [1].

§2. The results of the first two sections are essentially due to J.J. Moreau [1], [2] and R. T. Rockafellar [1], [2], [6] (see also [5] for a complete bibliography on the subject). Theorem 2.2 is an easy extension of the earlier results of J.J. Moreau [1] and H. Brézis [7]. The results of the third section can be found in the memoires of H. Brézis [6], [7]. We have had to omit a number of results concerning the applications of convex analysis to the domain of elliptic variational inequalities, for which we refer the reader to the survey of G. Stamppachia [1], U. Mosco [1] and to the monograph of J.L. Lions [1], as well as the applications in the convex optimization (see R. T. Rockafellar [1], [5]).

§3. The problems developed in this paragraph represent a synthesis of the results of T. Kato [1], [2], M. G. Crandall and A. Pazy [1], [2] and H. Brézis [4], obtained in connection with semigroups of nonlinear contractions. Note that the last part of Theorem 3.1 is due to G.J. Minty [1]. Theorems 3.3 and 3.4 were first established by M.G. Crandall and A. Pazy [1], [2] for uniformly convex Banach spaces and were extended by H. Brézis [4] to the general case that we have presented here. Theorem 3.5 is due to M.G. Crandall and A. Pazy [1]. It was extended by T. Kato [2]. Theorem 3.7 together with interesting applications in the partial differential equations can be found in Brézis' paper [7]. Theorem 3.8 was established in author's paper [1]. Proposition 3.7 was obtained by Y. Konishi [1] (see also H. Brézis, M.G. Crandall and A. Pazy [1] for similar results).

Other perturbation results for nonlinear m-dissipative operators were established by G. F. Webb [2] and by B. Calvert and K. Gustafson [1] under conditions of relative boundedness. The work of H.M. Riedl and G. F. Webb [1] contains numerous interesting results in connection with the stability under perturbations of closedness, resolvent and spectrum of nonlinear operators. Along these lines we must also mention the paper by J. Neuberger [1].

Chapter III

Differential equations in Banach spaces

We shall be concerned in this chapter with some classes of nonlinear differential equations associated with dissipative operators – placing special emphasis on existence results for the Cauchy problem. The theory of semigroups of nonlinear contractions, as well as its extensions and applications, constitute the core of the present chapter.

The main feature that distinguishes the subsequent results from the classical ones is the fact that one makes unusually weak assumptions concerning the nonlinear operators involved in our considerations. No compactness or Lipschitz condition is assumed whatsoever, although, of course these properties could be exploited in analysing these conditions in special cases.

The "monotonicity" method combined with some ideas and techniques for the linear theory of semigroups is, of course, the tool which makes it possible to avoid these strong assumptions.

§ 1. Semigroups of nonlinear contractions in Banach spaces

1.1. General properties of nonlinear semigroups

Let X be a real Banach space with its norm denoted $\|\cdot\|$ and let X^* be its dual space.

DEFINITION 1.1. Let C be a closed subset of X. A *semigroup of contractions* on C is a function S with domain $[0, \infty[\times C$ and range in C satisfying the following conditions:

a) $S(t+s)x = S(t)S(s)x, \quad \forall x \in C,\ t, s \geqslant 0,$

b) $S(0)x = x, \quad \forall x \in C,$

c) For every $x \in C$, $S(t)x$ is continuous in $t \geqslant 0$.

d) $\|S(t)x - S(t)y\| \leqslant \|x - y\|, \quad \forall t > 0, \quad \forall x, y \in C.$

Let S be a semigroup of contractions on C. For every $h > 0$ let

$$A_h x = (S(h)x - x)/h, \qquad x \in C. \tag{1.1}$$

Let $D(A_s)$ (respectively $D(A_w)$) be the set of all $x \in C$ for which $\lim_{h\to 0} A_h x$ (respectively $w\text{-}\lim_{h\to 0} A_h x$) exists. Define

$$A_s x = \lim_{h\to 0} A_h x, \qquad x \in D(A_s), \tag{1.2}$$

respectively

$$A_w x = w\text{ - }\lim_{h\to 0} A_h x, \qquad x \in D(A_w). \tag{1.3}$$

The operator A_s (respectively A_w) is called *the strong generator* (respectively *weak generator*) of S. Obviously, $D(A_s) \subset D(A_w)$ and $A_s x = A_w x$ for every $x \in D(A_s)$.

If S is a semigroup on C, we define

$$D_s = \{x \in C;\ \liminf_{t\to 0} \|S(t)x - x\|/t < \infty\} \tag{1.4}$$

and prove the following lemma.

LEMMA 1.1. *Let S be a semigroup of nonlinear contractions on C and let $x \in D_S$. Then $S(t)x$ is Lipschitz on every interval* $[0, T]$, $T > 0$.

Proof. Let $\{t_k\}$ be a monotonically decreasing sequence of positive numbers tending to zero and let $L > 0$ be such that

$$\|S(t_k)x - x\| \leqslant Lt_k.$$

Let $h > 0$ and let n_k be a nonnegative integer such that $0 \leqslant h - n_k t_k < t_k$. Then by (a) and (d) we have

$$\|S(t+h)x - S(t)x\| \leqslant \|S(h)x - x\| = \|S(h - n_k t_k + n_k t_k)x - x\| \leqslant$$

$$\leqslant \|S(h - n_k t_k)x - x\| + Lt_k n_k$$

for every $t > 0$. Consequently

$$\|S(t+h)x - S(t)x\| \leqslant \|S(h - n_k t_k)x - x\| + Lh.$$

But by the strong continuity of $S(t)x$, $\lim_{k\to\infty} \|S(h - n_k t_k)x - x\| = 0$, and the result follows.

In the next theorem we shall collect some elementary properties of the generators A_s and A_w defined above.

THEOREM 1.1. *Let S be a semigroup of nonlinear contractions defined on a closed subset C of X. Then*

(i) $(A_w x - A_w y, f) \leqslant 0$, *for all* $x, y \in D(A_w)$ *and every* $f \in F(x - y)$ (i.e. A_w *and* A_s *are dissipative operators in* X).

(ii) *If X is reflexive, then* $\overline{D(A_s)} = \overline{D(A_w)} = D_S$.

(iii) *If X is reflexive and strictly convex, then* $D(A_w) = D_S$. *In addition, if X is uniformly convex, then* $D(A_s) = D(A_w) = D_S$ *and* $A_w = A_s$.

We start with the following lemma:

LEMMA 1.2. *Let $u(t)$ be an X-valued function on an interval of real numbers. Suppose that $u(t)$ has a weak derivative $u'(s) \in X$ at $t = s$* $\left(\text{i.e. } \frac{\mathrm{d}}{\mathrm{d}t}(u(t), x^*)|_{t=s} = (u'(s), x^*), \text{ for all } x^* \in X^*\right)$. *Assume also that $t \to \|u(t)\|$ is differentiable at $t = s$. Then*

$$\|u(s)\| \frac{\mathrm{d}}{\mathrm{d}s} \|u(s)\| = (u'(s), f), \text{ for every } f \in F(u(s)). \tag{1.5}$$

Proof. Let $h > 0$. We have

$$(u(s+h) - u(s), f) \leqslant (\|u(s+h)\| - \|u(s)\|) \|f\|, \qquad \forall\, f \in F(u(s)).$$

As $u(t)$ is weakly differentiable at $t = s$, this yields

$$(u'(s), f) \leqslant \|u(s)\| \frac{\mathrm{d}}{\mathrm{d}s} \|u(s)\|.$$

The converse inequality follows much similarly.

Proof of Theorem 1.1. (i) We have

$$(A_h x - A_h y, f) = h^{-1}((S(h)x - S(h)y, f) - \|x - y\|^2) \leqslant$$

$$\leqslant h^{-1}(\|S(h)x - S(h)y\| \; \|x - y\| - \|x - y\|^2) \leqslant 0$$

for all $h > 0$ and $f \in F(x - y)$. Letting h tend to zero we obtain the desired result.

(ii) Obviously, $D(A_s) \subset \overline{D(A_w)} \subset D_S$. Let x be any fixed element of D_S. We shall show that $x \in \overline{D(A_s)}$. From Lemma 1.1 it follows that the function $t \to S(t)x$ is Lipschitz on $[0, \infty[$. Since X is a reflexive Banach space this implies that $S(t)x$ is a.e. differentiable on $]0, \infty[$ (see Theorem. 2.1, Chapter I). But $\frac{\mathrm{d}}{\mathrm{d}t} S(t)x = A_s S(t)x$ whenever $\frac{\mathrm{d}}{\mathrm{d}t} S(t)x$ exists. Hence $S(t)x \in D(A_s)$ a.e. $t > 0$, and the strong continuity of $S(t)x$ then implies that $x \in \overline{D(A_s)}$ (the strong closure of $D(A_s)$ in X).

(iii) Now, assume that X is reflexive and strictly convex. Let $x_0 \in D_S$ and let Y be the set of all weak cluster points of $\{(S(t)x_0 - x_0)/t, t \to 0\}$. Let A_1 be a subset of $X \times X$ defined by

$$A_1 = A_s \cup [x_0, \overline{\text{conv}\ Y}], \ D(A_1) = D(A_s) \cup \{x_0\}$$

where $\overline{\text{conv}}$ denotes the closure of the convex hull. We note that

$$(A_s x - A_s y, f) \leqslant 0, \qquad \forall f \in F(x - y), \ x, y \in D(A_s)$$

and for every $u \in Y$

$$(A_s x - u, f) \leqslant 0, \qquad \forall f \in F(x - x_0), \ x \in D(A_s).$$

This implies that A_1 is a dissipative subset of $X \times X$. On the other hand, X being reflexive, the function $t \to S(t)x_0$ is a.e. differentiable on $]0, \infty[$ and

$$\frac{\mathrm{d}}{\mathrm{d}t} S(t)x_0 = A_s S(t)x_0 \in A_1 S(t)x_0, \qquad \text{a.e.} \quad t > 0.$$

Next, by the dissipativeness of A_1, it follows that

$$\left(\frac{\mathrm{d}}{\mathrm{d}t}(S(t)x_0 - x_0), f\right) \leqslant (y, f), \qquad \text{a.e. } t > 0, \qquad \forall y \in A_1 x_0,$$

where $f \in F(S(t)x_0 - x_0)$. Thus from Lemma 1.2 it follows that

$$\|S(t)x_0 - x_0\| \frac{\mathrm{d}}{\mathrm{d}t} \|S(t)x_0 - x_0\| \leqslant \|y\| \ \|S(t)x_0 - x_0\|, \quad \text{a.e.} \quad t > 0$$

and therefore

$$\|S(t)x_0 - x_0\| \leqslant t|A_1 x_0|, \qquad \forall\ t > 0, \tag{1.6}$$

where $|A_1 x_0| = \inf\{\|y\|; y \in A_1 x_0\}$. But $A_1 x_0 = \overline{\text{conv}\ Y}$ is a closed convex subset of X which is reflexive and strictly convex. Therefore, there exists a unique element $y_0 \in A_1 x_0$ such that $\|y_0\| = |A_1 x_0|$. Then inequality (1.6) implies that $\overline{\text{conv}\ Y} = y_0 = A_w x_0$, i.e., $x_0 \in D(A_w)$. Thus we have shown that $D_S = D(A_w)$. Next assume that X is uniformly convex and let $x_0 \in D(A_w) = D_S$. Then

$$\frac{S(t)x_0 - x_0}{t} \rightharpoonup y_0 \qquad \text{as } t \to 0,$$

and by (1.6)

$$\left\|\frac{S(t)x_0 - x_0}{t}\right\| \leqslant |A_1 x_0| = \|y_0\|, \qquad \forall\ t > 0.$$

Since X is uniformly convex, we deduce that

$$\lim_{t\to 0} \frac{S(t)x_0 - x_0}{t} = y_0.$$

Hence $x_0 \in D(A_s)$ which concludes the proof.

Remark. The previous theorem shows that if X is uniformly convex then the strong generator A_s of the semigroup of nonlinear contractions S coincides with the weak generator A_w. In this case the operator $A_s = A_w$ is denoted by A_0 and is called *the generator* of S.

THEOREM 1.2. *Let X and X^* be uniformly convex Banach spaces. Let S be a semigroup of nonlinear contractions defined on a closed subset C of X and let A_0 be the generator of S. If x is any fixed element of $D(A_0)$, then*

(i) *$S(t)x \in D(A_0)$ for all $t \geqslant 0$ and the function $t \to A_0S(t)x$ is continuous from the right on* $[0, \infty[$.

(ii) *$S(t)x$ has a right derivative $\frac{\mathrm{d}}{\mathrm{d}t}S(t)x$ at every $t \geqslant 0$ and*

$$\frac{\mathrm{d}^+}{\mathrm{d}t} S(t)x = A_0S(t)x, \qquad \textit{for all} \quad t \geqslant 0.$$

(iii) $\frac{\mathrm{d}}{\mathrm{d}t} S(t)x = A_0S(t)x$ *exists and is continuous except at a countable number of values $t \geqslant 0$.*

Proof. Let x be any fixed element of $D(A_0)$. By Theorem 1.1, $D(A_0) = D_S$. Hence $S(t)x \in D(A_0)$ for all $t \geqslant 0$ and

$$\frac{\mathrm{d}^+}{\mathrm{d}t} S(t)x = A_0S(t)x \quad \text{for every} \quad t \geqslant 0. \tag{1.7}$$

Moreover, $t \to S(t)x$ is a.e. differentiable on $]0, \infty[$ and

$$\frac{\mathrm{d}}{\mathrm{d}t} S(t)x = A_0S(t)x, \qquad \text{a.e. } t > 0. \tag{1.8}$$

We shall prove that $A_0S(t)x$ is right continuous in t on $[0, \infty[$. Let $h > 0$. Then we have

$$\frac{\mathrm{d}}{\mathrm{d}t}(S(t+h)x - S(t)x) = AS(t+h)x - AS(t)x, \quad \text{a.e.} \quad t > 0.$$

Next, by Lemma 1.2 we get

$$\frac{\mathrm{d}}{\mathrm{d}t}\|S(t+h)x - S(t)x\| \leqslant 0, \quad \text{a.e.} \quad t > 0; \quad h > 0$$

because A_0 is dissipative. Integrating over $[s, T]$ this yields

$$\|S(t+h)x - S(t)x\| \leqslant \|S(s+h)x - S(s)x\|, \quad 0 \leqslant s \leqslant t,$$

and therefore

$$\left\| \frac{d^+}{dt} S(t)x \right\| \leqslant \left\| \frac{d^+}{ds} S(s)x \right\|, \qquad 0 \leqslant s \leqslant t.$$

Comparing this with formula (1.7) we conclude that $\|A_0 S(t)x\|$ is a monotone non-increasing function of t. This shows that $\|A_0 S(t)x\|$ is right continuous in t on $[0, \infty[$. Let $t_0 \geqslant 0$ and let $\{t_k\}$ be a decreasing sequence of positive numbers $t_k \to t_0$. Extracting a subsequence, if necessary, we may assume that $w\text{-}\lim_{k\to\infty} A_0 S(t_k)x = y_0$. The right continuity of $\|A_0 S(t)x\|$, at $t = t_0$, then implies that

$$\|y_0\| \leqslant \|A_0 S(t_0)x\|. \tag{1.9}$$

Let $\tilde{A}$ be a maximal dissipative extension of A_0 (such an $\tilde{A}$ exists in virtue of Zorn's lemma). Since $\tilde{A}$ is a demiclosed subset of $X \times X$ (see Proposition 3.5, Chapter II) we find that $y_0 \in \tilde{A}S(t_0)x$. On the other hand, for any $x \in D(A_0)$ and $u \in \tilde{A}x$, we have

$$\left(\frac{d}{dt}(S(t)x - x),\ F(S(t)x - x)\right) \leqslant (u, F(S(t)x - x)), \text{ a.e.} \quad t > 0$$

because $\tilde{A}$ is dissipative and $\dfrac{d}{dt} S(t)x = A_0 S(t)x \in \tilde{A}S(t)x$ a.e. $t > 0$. By Lemma 1.2 this yields

$$\|S(t)x - x\|/t \leqslant |\tilde{A}x| = \|\tilde{A}^0 x\|, \quad \text{for every} \quad t > 0.$$

Hence $A_0 x = \tilde{A}^0 x$. (We recall that $\tilde{A}^0$ denotes the minimal section of $\tilde{A}$.) Then from (1.9) it follows that $y_0 = A_0 S(t_0)x$ and

$$\lim_{k\to\infty} A_0 S(t_k)x = y_0,$$

because X is uniformly convex (see Proposition 1.4, Chapter I). Thus, we have proved the right continuity of $A_0 S(t)x$ at $t = t_0$. In order to prove (iii) we proceed as follows. Integrating (1.8) from t to $t + h$, we obtain

$$S(t+h)x - S(t)x = \int_t^{t+h} A_0 S(\tau)x \, d\tau, \qquad t, h \geqslant 0.$$

This equality shows that it suffices to prove that the function $t \to A_0 S(t)x$ is continuous except at a countable number of $t > 0$. Again using the previous argument one easily deduces that every continuity point t of $\|A_0S(t)x\|$ is also a continuity point of $t \to A_0S(t)x$. But since $\|A_0S(t)x\|$ is monotonic in t, it has at most countably many discontinuities. This completes the proof.

Remarks 1°. Incidentally, we have also proved that the generator A_0 of S can be expressed as $A_0x = A^0x$ for every $x \in D(A_0)$, where A is a maximal dissipative subset of $X \times X$.

2°. A function $S(t) : [0, \infty[\times C \to C$ satisfying (a), (b), (c) and

$$\|S(t)x - S(t)y\| \leqslant e^{\omega t} \|x - y\|, \tag{1.10}$$

for every $t \geqslant 0$ and $x, y \in C$ is called *semigroup of type* ω. In particular, a semigroup of type 0 is a semigroup of contractions. The extension of the previous results to the general case ω is straightforward. Note only that generator A (weak or strong) of a semigroup of type ω is ω-dissipative, i.e., $A - \omega I$ is dissipative.

1.2. The exponential formula

The main result, which shall be proved in this section, is the following.

THEOREM 1.3. *Let X be a real Banach space and A be a dissipative subset of $X \times X$ such that*

$$\overline{D(A)} \subset R(1 - \lambda A), \tag{1.11}$$

for all sufficiently small positive λ. Then

$$S(t)x = \lim_{n\to\infty} \left(1 - \frac{t}{n} A\right)^{-n} x \tag{1.12}$$

exists for $x \in \overline{D(A)}$, uniformly in t on every compact interval of $[0, \infty[$. *Moreover, $S(t)$ defined by formula* (1.12) *is a semigroup of contractions on $\overline{D(A)}$.*

The proof of this theorem will be divided into several steps. We set

$$J_\lambda = (1 - \lambda A)^{-1}, \quad \lambda > 0$$

and denote by D_λ the domain of J_λ. We note that by condition (1.11) $D_\lambda \supset \overline{D(A)}$ for all sufficiently small λ.

LEMMA 1.3. *Let $\lambda > 0$. Then the following statements hold*

(i) *J_λ is single-valued, and for $x, y \in D_\lambda$*

$$\|J_\lambda x - J_\lambda y\| \leqslant \|x - y\|$$

$$\|J_\lambda x - x\| \leqslant \lambda|Ax|, \quad \forall\ x \in D_\lambda \cap D(A).$$

(ii) $\|J_\lambda^n x - x\| \leqslant n \, \| J_\lambda x - x\|, \qquad \forall \, x \in D_\lambda, \; n = 1, 2, \ldots$

(iii) *If* $x \in D_\lambda$ *and* $\mu > 0$, *then* $\dfrac{\mu}{\lambda} x + \dfrac{\lambda - \mu}{\lambda} J_\lambda x \in D_\mu$ *and*

$$J_\lambda x = J_\mu \left(\frac{\mu}{\lambda} x + \frac{\lambda - \mu}{\lambda} J_\lambda x \right). \tag{1.13}$$

Proof. From Proposition 3.2 (Chapter II) we obtain (i) and (ii). To prove (iii), we note that (i) implies that

$$\|J_\lambda^n x - x\| = \left\| \sum_{i=0}^{n-1} (J_\lambda^{n-i} x - J_\lambda^{n-(i+1)} x) \right\| \leqslant \sum_{i=0}^{n-1} \|J_\lambda x - x\| =$$

$$= n \|J_\lambda x - x\|.$$

Now we shall prove (1.13) which is the nonlinear version of the resolvent formula. If $x \in D_\lambda$, then there is $[x_0, y_0] \in A$ such that $x_0 - \lambda y_0 = x$. We can write

$$\frac{\mu}{\lambda} x + \frac{\lambda - \mu}{\lambda} J_\lambda x = \frac{\mu}{\lambda} (x_0 - \lambda y_0) + \frac{\lambda - \mu}{\lambda} x_0.$$

and therefore

$$\frac{\mu}{\lambda} x + \frac{\lambda - \mu}{\lambda} J_\lambda x \in (1 - \mu A) \, x_0 .$$

This means that

$$J_\mu \left(\frac{\mu}{\lambda} x + \frac{\lambda - \mu}{\lambda} J_\lambda x \right) = x_0 = J_\lambda x$$

as claimed.

LEMMA 1.4. *Let* $\lambda \geqslant \mu > 0$ *and* $x \in D(J_\lambda^m) \cap D(J_\mu^n)$ *where* m *and* n *are positive integers satisfying* $n \geqslant m$. *Then*

$$\|J_\mu^n x - J_\lambda^m x\| \leqslant \sum_{j=0}^{m} \alpha^j \beta^{n-j} \binom{n}{j} \|J_\lambda^{m-j} x - x\| +$$

$$+ \sum_{j=m}^{n} \alpha^m \beta^{j-m} \binom{j-1}{m-1} \|J_\mu^{n-j} x - x\| \tag{1.14}$$

where $\alpha = \mu/\lambda$ *and* $\beta = (\lambda - \mu)/\lambda$.

Proof. For the positive integers j and k satisfying $0 \leqslant j \leqslant n$ and $0 \leqslant k \leqslant m$, set

$$a_{k,j} = \|J_\mu^j x - J_\lambda^k x\|.$$

Using the properties of J_λ listed in Lemma 1.3 we obtain

$$a_{k,j} = \left\| J_\mu^j x - J_\mu \left(\frac{\mu}{\lambda} J_\lambda^{k-1} x + \frac{\lambda - \mu}{\lambda} J_\lambda^k x \right) \right\| \leqslant$$

$$\leqslant \frac{\mu}{\lambda} \|J_\mu^{j-1} x - J_\lambda^{k-1} x\| + \frac{\lambda - \mu}{\lambda} \|J_\mu^{j-1} x - J_\lambda^k x\| \leqslant \alpha a_{k-1,j-1} + \beta a_{k,j-1}.$$

By solving the inequalities

$$a_{k,j} \leqslant \alpha a_{k-1,j-1} + \beta a_{k,j-1}$$

one obtains the desired estimate.

We also need the following combinatorial lemma.

LEMMA 1.5. *Let $n \geqslant m > 0$ be integers, and α, β be positive numbers satisfying $\alpha + \beta = 1$. Then*

$$\sum_{j=0}^{m} \binom{n}{j} \alpha^j \beta^{n-j} (m - j) \leqslant ((n\alpha - m)^2 + n\alpha\beta)^{1/2}, \tag{1.15}$$

$$\sum_{m=j}^{n} \binom{j-1}{m-1} \alpha^m \beta^{j-m} (n - j) \leqslant \left(\frac{m\beta}{\alpha^2} + \left(\frac{m\beta}{\alpha} + m - n \right)^2 \right)^{1/2}. \tag{1.16}$$

Proof. In order to prove (i), note first that

$$\sum_{j=0}^{m} \binom{n}{j} \alpha^j \beta^{n-j} (m - j) \leqslant \sum_{j=0}^{n} \binom{n}{j} \alpha^j \beta^{n-j} (m - j) \leqslant$$

$$\leqslant \left(\sum_{j=0}^{n} \binom{n}{j} \alpha^j \beta^{n-j} \right)^{1/2} \left(\sum_{j=0}^{m} \binom{n}{j} \alpha^j \beta^{n-j} (m - j)^2 \right)^{1/2}.$$

Combining this with the following relations

$$\sum_{j=0}^{n} \binom{n}{j} \alpha^j \beta^{n-j} = 2^n$$

$$\sum_{j=0}^{n} j \binom{n}{j} \alpha^j \beta^{n-j} = 2^{n-1} n\alpha$$

$$\sum_{j=0}^{n} j^2 \binom{n}{j} \alpha^j \beta^{n-j} = 2^{n-2} n\alpha((n - 1)\alpha + 2)$$

one obtains the desired estimate. To obtain (ii) one starts from the inequality

$$\sum_{j=m}^{n}\binom{j-1}{m-1}\alpha^m\beta^{j-m}(n-j) \leqslant \sum_{j=m}^{\infty}\binom{j-1}{m-1}\alpha^m\beta^{j-m}|n-j| \leqslant$$
$$\leqslant \left(\sum_{j=m}^{\infty}\binom{j-1}{m-1}\alpha^m\beta^{j-m}\right)^{1/2}\left(\sum_{j=m}^{\infty}\binom{j-1}{m-1}\alpha^m\beta^{j-m}(n-j)^2\right)^{1/2}$$

and one uses the equality

$$\sum_{j=m}^{\infty}\binom{j-1}{m-1}\beta^{j-m} = \frac{1}{(1-\beta)^m}, \qquad |\beta| < 1$$

to evaluate the expression from the right-hand side.

Proof of Theorem 1.3. Let $x \in D(A)$ and let $\lambda \geqslant \mu > 0$. Let n and m be positive integers such that $n \geqslant m$. By assumption, $x \in D(J_\lambda^m) \cap D(J_\mu^n)$. Then we combine (1.14), (1.15) and (1.16) to obtain

$$\|J_\mu^n x - J_\lambda^m x\| \leqslant \{[n\mu - m\lambda)^2 + n\mu(\lambda-\mu)]^{1/2} +$$
$$+ [m\lambda(\lambda-\mu) + (m\lambda - n\mu)^2]^{1/2}\}\, |Ax|. \tag{1.17}$$

Taking $\mu = \dfrac{t}{n}$ and $\lambda = \dfrac{t}{m}$ in (1.17), we obtain

$$\|J_{\frac{t}{n}}^n x - J_{\frac{t}{m}}^m x\| \leqslant 2t\left(\frac{1}{m} - \frac{1}{n}\right)^{1/2} |Ax|. \tag{1.18}$$

Therefore, $\lim\limits_{n\to\infty} J_{\frac{t}{n}}^n x = S(t)x$ exists uniformly in t on every compact subset of $[0, \infty[$. Since $J_{\frac{t}{n}}^n$ is nonexpansive on $D(J_{\frac{t}{n}}^n)$, we find that

$$\|S(t)x - S(t)y\| \leqslant \|x-y\| \text{ for every } t \geqslant 0, \text{ and } x, y \in D(A).$$

This implies that $S(t)x = \lim\limits_{n\to\infty} J_{\frac{t}{n}}^n x$ exists for $x \in \overline{D(A)}$ and $S(t)$ is nonexpansive on $\overline{D(A)}$. Moreover, for every $x \in D(A)$, $S(t)x$ is Lipschitz continuous in t on $[0, \infty[$. Indeed, taking the limit in inequality (1.17) with $n = m$, $\mu = \dfrac{t}{n}$ and $\lambda = \dfrac{s}{n}$, where $0 \leqslant t \leqslant s$, we obtain

$$\|S(t)x - S(s)x\| \leqslant 2|t-s|\; |Ax|, \quad x \in D(A). \tag{1.19}$$

In particular, this shows that $S(t)x$ is continuous in t for every $x \in D(A)$. In order to complete the proof, we shall verify the semigroup property $S(t+s) = S(t)S(s)$. We note that the strong convergence, $S(t) = \lim_{n\to\infty} J^n_{\frac{t}{n}}$ together with (1.18), implies that

$$[S(t)]^m x = \lim_{n\to\infty} [J^n_{\frac{t}{n}}]^m x = \lim_{n\to\infty} [J^m_{\frac{t}{n}}]^n x.$$

Hence

$$S(mt)x = \lim_{n\to\infty} J^n_{\frac{mt}{n}} x = \lim_{k\to\infty} J^{mk}_{\frac{mt}{mk}} x = \lim_{k\to\infty} [J^m_{\frac{t}{k}}]^k x = [S(t)]^m x.$$

Then by using a standard argument one proves that $S(t+s) = S(t)S(s)$ if t, s are rational. Since $S(t)x$ is continuous in t and Lipschitz continuous in x this implies $S(t+s) = S(t)S(s)$ for t, s real. This completes the proof.

Remark. The previous proof can be easily modified to obtain the exponential formula (1.12) under the following weaker assumption on A: There is a real number ω such that $A - \omega I$ is dissipative and $\overline{D(A)} \subset R(1 - \lambda A)$ for all sufficiently small positive λ. In this case the semigroup $S(t)$ defined by the exponential formula (1.12) is of type ω.

1.3. Convergence theorems

We start with an existence result for the solutions of the differential equations associated with Lipschitz operators.

PROPOSITION 1.1. *Let C be a closed convex subset of a Banach space X and let T be a nonexpansive operator which maps C into itself. Then for every $x \in C$ the initial value problem*

$$\begin{aligned} \frac{du}{dt} &= Tu - u, \quad 0 \leqslant t < \infty \\ u(0) &= x \end{aligned} \tag{1.20}$$

has a unique solution $u \in C^1(0, \infty; X)$ such that $u(t) \in C$ for all $t \geqslant 0$. Moreover, the following inequality holds

$$\|u(n) - T^n x\| \leqslant \sqrt{n}\, \|x - Tx\|, \quad n = 1, 2, \ldots \tag{1.21}$$

Proof. The initial value problem (1.20) is equivalent to

$$u(t) = e^{-t}x + \int_0^t e^{s-t} Tu(s)\, ds, \quad 0 \leqslant t < \infty. \tag{1.22}$$

Let T_0 be any fixed positive number. Equation (1.22) can be solved by the Banach fixed-point theorem noting that the mapping

$$\Phi(u)(t) = \mathrm{e}^{-t}x + \int_0^t \mathrm{e}^{s-t}T(u(s))\,\mathrm{d}s$$

maps the closed convex set

$$\{u \in C(0, T_0; X);\ u(t) \in C,\ 0 \leqslant t \leqslant T_0\}$$

into itself and Φ is Lipschitz with constant $(1 - \mathrm{e}^{-T_0})$.

Since $T - 1$ is dissipative, it follows from (1.20) that

$$\frac{\mathrm{d}}{\mathrm{d}t}\,\|u(t) - x\| \leqslant \|Tx - x\|,\ \text{a.e. } t > 0$$

and hence

$$\|u(t) - x\| \leqslant t\|x - Tx\|, \quad 0 \leqslant t < \infty. \tag{1.23}$$

We notice that

$$u(t) - T^n(x) = \mathrm{e}^{-t}(x - T^n x) + \int_0^t \mathrm{e}^{s-t}(Tu(s) - T^n x)\,\mathrm{d}s, \quad 0 \leqslant t < \infty$$

and

$$\|x - T^n x\| \leqslant \sum_{k=1}^{n} \|T^{k-1}x - T^k x\| \leqslant n\|x - Tx\|.$$

Therefore

$$\|u(t) - T^n x\| \leqslant n\mathrm{e}^{-t}\|x - Tx\| + \int_0^t \mathrm{e}^{s-t}\|u(s) - T^{n-1}x\|\,\mathrm{d}s. \tag{1.24}$$

Then estimate (1.21) follows from the following simple lemma.

LEMMA 1.6. *Let $\varphi_n(t)$ be a sequence of locally integrable functions on* $]0, \infty[$ *satisfying*

$$\varphi_n(t) \leqslant n\mathrm{e}^{-t} + \int_0^t \mathrm{e}^{s-t}\varphi_{n-1}(s)\,\mathrm{d}s, \quad n = 1, 2, \ldots;\ \varphi_0(t) \leqslant t. \tag{1.25}$$

Then

$$\varphi_n(t) \leqslant ((n - t)^2 + t)^{1/2}, \quad 0 < t < \infty. \tag{1.26}$$

Proof. Obviously, relation (1.26) holds for $n = 0$. Suppose that relation (1.26) also holds for $n - 1$, i.e.

$$\varphi_n(t) \leqslant n e^{-t} + \int_0^t e^{s-t}[(n-1-s)^2 + s]^{1/2}\, ds = \psi_n(t) e^{-t}.$$

To conclude the proof we must verify that $\psi_n(t) \leqslant e^t[(n-t)^2 + t]^{1/2}$. Since $\psi_n(0) = 0$, it suffices to show that

$$\psi_n'(t) \leqslant e^t[(n-t)^2 + t]^{1/2} + 1/2\, e^t[(n-t)^2 + t]^{-1/2}(1 - 2n + 2t)$$

which follows by an elementary calculation. This completes the proof.

Consider now the initial value problem

$$\begin{aligned} &\frac{du}{dt}(t) \in Au(t), \quad 0 < t < \infty \\ &u(0) = x \end{aligned} \tag{E_0}$$

where A is a given dissipative subset of $X \times X$.

Definition 1.2. A function $u(t)$ defined on $[0, \infty[$ with values in X is said to be *a solution* (*strong solution* in other terminology) of (E_0) if $u(t)$ is continuous in t on $[0, \infty[$ and Lipschitz on every compact interval of $]0, \infty[$, $u(0) = x$, $u(t)$ is differentiable a.e. on $]0, \infty[$, $u(t) \in D(A)$ a.e. $t > 0$ and $\frac{du}{dt}(t) \in Au(t)$ a.e. $t > 0$.

Proposition 1.2. *Let A be a dissipative set in $X \times X$. Suppose that for every $x \in D(A)$, (E_0) has at least a solution u_x. Then*

(i) $\|u_x(t) - u_y(t)\| \leqslant \|x - y\|$, *for every* $t \geqslant 0$ *and* $x, y \in D(A)$.

(ii) $\left\|\frac{du_x}{dt}(t)\right\| = |Au_x(t)| \leqslant |Ax|$, a.e. $t \in]0, \infty[$.

(iii) $\|u_x(t) - x\| \leqslant t|Ax|$, *for every* $t \geqslant 0$, $x \in D(A)$.

(iv) *For every* $t \geqslant 0$ *define* $S(t)x = u_x(t)$ *and again denote by* $S(t)$ *the extension of* $S(t)$ *on* $\overline{D(A)}$. *Then* $S(t)$ *is a semigroup of nonlinear contractions on* $\overline{D(A)}$.

Proof. Let $x, y \in D(A)$. Then we have

$$\frac{d}{dt}(u_x(t) - u_y(t)) \in Au_x(t) - Au_y(t) \quad \text{a.e. } t \in]0, \infty[.$$

Then the dissipativeness of A and Lemma 1.2 yields

$$\frac{d}{dt}\|u_x(t) - u_y(t)\| \leqslant 0, \quad \text{a.e. } t \in]0, \infty[$$

which implies (i). Let Ω be the set of all values of t for which u_x is differentiable, $u_x(t) \in D(A)$ and $\mathrm{d}u_x/\mathrm{d}t \in Au_x$. Let $s>0$ be fixed such that $u_x(s) \in D(A)$. A being dissipative, for every $y \in Au_x(s)$ there exists $f \in F(u_x(t) - u_x(s))$ such that

$$\frac{1}{2}\frac{\mathrm{d}}{\mathrm{d}t}\|u_x(t) - u_x(s)\|^2 \leqslant (y, f) \leqslant \|y\| \, \|u_x(t) - u_x(s)\|, \quad \text{a.e. } t > 0.$$

Hence

$$\|u_x(t) - u_x(s)\| \leqslant |Au_x(s)| \, (t - s) \qquad \text{for } t \geqslant s.$$

This implies that $\left\|\frac{\mathrm{d}u_x}{\mathrm{d}t}(t)\right\| = |Au_x(t)|$ for almost all $t > 0$. Let $h > 0$ and let $v(t) = u_x(t + h)$. Clearly $v(t)$ satisfies

$$\frac{\mathrm{d}v}{\mathrm{d}t} \in Av(t), \quad \text{a.e.} \quad t > 0.$$

Using again the dissipativeness of A and Lemma 1.2 we get

$$\frac{\mathrm{d}}{\mathrm{d}t}\|u_x(t) - v(t)\| \leqslant 0, \quad \text{a.e. } t > 0$$

and hence

$$\|u_x(t + h) - u_x(t)\| \leqslant \|u_x(h) - x\|, \quad \text{for all } t \geqslant 0. \tag{1.27}$$

On the other hand, in much the same manner, the relation

$$\frac{\mathrm{d}}{\mathrm{d}h}(u_x(h) - x) \in Au_x(h), \quad \text{a.e. } h > 0$$

implies that $\|u_x(h) - x\| \leqslant h|Ax|$, for every $h>0$. Combined with relation (1.27) the last inequality gives $\left\|\frac{\mathrm{d}u_x}{\mathrm{d}t}(t)\right\| \leqslant |Ax|$ as claimed. Note that (iii) was already proved.

In particular, (i) shows that the solution u_x is unique. Define $S(t)x = u_x(t)$. The function $S(t)x$ is continuous in t and nonexpansive in x. Then for every $t \geqslant 0$, $S(t)$ can be extended to a nonexpansive mapping on $\overline{D(A)}$. Obviously, $S(t)$ maps $\overline{D(A)}$ into itself and

$$\|S(t)x - S(t)y\| \leqslant \|x - y\|, \text{ for every } t \geqslant 0, \ x, y \in \overline{D(A)}.$$

The semigroup property $S(t + s) = S(t)S(s)$ is an immediate consequence of the fact that the solution u_x exists and is unique for every $x \in D(A)$.

In order to obtain existence and approximation results for the problem (E_0) associated with a dissipative set A, we have to impose further conditions on A. More precisely we shall need the following stronger version of condition (1.11), namely

$$\overline{\text{conv } D(A)} \subset \bigcap_{\lambda > 0} R(1 - \lambda A). \tag{1.28}$$

Consider the approximating equation

$$\frac{\mathrm{d}u_\lambda}{\mathrm{d}t} = A_\lambda u_\lambda, \quad 0 \leqslant t < \infty, \tag{1.29}$$

$$u_\lambda(0) = x,$$

where $A_\lambda = \lambda^{-1}(J_\lambda - 1)$ and $J_\lambda = (1 - \lambda A)^{-1}$, $\lambda > 0$. We can now state the main result of this section.

THEOREM 1.4. *Let X be a real Banach space and let A be a dissipative set in $X \times X$ satisfying condition* (1.28). *Let $x \in \overline{D(A)}$. Then*

(i) *For every $\lambda > 0$ the initial value problem* (1.29) *has a unique solution* $u_\lambda \in C^1(0, \infty; X)$.

(*ii*) $\lim\limits_{\lambda \to 0} u_\lambda(t) = S(t)\, x = \lim\limits_{n \to \infty} \left(1 - \dfrac{t}{n} A\right)^{-n} x,$

and the limit is uniform on bounded intervals.

(iii) *If $x \in D(A)$ and if the initial value problem (E_0) has a solution $u(t)$ then*

$$u(t) = \lim_{\lambda \to 0} u_\lambda(t) = \lim_{n \to \infty} \left(1 - \frac{t}{n} A\right)^{-n} x, \ t \geqslant 0.$$

Proof. The operator J_λ restricted to $C = \overline{\text{conv } D(A)}$ is a contraction defined on C into C. Then by Proposition 1.1, the equation

$$\frac{\mathrm{d}v_\lambda}{\mathrm{d}t} = (J_\lambda - 1)v_\lambda, \quad 0 \leqslant t < \infty$$

$$v_\lambda(0) = x$$

has a solution $v_\lambda(t)$ which satisfies

$$\| v_\lambda(n) - J_\lambda^n x\| \leqslant \sqrt{n} \|J_\lambda x - x\| = \sqrt{n\lambda} \|A_\lambda x\|.$$

Note that $u_\lambda(t) = v_\lambda(t/\lambda)$ is a solution of Eq. (1.29) and satisfies

$$\|u_\lambda(n\lambda) - J_\lambda^n x\| \leqslant \sqrt{n\lambda} \|A_\lambda x\|, \quad n = 1, 2, \ldots \tag{1.30}$$

We set $S_\lambda(t)x = u_\lambda(t)$. A_λ being dissipative, it follows that $\|S_\lambda(t)x - S_\lambda(t)y\| \leqslant \|x - y\|$ for every $t \geqslant 0$ and $x, y \in \overline{D(A)}$. Since $S(t)$ also has this property we can confine ourselves to verify (ii) in the case $x \in D(A)$.

Let $t \geqslant 0$ and let n be a positive integer such that $t = n\lambda - \mu$, $0 \leqslant \mu \leqslant \lambda$. Since the operator A_λ is dissipative, we have by Proposition 1.2, part (ii), that

$$\left\| \frac{\mathrm{d}u_\lambda}{\mathrm{d}t}(t) \right\| \leqslant \|A_\lambda x\|, \quad 0 \leqslant t < \infty,$$

and therefore

$$\|u_\lambda(t) - u_\lambda(n\lambda)\| \leqslant \mu\|A_\lambda x\|. \tag{1.31}$$

On the other hand, from relation (1.18) we find

$$\|J_\lambda^n x - S(n\lambda)x\| \leqslant (2n\lambda/\sqrt{n})\,|Ax|. \tag{1.32}$$

Next by (1.19)

$$\|S(n\lambda)x - S(t)x\| \leqslant 2\mu|Ax|. \tag{1.33}$$

Since $\|A_\lambda x\| \leqslant |Ax|$, it follows from relations (1.30), (1.31), (1.32) and (1.33) that

$$\|u_\lambda(t) - S(t)x\| \leqslant 3(\lambda + \sqrt{t\lambda})\,|Ax|;\ \lambda, t > 0. \tag{1.34}$$

Thus $u_\lambda(t) \to S(t)x$ as $\lambda \to 0$.

To prove (iii) we proceed as follows. Consider the family of step functions $u_\varepsilon(t)$ defined by

$$u_\varepsilon(t) = (1 - \varepsilon A)^{-j}x \qquad \text{for } (j-1)\varepsilon \leqslant t \leqslant j\varepsilon,\ j = 1, 2, \ldots$$

Clearly

$$u_\varepsilon(t) = (1 - \varepsilon A)^{-1}u_\varepsilon(t - \varepsilon), \qquad \text{for } \varepsilon \leqslant t < \infty. \tag{1.35}$$

Let u(t) be the solution of the problem (E_0). We extend $u(t)$ as x for $t < 0$ and define

$$y_\varepsilon(t) = \frac{\mathrm{d}u}{\mathrm{d}t}(t) - \frac{u(t) - u(t-\varepsilon)}{\varepsilon}.$$

We have

$$\frac{\mathrm{d}}{\mathrm{d}t}(u(t) - u(t-h)) \in Au(t) - Au(t-h), \qquad \text{a.e. } t, t-h > 0.$$

Then the dissipativeness of A and Lemma 1.2 imply that

$$\frac{\mathrm{d}}{\mathrm{d}t}\,\|u(t)-u(t-h)\| \leqslant 0, \quad \text{a.e. } t,\, t-h>0$$

or

$$\|u(t)-u(t-h)\| \leqslant \|u(h)-x\| \leqslant h|Ax|, \qquad t \geqslant h > 0.$$

In particular, this inequality implies that

$$\|y_\varepsilon(t)\| \leqslant 2\,|Ax|, \text{ a.e. on }]0, \infty[. \tag{1.36}$$

On the other hand, one sees that

$$u(t) = (1-\varepsilon A)^{-1}(u(t-\varepsilon) - \varepsilon y_\varepsilon(t)) \quad \text{a.e. on }]0, \infty[$$

and hence

$$\|u(t)-u_\varepsilon(t)\| \leqslant \varepsilon\|y_\varepsilon(t)\| + \|u(t-\varepsilon)-u_\varepsilon(t-\varepsilon)\|, \tag{1.37}$$

because $(1-\varepsilon A)^{-1}$ is nonexpansive. Integrating (1.37) on $]\varepsilon, .T[$ with $T>\varepsilon$ we obtain

$$\varepsilon^{-1}\int_{T-\varepsilon}^{T} \|u(t)-u_\varepsilon(t)\|\,\mathrm{d}t \leqslant \int_{\varepsilon}^{T}\|y_\varepsilon(t)\|\,\mathrm{d}t + \varepsilon^{-1}\int_0^{\varepsilon}\|u(t)-u_\varepsilon(t)\|\,\mathrm{d}t.$$

But $u_\varepsilon(t) = (1-\varepsilon A)^{-1}x$ on $[0, \varepsilon]$ and $\|u(t)-x\| \leqslant t|Ax|$. In view of Proposition 1.2, the last inequality yields

$$\varepsilon^{-1}\int_{T-\varepsilon}^{T}\|u(t)-u_\varepsilon(t)\|\,\mathrm{d}t \leqslant \int_0^T\|y_\varepsilon(t)\|\,\mathrm{d}t + \varepsilon\,|Ax|. \tag{1.38}$$

In formula (1.38) we take $\varepsilon = T/n$. Since $u_\varepsilon(t) = \left(1-\frac{T}{n}A\right)^{-n}x$ whenever $T-\varepsilon \leqslant \leqslant t \leqslant T$, we obtain

$$\frac{n}{T}\int_{\left(1-\frac{1}{n}\right)T}^{T}\left\|u(t)-\left(1-\frac{T}{n}A\right)^{-n}x\right\|\mathrm{d}t \leqslant \int_0^T\|y_\varepsilon(t)\|\,\mathrm{d}t + \frac{T}{n}\,|Ax|. \tag{1.39}$$

Since $\lim\limits_{\varepsilon\to 0} y_\varepsilon(t) = 0$ a.e. $t \in]0, T[$, estimate (1.36) and the Lebesgue dominated convergence theorem imply that $\lim\limits_{\varepsilon\to 0} y_\varepsilon = 0$ in $L^1(0, T; X)$. Then, if in relation (1.39) we make $n \to \infty$, we obtain

$$u(T) = \lim_{n\to\infty}\left(1-\frac{T}{n}A\right)^{-n}x,$$

which completes the proof.

1.4. Generation of nonlinear semigroups

The exponential formula (1.12) shows that $u(t) = S(t)x$ can be interpreted as a "generalized solution" of the equation (E_0). In fact $u(t)$ is the limit of solutions $u_\varepsilon(t)$ of the approximate Cauchy problem

$$\frac{1}{\varepsilon}(u_\varepsilon(t) - u_\varepsilon(t-\varepsilon)) \in Au_\varepsilon(t), \quad \text{for } t \geqslant \varepsilon$$

$$u_\varepsilon(t) = x. \qquad \text{for } -\varepsilon \leqslant t \leqslant 0.$$

The main goal pursued in this section is to find sufficient conditions to assure that $S(t)x$ is a strong solution of (E_0). We shall begin with the following general result.

THEOREM 1.5. *Let X be a real Banach space and let A be a closed and dissipative subset of $X \times X$ satisfying condition (1.28). Let x be any fixed element of $\overline{D(A)}$ and let*

$$S(t)x = \lim_{n\to\infty}\left(1 - \frac{t}{n}A\right)^{-n} x, \quad t \geqslant 0.$$

If $S(t)x$ is differentiable a.e. on $]0, \infty[$, then $u(t) = S(t)x$ is a solution of (E_0).

Let $\langle , \rangle_s : X \times X \to R$ be the function defined by

$$\langle y, x\rangle_s = \sup\{(y, x^*);\ x^* \in F(x)\}.$$

We recall that $\langle , \rangle_s$ is upper-semicontinuous on $X \times X$ (see Proposition 1.2, Chapter I). In the proof of Theorem 1.5 we shall use the following lemma which also presents interest in other problems.

LEMMA 1.7. *Let the assumptions of Theorem 1.5 be satisfied. If $x_0 \in \overline{D(A)}$ and $[x, y] \in A$ then*

$$\overline{\lim_{t\to 0}}\left\langle \frac{S(t)x_0 - x_0}{t},\quad x_0 - x\right\rangle_s \leqslant \langle y, x_0 - x\rangle_s. \tag{1.40}$$

Proof. For every $\lambda > 0$ consider the solution u_λ of Eq. (1.29) with an initial value $u_\lambda(0) = x_0$. Let $[x, y] \in A$ and let $x_\lambda = x - \lambda y$. Clearly, $x = J_\lambda x_\lambda$ and $y = A_\lambda x_\lambda$. We have

$$\frac{d}{dt}(u_\lambda(t) - x_\lambda) = A_\lambda u_\lambda(t), \quad 0 \leqslant t < \infty.$$

Next by Lemma 1.2

$$\frac{1}{2}\frac{d}{dt}\|u_\lambda(t) - x_\lambda\|^2 \leqslant (A_\lambda x_\lambda, f_\lambda), \text{ a.e. } t > 0,$$

where f_λ is any element of $F(u_\lambda(t) - x_\lambda)$. To obtain this we have used the obvious inequality

$$(A_\lambda x - A_\lambda \bar{x}, f) \leqslant 0, \quad \forall\, f \in F(x - \bar{x}).$$

Consequently

$$\frac{1}{2}(\|u_\lambda(t) - x_\lambda\|^2 - \|x_0 - x_\lambda\|^2) \leqslant \int_0^t (y, f_\lambda(\tau))\,\mathrm{d}\tau,$$

and therefore

$$\frac{1}{2}(\|u_\lambda(t) - x_\lambda\|^2 - \|x_0 - x_\lambda\|^2) \leqslant \int_0^t \langle y, u_\lambda(\tau) - x_\lambda \rangle_s\,\mathrm{d}\tau, \quad \text{for} \quad t \geqslant 0.$$

Inasmuch as $\lim_{\lambda\to 0} u_\lambda(t) = S(t)x_0$, $\lim_{\lambda\to 0} x_\lambda = x$ and $\langle , \rangle_s$ is upper-semicontinuous we infer that

$$\frac{1}{2}(\|u(t) - x\|^2 - \|x_0 - x\|^2) \leqslant \int_0^t \langle y, u(\tau) - x\rangle_s\,\mathrm{d}\tau.$$

But

$$\langle u(t) - x_0, x_0 - x\rangle_s \leqslant \frac{1}{2}[\|u(t) - x\|^2 - \|x_0 - x\|^2].$$

because

$$\langle u - x, x - v\rangle_s \leqslant \frac{1}{2}[\|u - v\|^2 - \|x - v\|^2], \text{ for all } u, x, v \in X. \tag{1.41}$$

Thus

$$\left\langle \frac{u(t) - x_0}{t}, x_0 - x \right\rangle_s \leqslant \int_0^1 \langle y, u(t\tau) - x\rangle_s\, d\tau$$

and again by the upper-semicontinuity of $\langle , \rangle_s$ we obtain the desired estimate (1.40).

Proof of Theorem. 1.5 Let $x \in D(A)$ and $t_0 > 0$ be such that $\frac{\mathrm{d}}{\mathrm{d}t} S(t)x$ exists at $t = t_0$. We have to prove that

$$\left[S(t_0)x, \frac{\mathrm{d}}{\mathrm{d}t} S(t_0)x \right] \in A.$$

We can write

$$S(t_0 - h)x = S(t_0)x - h\frac{\mathrm{d}}{\mathrm{d}t} S(t_0)x + g(h), \quad 0 < h < t_0,$$

where $\lim_{h\to 0} \|g(h)\|/h = 0$. Since $S(t_0 - h)x \in \overline{D(A)}$, it follows from relation (1.28) that there exist $[x_h, y_h] \in A$ such that

$$S(t_0 - h)x = S(t_0)x - h \frac{\mathrm{d}}{\mathrm{d}t} S(t_0)x + g(h) = x_h - hy_h,$$

or

$$S(t_0)x - x_h = h\left(\frac{\mathrm{d}}{\mathrm{d}t} S(t_0)x - y_h\right) - g(h). \tag{1.42}$$

Now, applying Lemma 1.7 with $x = x_h$, $y = y_h$ and $x_0 = S(t_0)x$, one finds that

$$\left\langle \frac{\mathrm{d}}{\mathrm{d}t} S(t_0)x,\ S(t_0)x - x_h \right\rangle_s \leqslant \langle y_h, S(t_0)x - x_h\rangle_s.$$

Let $f_h \in F(S(t_0)x - x_h)$ be such that $\langle y_h, S(t_0)x - x_h \rangle_s = (y_h, f_h)$. (Since $F(S(t_0)x - x_h)$ is w^*-compact in X^* such f_h always exists.) This permits us to write

$$\left(\frac{\mathrm{d}}{\mathrm{d}t} S(t_0)x - y_h,\ f_h\right) \leqslant 0$$

and by (1.42), it follows that

$$\| S(t_0)x - x_h\| \leqslant \|g(h)\|,$$

which implies that $\lim_{h\to 0} x_h = S(t_0)x_0$ and $\lim_{h\to 0} y_h = \frac{\mathrm{d}}{\mathrm{d}t} S(t_0)x$. Since $[x_h, y_h] \in A$ and A is closed in $X \times X$, we find that

$$\frac{\mathrm{d}}{\mathrm{d}t} S(t_0)x \in A\,S(t_0)x,$$

as claimed. The theorem is proven since t_0 is arbitrary.

We have already seen in the proof of Theorem 1.3 that for every $x \in D(A)$ the function $S(t)x$ is Lipschitz continuous on $[0, \infty[$, namely

$$\| S(t)x - S(s)x\| \leqslant 2\,|t - s|\ |Ax|, \text{ for all } t, s \geqslant 0.$$

If X has the property that each Lipschitz-continuous X-valued function is differentiable almost everywhere, then in view of the previous theorem for every $x \in D(A)$, $u(t) = S(t)x$ is a solution of the problem (E_0). In particular, this is the case if X is reflexive (see Theorem 2.1, Chapter I).

COROLLARY 1.1. *Let X be reflexive and let A be a closed and dissipative subset of $X \times X$ satisfying* (1.28). *Then for each $x \in D(A)$ the initial value problem* (E_0) *has a unique solution $u : [0, \infty[\to X$ given by the exponential formula*

$$u(t) = S(t)x = \lim_{n\to\infty} \left(1 - \frac{t}{n} A\right)^{-n} x, \quad t \geqslant 0.$$

Moreover,

$$\left\| \frac{\mathrm{d}u}{\mathrm{d}t}(t) \right\| = |Au(t)| \leqslant |Ax| \qquad \text{a.e.} \quad t > 0. \tag{1.43}$$

If A is an m-dissipative set in $X \times X$ then it is closed and (1.28) is trivially satisfied. Corollary 1.1 may thus be applied. If X also happens to be strictly convex, then $Au(t)$ has a unique element of minimum norm $A^0u(t)$ (see Section 3.1, Chapter II) and (1.43) implies that

$$\frac{\mathrm{d}u}{\mathrm{d}t}(t) = A^0u(t), \qquad \text{a.e. } t > 0.$$

Our next theorem shows that if X and X^* are "sufficiently smooth" this also occurs in the general case considered before. More precisely, we have the folloving theorem

THEOREM 1.6. *Let X and X^* be uniformly convex and let $A \subset X \times X$ be closed, dissipative and satisfying condition* (1.28). *Then for every $x \in D(A)$*

(i) *The set Ax has a unique element of minimum norm which is denoted by A^0x.*

(ii) *There exists a unique function $u(t) \in D(A)$ for every $t \geqslant 0$ which is Lipschitz continuous on $[0, \infty[$ and everywhere differentiable from the right, satisfying*

$$\begin{aligned} &\frac{\mathrm{d}^+u}{\mathrm{d}t}(t) = A^0u(t), \quad 0 \leqslant t < \infty, \\ &u(0) = x \\ &\left\| \frac{\mathrm{d}^+u}{\mathrm{d}t}(t) \right\| \leqslant \|A^0x\| = |Ax|, \; 0 \leqslant t < \infty. \end{aligned} \tag{1.44}$$

Proof. We assume first that A is demiclosed (i.e. it is closed in $X \times X_w$, where X_w denotes the space X endowed with weak topology). Define a set B in $X \times X$ as follows

$$Bx = \overline{\text{conv}\; Ax}, \qquad x \in D(B) = D(A).$$

A being dissipative, we have

$$(y_1 - y_2, F(x_1 - x_2)) \leqslant 0, \qquad \text{for every } [x_i, y_i] \in A, \; i = 1, 2,$$

and this clearly implies that

$$(y_1 - y_2, F(x_1 - x_2)) \leqslant 0, \text{ for every } x_1 \in D(A),\ y_1 \in Ax_1,\ y_2 \in \overline{\text{conv }Ax_2}.$$

Therefore, B is dissipative. Furthermore, it follows from (1.28) that

$$D(A) \subset \bigcap_{\lambda>0} R(1 - \lambda B).$$

Let $x \in D(A)$. Then $x_\lambda = (1 - \lambda A)^{-1}x$ and $y_\lambda = A_\lambda x$ are well defined for every $\lambda > 0$. Moreover $\|A_\lambda x\| \leqslant |Ax|$ and $\lim_{\lambda\to 0} x_\lambda = x$ (see Proposition 3.2, Chapter II). Let $\lambda_n \to 0$ be a sequence for which $A_{\lambda_n}x \to y$. Since $A_{\lambda_n} x \in Ax_{\lambda_n}$, the demiclosedness of A implies that $y \in Ax$. But $\|A_\lambda x\| = \|B_\lambda x\| \leqslant |Bx| = \|B^0 x\|$ where B^0x is the element of minimum norm in Bx which exists and is unique because in particular X^* is strictly convex. We may conclude therefore that $y = B^0x \in Ax$. Consequently Ax has a unique element of minimum norm $A^0x = B^0x$. Next, by Corollary 1.1,

$$\frac{\mathrm{d}}{\mathrm{d}t} S(t)x = A^0 S(t)x, \quad \text{a.e.} \quad t > 0$$

and

$$\| S(t)x - x\| \leqslant t|Ax|, \quad \text{for every } t \geqslant 0. \tag{1.45}$$

Let A_0 be the generator of the semigroup $S(t)$. Then (1.45) implies that $x \in D(A_0)$ (see Theorem 1.1) and by Theorem 1.2

$$\frac{\mathrm{d}^+}{\mathrm{d}t} S(t)x = A_0 S(t)x \quad \text{for every } t \geqslant 0.$$

Since $A_0S(t)x$ is right continuous in t and A is closed we find that

$$A_0x = \lim_{t\to 0} A_0S(t)x \in Ax.$$

Thus inequality (1.45) implies that $A_0x = A^0x$. Therefore we have proved that $u(t) = S(t)x$ satisfies condition (ii).

To complete the proof, we turn to the case when A is only closed. Let $\tilde{A}$ be the closure of A in $X \times X_w$, i.e. the smallest demiclosed extension of A. Clearly $D(A) \subset D(\tilde{A}) \subset \overline{D(A)}$ and

$$\overline{\text{conv } D(\tilde{A})} \subset \bigcap_{\lambda>0} R(1 - \lambda\tilde{A}).$$

The uniform convexity of X^* implies in particular that the duality map F of X is continuous on X (see Proposition 1.5, Chapter I). This shows that $\tilde{A}$ is dissipative and according to the first part of the proof, the initial value problem

$$\frac{d^+u}{dt} = (\tilde{A})^0(u), \quad t \geqslant 0$$

$$u(0) = x, \ x \in D(\tilde{A})$$

has a unique solution $u(t)$ satisfying (ii). To conclude the proof it suffices to show that $D(\tilde{A}) = D(A)$ and $\tilde{A}^0 = A^0$. Let $x \in D(\tilde{A})$.

By assumption, there is $[x_\lambda, y_\lambda] \in A \subset \tilde{A}$ such that $x = x_\lambda - \lambda y_\lambda$. We note that $x_\lambda = (1 - \lambda A)^{-1} x = (1 - \lambda \tilde{A})^{-1} x$ and $y_\lambda = A_\lambda x = \tilde{A}_\lambda x$. Since $x \in D(\tilde{A})$, $x_\lambda \to x$ and $\|y_\lambda\| \leqslant |\tilde{A}x| = \|\tilde{A}^0 x\|$. $\tilde{A}$ being demiclosed and X uniformly convex, this implies by a standard device that $\lim_{\lambda \to 0} y_\lambda = \tilde{A}^0 x$. From the closedness of A we deduce that $x \in D(A)$ and $\tilde{A}^0 x \in Ax$. Therefore $D(\tilde{A}) = D(A)$ and for any $x \in D(A)$, Ax has a unique element of minimum norm $A^0 x = \tilde{A}^0 x$. This concludes the proof of Theorem 1.6.

The following simple consequence of Theorem 1.6 can be regarded as a partial extension of the Hille-Yosida Theorem (see Theorem 3.1, Chapter I).

Corollary 1.2. *Let X and X^* be uniformly convex and let A be a dissipative and closed subset of $X \times X$ satisfying condition* (1.28). *Let A^0 be the minimal section of A. Then $D(A^0) = D(A)$ and A^0 is single-valued. Moreover, there exists a semigroup of contractions on $\overline{D(A)}$ whose generator coincides with A^0.*

Proof. Let $S(t)$: $[0, \infty[\times \overline{D(A)} \to \overline{D(A)}$ be the semigroup defined by the exponential formula (1.12). Let A_0 be the generator of S. In view of Theorem 1.6, $D(A) \subset D(A_0)$ and $A_0 x = A^0 x$ for each $x \in D(A)$. We shall prove that $D(A_0) = D(A)$. Let $x \in D(A_0)$. Then $S(t)x$ is right-differentiable everywhere on $[0, \infty[$ and $d^+/dt S(t)x = A_0 S(t)x$. Next from Theorem 1.5 it follows that $A_0 S(t)x = A^0 S(t)x$ a.e. $t > 0$. Since $A_0 S(t)x$ is right-continuous and A is closed, this implies that $x \in D(A)$.

Definition 1.3. Let S be a semigroup of contractions defined on a closed subset C of a Banach space X. Let A be a dissipative set in $X \times X$. One says that S is *generated* by A if the following conditions hold

(a) $\overline{D(A)} = C$.

(b) For every $x \in D(A)$, function $u(t) = S(t)x$ is a solution (in the sense of Definition 1.2) of the initial value problem (E_0).

We observe that if S is generated by some dissipative set, then the generator A_0 of S is densely defined on the domain of S.

We have already seen (Corollary 1.1) that if X is reflexive, then every closed and dissipative subset A of $X \times X$ which satisfies condition (1.28) (in particular every m-dissipative subset) generates a unique semigroup of contractions on $\overline{D(A)}$. In this context, one might ask whether there is at most one A which generates a given semigroup S.

Our next theorem gives a partial answer to this question.

THEOREM 1.7. *Let X^* be uniformly convex and let A and B be m-dissipative sets in $X \times X$. Let S_A and S_B be the semigroups generated by A and B on $\overline{D(A)}$ and $\overline{D(B)}$, respectively. If $S_A(t) = S_B(t)$ for all $t \geqslant 0$, then $A = B$.*

Proof. First we shall prove that $D(A) = D(B)$. Let us denote $S = S_A = S_B$ and let x_0 be any fixed element of $D(B)$. Applying Lemma 1.7 we get

$$\limsup_{t \to 0} \left(\frac{S(t)x_0 - x_0}{t} - y,\ F(x_0 - x) \right) \leqslant 0, \quad \forall [x, y] \in A. \tag{1.46}$$

On the other hand it follows from Proposition 1.2 that

$$\left\| \frac{S(t)x_0 - x_0}{t} \right\| \leqslant |Bx_0|, \qquad \forall\, t \geqslant 0.$$

Thus we may assume without loss of generality that $\dfrac{S(t)x_0 - x_0}{t} \to y_0$ as $t \to 0$. Using (1.46) we obtain

$$(y_0 - y, F(x_0 - x)) \leqslant 0 \quad \forall\, [x, y] \in A.$$

Since A is maximal dissipative, this implies that $[x_0, y_0] \in A$. Hence $D(B) \subset D(A)$ and similarly one proves that $D(A) \subset D(B)$. Let L_0 denote the (strong) generator of S. Then for every $x \in D(A) = D(B)$ we have

$$\frac{d}{dt} S(t)x = L_0 S(t)x \in (AS(t)x) \cap (BS(t)x), \qquad \text{a.e.} \quad t > 0$$

and again using Proposition 1.2 we deduce that $L_0 S(t)x \in A^0 S(t)x \cap B^0 S(t)x$ a.e. $t > 0$ and

$$\|L_0 S(t)x\| \leqslant \inf(|Ax|, |Bx|).$$

Let $\{t_n\}$ be a sequence of positive numbers such that $t_n \to 0$ as $n \to \infty$ and $L_0 S(t_n)x \rightharpoonup y$ as $n \to \infty$. Inasmuch as A and B are demiclosed it follows that $y \in A^0 x \cap B^0 x$. Therefore, we have proved that $A^0 x \cap B^0 x \neq \emptyset$ for every $x \in D(A) = D(B)$. Thus from Corollary 1.3, Chapter II, $A = B$ and the proof of Theorem 1.7 is complete.

We are now able to prove an important theorem stated in Chapter II (Theorem 3.4), namely: *If X^* is uniformly convex and if A is an m-dissipative set in $X \times X$ then A^0 is a principal section of A*. Let $y_0 \in X$ and $x_0 \in \overline{D(A)}$ be such that

$$(y_0 - y, F(x_0 - x)) \leqslant 0 \quad \text{for every } [x, y] \in A^0. \tag{1.47}$$

Let us denote by $S(t)$ the semigroup generated by A on $\overline{D(A)}$ i.e.

$$\frac{\mathrm{d}}{\mathrm{d}t} S(t)\,x \in A^0 S(t)x, \text{ for every } x \in D(A) \text{ and a.e. } t > 0.$$

Taking in (1.47), $x = S(t)u$ with $u \in D(A)$, we obtain

$$\left(y_0 - \frac{\mathrm{d}}{\mathrm{d}t} S(t)u,\ F(x_0 - S(t)u)\right) \leqslant 0,$$

or

$$\frac{1}{2}\ \frac{\mathrm{d}}{\mathrm{d}t} \| S(t)u - x_0\|^2 \leqslant (y_0, F(S(t)u - x_0)), \text{ a.e. on }]0, \infty[.$$

Therefore

$$\| S(t)u - x_0\| \leqslant \|u - x_0\| + t\|y_0\|, \qquad \forall\, t \geqslant 0.$$

In particular, this implies that

$$\| S(t)x_0 - x_0\| \leqslant t\|y_0\|, \qquad \forall\, t \geqslant 0.$$

Thus by the same argument as that used in the proof of Theorem 1.7, one deduces that $x_0 \in D(A)$. Therefore, we are in the situation of Theorem 3.3 (Chapter II). Hence $[x_0, y_0] \in A$, which concludes the proof of Theorem 3.4.

Another question which arises is the following. If S is a semigroup of nonlinear contractions, defined on a closed subset $C \subset X$, is there a dissipative set $A \subset X \times X$ which generates S? In this context, one might ask whether the generator of a semigroup of nonlinear contractions S is densely defined on the domain of S.

As will be seen in the next chapter, these questions have an affirmative answer if X is a Hilbert space. The following example due to G. F. Webb shows that this is not the case in general Banach spaces. Let $X = C(0,1)$ be the Banach space of all real-valued continuous functions defined on the compact [0,1]. Consider the function

$$f(r) = \begin{cases} r & \text{if } r > 0 \\ 2r & \text{if } r \leqslant 0 \end{cases}$$

and define

$$(S(t)x)\,(s) = f(t + f^{-1}(x(s)),\ x \in X.$$

It is easy to see that $S(t)\colon X \to X$ is a semigroup of nonlinear contractions on $X = C(0,1)$. Moreover,

$$\frac{(S(t)x)\,(s) - x(s)}{t} = \begin{cases} 1 & \text{if } x(s) \geqslant 0 \\ 1 - \dfrac{1}{2t}\,x(s) & \text{if } x(s) < 0. \end{cases}$$

This implies that $\dfrac{S(t)x - x}{t}$ has no limit as $t \to 0$ in X unless x is positive on [0,1]. In other words, the domain of the generator A_0 of S coincides with the set of all continuous functions $x(s)$ which are positive on [0,1]. Therefore $D(A_0)$ is not dense in X. In particular, this implies that the given semigroup is not generated.

The situation is even worse than this. There are semigroups of contractions $S(t)$ defined on $X = C(0,1)$ for which $(S(t)x - x)/t$ do not converge in any sense as $t \to 0$, to an element of X for any $x \in X$. (See M.G. Crandall and T. Liggett [1] for examples of this kind.) In other words, there are semigroups of contractions on some Banach spaces which have no generators in any sense.

§2. Quasi-autonomous differential equations

This paragraph is mainly concerned with the initial value problem

$$\frac{du}{dt}(t) \in Au(t) + f(t), \qquad 0 < t < T \tag{2.1}$$

$$u(0) = u_0$$

where A is a given dissipative set in $X \times X$ and $f(t)$ is an X-valued function on $[0, T]$. We shall usually assume that $f \in L^1(0, T; X)$, and A satisfies condition (1.28). The arguments and results generalize those of Section 1 in which the autonomous case was treated.

2.1. Existence theorems

As in the autonomous case (see Definition 1.2), a function $u(t)\colon [0, T] \to X$ is called *solution* (or *strong solution*) of the Cauchy problem 2.1, if $u(t)$ is continuous on $[0, T]$ and Lipschitz on every compact of $]0, T[$, $u(t)$ is differentiable a.e. on $]0,T[$ and $\dfrac{du}{dt}(t) \in Au(t) + f(t)$, a.e. on $]0, T[$.

We also need another definition in order to state an important existence theorem concerning problem (2.1).

Definition 2.1. The function $u(t)\colon [0, T] \to X$ is called an *integral solution* of the initial value problem (2.1) if $u(t)$ is continuous on $[0, T]$, $u(0) = u_0$ and the inequality

$$\frac{1}{2}\|u(t) - x\|^2 \leqslant \frac{1}{2}\|u(s) - x\|^2 + \int_s^t \langle f(\tau) + y, u(\tau) - x\rangle_s \, d\tau. \tag{2.2}$$

holds for each $[x, y] \in A$ and $0 \leqslant s \leqslant t \leqslant T$.

Clearly, every strong solution is also an integral solution. We observe that if A is a closed dissipative set in $X \times X$, satisfying condition (1.28), then $u(t) = S(t)x = \lim_{n\to\infty} \left(1 - \frac{t}{n} A\right)^{-n} x$ provided by Theorem 1.4 is an integral solution of the autonomous equation (E_0). Indeed if we take in the proof of Lemma 1.7, $x_0 = x(s)$ where $0 \leqslant s \leqslant t \leqslant T$, we obtain (see (1.41))

$$\frac{1}{2}(\|u_\lambda(t) - x_\lambda\|^2 - \|u_\lambda(s) - x_\lambda\|^2) \leqslant \int_s^t \langle y, u_\lambda(\tau) - x_\lambda \rangle_s \, d\tau.$$

Letting $\lambda \to 0$ we find (2.2).

THEOREM 2.1. *Let X be a real Banach space, C a closed convex cone of X and A a closed dissipative set in $X \times X$ such that*

$$D(A) \subset C \quad \text{and} \quad R(1 - \lambda A) \supset C, \text{ for every } \lambda > 0. \tag{2.3}$$

Let $u_0 \in \overline{D(A)}$ and $f \in L^1(0, T; X)$ be such that $f(t) \in C$ a.e. on $]0, T[$. Then Eq. (2.1) *has a unique integral solution $u(t)$ such that $u(t) \in \overline{D(A)}$* a.e. $t \in]0, T[$. *Moreover, if u and v are integral solutions of $\frac{du}{dt} \in Au + f$ and $\frac{dv}{dt} \in Av + g$, respectively, with the same initial value, then*

$$\|u(t) - v(t)\|^2 \leqslant \|u(s) - v(s)\|^2 + 2\int_s^t \langle f(\tau) - g(\tau), u(\tau) - v(\tau) \rangle_s \, d\tau \tag{2.4}$$

for $0 \leqslant s \leqslant t \leqslant T$.

The proof of Theorem 2.1 is based on the following lemma.

LEMMA 2.1. *Let the assumptions of Theorem 2.1 hold and let $\lambda > 0$. Let $u_\lambda(t)$ be the solution of the approximate problem*

$$\frac{du_\lambda}{dt} = A_\lambda u_\lambda + f(t), \text{ a.e. } 0 < t < T, \tag{2.5}$$

$$u_\lambda(0) = u_0.$$

Then $u(t) = \lim_{\lambda\to 0} u_\lambda(t)$ exists and the limit is uniform in t on $[0, T]$. Moreover, $u(t)$ is an integral solution of (2.1).

Proof. The proof of this lemma is more complicated and therefore we shall divide it into several steps.

(a) First we assume that $f(t) = y_0$ on $[0, T]$ where y_0 is any fixed element of C. Let us denote by A^1 the dissipative set $u \to Au + y_0$. From (2.3) it follows that $C \subset R(1 + \lambda A^1)$ for every $\lambda > 0$, and we easily deduce that $A^1_\lambda u = A_\lambda(u + \lambda y_0) + y_0$ for all $u \in C$. Consider the approximate equations

$$\frac{dv_\lambda}{dt} = A_\lambda(v_\lambda + \lambda y_0) + y_0, \quad 0 \leqslant t \leqslant T, \tag{2.6}$$

$$v_\lambda(0) = u_0.$$

By Theorem 1.4, $\lim_{\lambda \to 0} v_\lambda(t) = S_1(t)\, u_0$ where $S_1(t)$ is the semigroup generated by A^1, i.e.,

$$S_1(t)u_0 = \lim_{n \to \infty} \left(1 - \frac{t}{n} A^1\right)^{-n} u_0, \quad t \geqslant 0.$$

As we have already remarked $S_1(t)u_0 = u(t)$ is an integral solution of (2.1) with $f(t) = y_0$. Let u_λ be the unique solution of Eq. (2.5) with $f(t) = y_0$; then

$$\frac{d}{dt}(u_\lambda(t) - v_\lambda(t)) = A_\lambda u_\lambda(t) - A_\lambda(v_\lambda(t) + \lambda y_0), \text{ a.e. } t \in]0, T[.$$

Fix $f_\lambda \in F(u_\lambda(t) - v_\lambda(t) - \lambda y_0)$ and multiply the last equality by f_λ. Since A_λ is dissipative, we obtain

$$\frac{d}{dt} \|u_\lambda(t) - v_\lambda(t) - \lambda y_0\| \leqslant 0, \quad \text{a.e.} \quad t \in]0, T[,$$

and therefore

$$\|u_\lambda(t) - v_\lambda(t) - \lambda y_0\| \leqslant \lambda \|y_0\|, \quad 0 \leqslant t \leqslant T.$$

Consequently, $\lim_{\lambda \to 0} u_\lambda(t) = u(t) = S_1(t)u_0$ uniformly in t on $[0, T]$.

(b) Suppose now that $f(t)$ is a step function on $[0, T]$, namely

$$f(t) = y_i \quad \text{for } t_{i-1} \leqslant t < t_i, \ i = 1, 2, \ldots, n$$

where $t_0 = 0$, $t_n = T$ and $y_i \in C$ for all i. Clearly, the solution u_λ of the problem (2.5) can be represented as

$$u_\lambda(t) = v^i_\lambda(t - t_{i-1}) \quad \text{for } t_{i-1} \leqslant t < t_i, \ i = 1, \ldots, n$$

where v_λ^i is the solution of the following initial value problem

$$\frac{dv_\lambda^i}{dt} = A_\lambda v_\lambda^i + y_i, \qquad 0 \leqslant t \leqslant t_i - t_{i-1},$$

$$v_\lambda^i(0) = u_\lambda(t_{i-1}).$$

Thus according to the first part of this proof $\lim_{\lambda\to 0} u_\lambda(t) = u(t)$ where

$$u(t) = S_i(t - t_{i-1})u(t_{i-1}), \quad \text{for} \quad t_{i-1} \leqslant t \leqslant t_i$$

while S_i is the semigroup generated by $A_i u = Au + y_i$, $i = 1, 2, \ldots, n$. Inasmuch as $u(t)$ satisfies (2.2) on every interval $[t_{i-1}, t_i]$, we infer that $u(t)$ is an integral solution of (2.1) on $[0, T]$.

(c) We shall now treat the general case $f \in L^1(0, T; X)$ and $f(t) \in C$ a.e. $t \in]0, T[$. Then there is a sequence of step functions $\{f_n\}$ such that $f_n \to f$ in $L^1(0, T; X)$ as $n \to \infty$ and $f_n(t) \in C$ a.e. $t \in]0, T[$. Let u_λ^n be the solution of the Cauchy problem

$$\frac{du_\lambda^n}{dt} = A_\lambda u_\lambda^n + f_n, \quad 0 < t < T \tag{2.7}$$

$$u_\lambda^n(0) = u_0$$

and let u_λ be the solution of (2.5). Subtracting (2.7) from (2.5) and multiplying by $f_\lambda^n \in F(u_\lambda(t) - u_\lambda^n(t))$, after integrating, we obtain

$$\|u_\lambda(t) - u_\lambda^n(t)\| \leqslant \int_0^T \|f(t) - f_n(t)\| \, dt, \qquad \forall t \in [0, T]. \tag{2.8}$$

Let $\lambda, \mu > 0$. Then

$$\|u_\lambda(t) - u_\mu(t)\| \leqslant \|u_\lambda(t) - u_\lambda^n(t)\| + \|u_\lambda^n(t) -$$

$$- u_\mu^n(t)\| + \|u_\mu(t) - u_\mu^n(t)\|.$$

According to the preceding point of the proof for each n, $\lim_{\lambda\to 0} u_\lambda^n$ exists and is uniform on $[0, T]$. Thus it follows from (2.8) that

$$\|u_\lambda(t) - u_\mu(t)\| \leqslant \varepsilon, \qquad \forall t \in [0, T],$$

whenever $0 < \lambda, \mu < \delta(\varepsilon)$. We set $u(t) = \lim_{\lambda\to 0} u_\lambda(t)$. Since $u_\lambda^n(t) \in \overline{D(A)}$ a.e. on $]0, T[$, we see that also $u(t) \in \overline{D(A)}$ a.e. on $]0, T[$. We shall prove now that u satisfies (2.2) for every $[x, y] \in A$ and $0 \leqslant s \leqslant t \leqslant T$. To this end we set

$u^n(t) = \lim_{\lambda\to 0} u^n_\lambda(t)$. As mentioned before, $u^n(t)$ is for every n, an integral solution of (2.1), i.e.,

$$\|u^n(t) - x\|^2 \leqslant \|u^n(s) - x\|^2 + 2\int_s^t \langle y + f_n(\tau),\ u^n(\tau) - x\rangle_s \, d\tau, \tag{2.9}$$

for any $[x, y] \in A$ and $0 \leqslant s \leqslant t \leqslant T$. It follows from (2.8) that $u_n(t) \to u(t)$ uniformly on $[0, T]$ as $n \to \infty$. Recalling that function $\langle\, , \rangle_s$ is upper-semicontinuous on $X \times X$ and letting $n \to \infty$ in (2.9) we obtain (2.2) as claimed. This concludes the proof of Lemma 2.1.

We shall continue with a rather technical result.

LEMMA 2.2. *Let* $[u,f]$, $[v, g] \in C(0, T; X) \times L^1(0, T; X)$ *be such that*

$$\int_\alpha^\beta \frac{1}{2}\|u(\sigma) - v(t)\|^2 \, d\sigma \leqslant \int_\alpha^\beta \frac{1}{2}\|u(\sigma) - v(s)\|^2 \, d\sigma +$$

$$+\int_\alpha^\beta d\sigma \int_s^t \langle f(\sigma) - g(\tau), u(\sigma) - v(\tau)\rangle_s \, d\tau$$

$$+\int_s^t \frac{1}{2}\|u(\alpha) - v(\tau)\|^2 \, d\tau - \int_s^t \frac{1}{2}\|u(\beta) - v(\tau)\|^2 \, d\tau, \tag{2.10}$$

for all $0 \leqslant s \leqslant t \leqslant T$ *and* $0 \leqslant \alpha \leqslant \beta \leqslant T$. *Then*

$$\frac{1}{2}\|u(t) - v(t)\|^2 \leqslant \frac{1}{2}\|u(s) - v(s)\|^2 + \int_s^t \langle f(\tau) - g(\tau), u(\tau) - v(\tau)\rangle_s d\tau$$

for all $0 \leqslant s \leqslant t \leqslant T$.

Proof. Define

$$\varphi(\sigma, \tau) = \frac{1}{2}\|u(\sigma) - v(\tau)\|^2, \quad \text{for } \sigma, \tau \in [0, T]$$

and

$$\psi(\sigma, \tau) = \langle f(\sigma) - g(\tau), u(\sigma) - v(\tau)\rangle_s, \quad \text{for } \sigma, \tau \in [0, T].$$

We now consider the regularizations φ_n and ψ_n:

$$\varphi_n(t, s) = \int_0^T\int_0^T \rho_n(t - \sigma, s - \tau)\, \varphi(\sigma, \tau) \, d\tau \, d\sigma,$$

$$\psi_n(t, s) = \int_0^T\int_0^T \rho_n(t - \sigma, s - \tau)\, \psi(\sigma, \tau) \, d\sigma \, d\tau$$

where

$$\rho_n(t, s) = n^2\rho(nt)\,\rho(ns), \ \ \rho \in \mathscr{D}(R^1), \ \ \rho \geqslant 0,$$

supp $\rho \subset [-1, +1]$, $\int \rho(t)\,\mathrm{d}t = 1$ and $\rho(t) = \rho(-t)$ for all t.

Let $\frac{1}{n} \leqslant \alpha \leqslant \beta \leqslant T$ and $\frac{1}{n} \leqslant s \leqslant t \leqslant T$. Then (2.10) implies that

$$\int_\alpha^\beta [\varphi_n(\sigma, t) - \varphi_n(\sigma, s)\,\mathrm{d}\sigma + \int_s^t [\varphi_n(\beta, \tau) - \varphi_n(\alpha, \tau)]\,\mathrm{d}\tau \leqslant$$

$$\leqslant \int_\alpha^\beta \mathrm{d}\sigma \int_s^t \psi_n(\sigma, \tau)\,\mathrm{d}\tau.$$

Hence

$$\frac{\partial\varphi_n}{\partial\tau}(\sigma, \tau) + \frac{\partial\varphi_n}{\partial\sigma}(\sigma, \tau) \leqslant \psi_n(\sigma, \tau) \text{ for } \frac{1}{n} \leqslant \sigma \leqslant T, \frac{1}{n} \leqslant \tau \leqslant T.$$

This shows that

$$\varphi_n(t, t) \leqslant \varphi_n(s, s) + \int_s^t \psi_n(\tau, \tau)\,\mathrm{d}\tau, \ \ \frac{1}{n} \leqslant s \leqslant t \leqslant T.$$

Thus, letting $n \to \infty$ we obtain

$$\varphi(t, t) \leqslant \varphi(s, s) + \int_s^t \psi(\tau, \tau)\,\mathrm{d}\tau, \ \text{ for } 0 \leqslant s \leqslant t \leqslant T$$

as claimed.

LEMMA 2.3. *In Lemma 2.1 assume that* $u_0 \in D(A)$ *and* $f \in W^{1,1}(0, T; X)$. *Then*

$$\left\| \frac{\mathrm{d}u_\lambda}{\mathrm{d}t}(t) \right\| \leqslant |Au_0| + \|f(0)\| + \int_0^T \left\| \frac{\mathrm{d}f}{\mathrm{d}t}(t) \right\| \mathrm{d}t, \ \ 0 \leqslant t \leqslant T.$$

Proof. From (2.5) it follows that

$$\frac{\mathrm{d}}{\mathrm{d}t}(u_\lambda(t + h) - u_\lambda(t)) = A_\lambda u_\lambda(t + h) - A_\lambda u_\lambda(t) + f(t + h) - f(t)$$

almost everywhere on $]0, T - h[$. Since A_λ is dissipative this implies that

$$\frac{\mathrm{d}}{\mathrm{d}t}\|u_\lambda(t + h) - u_\lambda(t)\| \leqslant \|f(t + h) - f(t)\|, \text{ a.e. on }]0, T - h[$$

and therefore

$$\|u_\lambda(t+h) - u_\lambda(t)\| \leqslant \|u_\lambda(h) - u_\lambda(0)\| + \int_0^t \|f(s+h) - f(s)\| \, ds. \qquad (2.11)$$

Similarly, one obtains

$$\|u_\lambda(h) - u_\lambda(0)\| \leqslant h\|A_\lambda u_\lambda(0)\| + \int_0^h \|f(t)\| \, dt \leqslant h|Au_0| + \int_0^h \|f(t)| \, dt.$$

Thus dividing (2.11) by h and letting $h \to 0$ we obtain the desired estimate

$$\left\| \frac{du_\lambda}{dt}(t) \right\| \leqslant |Au_0| + \|f(0)\| + \int_0^t \left\| \frac{df}{ds}(s) \right\| ds, \quad 0 \leqslant t \leqslant T$$

because $f \in W^{1,1}(0, T; X)$ (see Section 2.2, Chapter I).

Proof of Theorem 2.1. By Lemma 2.1 the initial value problem (2.1) has an integral solution $u(t) \in \overline{D(A)}$ a.e. on $]0, T[$ given by

$$u(t) = \lim_{\lambda \to 0} u_\lambda(t) \qquad 0 \leqslant t \leqslant T$$

and the limit is uniform on $[0, T]$. Let $g \in L^1(0, T; X)$ and let $v \in C(0, T; X)$ be any integral solution of the Cauchy problem $\dfrac{dv}{dt} \in Av + g$, $v(0) = u_0$, i.e.,

$$\|v(t) - x\|^2 \leqslant \|v(s) - x\|^2 +$$

$$+ 2\int_s^t \langle y + g(\tau), v(\tau) - x\rangle_s \, d\tau, \ 0 \leqslant s \leqslant t \leqslant T \qquad (2.12)$$

for every $[x, y] \in A$. To conclude the proof it suffices to prove inequality (2.4). To this purpose, we first assume that $u_0 \in D(A)$ and $f, g \in W^{1,1}(0, T; X)$. Setting $w_\lambda(t) = (1 - \lambda A)^{-1} u_\lambda(t)$ and taking $x = w_\lambda(\sigma)$, $0 \leqslant \sigma \leqslant T$, in (2.12), we obtain

$$\frac{1}{2}\|v(t) - w_\lambda(\sigma)\|^2 \leqslant \frac{1}{2}\|v(s) - w_\lambda(\sigma)\|^2 +$$

$$+ \int_s^t \left\langle \frac{du_\lambda}{ds}(\sigma) + g(\tau) - f(\sigma), v(\tau) - w_\lambda(\sigma) \right\rangle_s d\tau, \quad \text{for } 0 \leqslant s \leqslant t \leqslant T. \qquad (2.13)$$

Since $w_\lambda(t) - u_\lambda(t) = \lambda A_\lambda u_\lambda(t)$, by Lemmas 2.1 and 2.3 it follows that

$$\lim_{\lambda \to 0} \|u_\lambda(t) - w_\lambda(t)\| = 0 \quad \text{uniformly on } [0, T]. \tag{2.14}$$

Thus, the upper-semicontinuity of $\langle , \rangle_s$ implies

$$\limsup_{\lambda \to 0} \int_\alpha^\beta \int_s^t \left\langle \frac{du_\lambda}{d\sigma}(\sigma) + g(\tau) - f(\sigma), v(\tau) - w_\lambda(\sigma) \right\rangle_s d\tau\, d\sigma \geqslant$$

$$\lim_{\lambda \to 0} \Bigg[\int_s^t \int_\alpha^\beta \langle g(\tau) - f(\sigma), v(\tau) - w_\lambda(\sigma) \rangle_s\, d\tau\, d\sigma - \frac{1}{2} \int_s^t (\|u_\lambda(\beta) -$$

$$- v(\tau)\|^2 - \|u_\lambda(\alpha) - v(\tau)\|^2)\, d\tau \Bigg].$$

Here we have used the relation $\dfrac{1}{2}\dfrac{d}{d\sigma}\|u_\lambda(\sigma) - v\|^2 = \left\langle \dfrac{du_\lambda}{d\sigma}(\sigma), u_\lambda(\sigma) - v \right\rangle_s$ (see Lemma 1.2). Thus (2.13) and (2.14) yield

$$\frac{1}{2}\int_\alpha^\beta \|u(\sigma) - v(t)\|^2\, d\sigma \leqslant \frac{1}{2}\int_\alpha^\beta \|u(\sigma) - v(s)\|^2\, d\sigma +$$

$$+ \int_\alpha^\beta d\sigma \int_s^t \langle f(\sigma) - g(\tau), u(\sigma) - v(\tau) \rangle_s\, d\tau +$$

$$+ \frac{1}{2}\int_s^t \|u(\alpha) - v(\tau)\|^2\, d\tau - \frac{1}{2}\int_s^t \|u(\beta) - v(\tau)\|^2\, d\tau$$

where $0 \leqslant \alpha \leqslant \beta \leqslant T$ and $0 \leqslant s \leqslant t \leqslant T$. This implies using Lemma 2.2 that u and v satisfy inequality (2.4) as claimed.

Now we consider the general case $u_0 \in \overline{D(A)}$ and $f, g \in L^1(0, T; X)$. Let $\{u_n^0\} \subset D(A)$ and $\{f_n\}, \{g_n\} \subset W^{1,1}(0, T; X)$ be such that $u_n^0 \to u_0$ in X, $f_n \to f$ and $g_n \to g$ in $L^1(0, T; X)$ as $n \to \infty$. Let u_n and v_n be the corresponding integral solutions of equation (2.1).

Then we have

$$\|u_n(t) - v_n(t)\|^2 \leqslant \|u_n(s) - v_n(s)\|^2 +$$

$$+ 2\int_s^t \langle f_n(\tau) - g_n(\tau), u_n(\tau) - v_n(\tau) \rangle_s\, d\tau \tag{2.15}$$

for all $0 \leqslant s \leqslant t \leqslant T$. Without any loss of generality, we may assume that $u_n(t) \to u(t)$ and $v_n(t) \to v(t)$ a.e. on $]0, T[$. In (2.15) we take the limit as $n \to \infty$ and obtain (2.4) almost everywhere in s and t. Since $u(t)$ and $v(t)$ are continuous, the inequality extends to all $0 \leqslant s \leqslant t \leqslant T$. The proof is complete.

Remark. Under the assumptions in Theorem 2.1 suppose in addition that there exists some $\omega > 0$ such that $A + \omega I$ is dissipative. Then instead of (2.4) we can obtain the following stronger estimate

$$\|u(t) - v(t)\|^2 \leqslant \exp(-2\omega(t-s))\,\|u(s) - v(s)\|^2 +$$

$$+ 2\int_s^t \exp(-2\omega(t-\tau))\,\langle f(\tau) - g(\tau), u(\tau) - v(\tau)\rangle_s \,\mathrm{d}\tau. \tag{2.16}$$

We sketch here the proof. From Lemma 2.1, $u = \lim_{\lambda\to 0} u_\lambda$ and $v = \lim_{\lambda\to 0} v_\lambda$. Since $A + \omega I$ is dissipative, it follows that $(A_\lambda u - A_\lambda v, f) \leqslant -\omega\|u - v\|^2$ for every $\lambda > 0$, u and v in X and $f \in F(u - v)$. This implies (2.16) with u_λ instead of u and v_λ instead of v. So letting $\lambda \to 0$ we obtain the desired result. If X is reflexive, Theorem 2.1 can be strengthened, namely by the following.

THEOREM 2.2. *Let X be a reflexive Banach space and let C be a closed convex cone of X. Let A be a closed dissipative set in $X \times X$ such that $D(A) \subset C$ and $C \subset R(1 - \lambda A)$ for every $\lambda > 0$. Let $u_0 \in D(A)$ and $f \in W^{1,1}(0, T; X)$ be such that $f(t) \in C$ on* $[0, T]$.

Then the initial value problem (2.1) *has a unique solution $u(t)$ which satisfies*

$$u \in W^{1,\infty}(0, T; X),\ u(t) \in D(A),\ \text{a.e. on }]0, T[. \tag{2.17}$$

$$\left\|\frac{\mathrm{d}u}{\mathrm{d}t}(t)\right\| = |Au(t) + f(t)| \leqslant |Au_0 + f(0)| + \int_0^t \left\|\frac{\mathrm{d}f}{\mathrm{d}s}(s)\right\| \mathrm{d}s,$$

$$\text{a.e. } on\ ]0, T[. \tag{2.18}$$

Proof. According to Theorem 2.1 the Cauchy problem (2.1) has a unique integral solution $u(t)$. We shall prove that $u(t)$ is in fact a (strong) solution of (2.1). First we shall show that $u \in W^{1,\infty}(0, T; X)$, i.e., $u(t)$ is absolutely continuous on $[0, T]$ and $\dfrac{\mathrm{d}u}{\mathrm{d}t} \in L^\infty(0, T; X)$ (see Section 2.2, Chapter I). In (2.2) we have taken $x = u_0$ and y to be any element of Au_0. We obtain

$$\frac{1}{2}\|u(h) - u_0\|^2 \leqslant \int_0^h \|f(\tau) + y\|\,\|u(\tau) - u_0\|\,\mathrm{d}\tau, \quad 0 < h < T.$$

Next by (2.4)

$$\frac{1}{2}\|u(t+h)-u(t)\|^2 \leqslant \frac{1}{2}\|u(h)-u(0)\|^2 +$$

$$+\int_0^t \|f(\tau+h)-f(\tau)\|\,\|u(\tau+h)-u(\tau)\|\,\mathrm{d}\tau$$

Thus by a variant of Gronwal's lemma we find that

$$\|u(h)-u_0\| \leqslant \int_0^h |f(s)+Au_0|\,\mathrm{d}s, \quad 0 \leqslant h \leqslant T,$$

respectively

$$\|u(t+h)-u(t)\| \leqslant \int_0^h |f(s)+Au_0|\,\mathrm{d}s + \int_0^t \|f(s+h)-f(s)\|\,\mathrm{d}s. \tag{2.19}$$

Since $f \in W^{1,1}(0, T; X)$ this implies that $u(t)$ is absolutely continuous on $[0, T]$. Inasmuch as X is reflexive the function $u(t)$ is also almost everywhere differentiable on $]0, T[$. So, dividing (2.19) by h yields, after passing to limit as $h \to 0$,

$$\left\|\frac{\mathrm{d}u}{\mathrm{d}t}(t)\right\| \leqslant |Au_0+f(0)| + \int_0^t \left\|\frac{\mathrm{d}f}{\mathrm{d}s}(s)\right\|\mathrm{d}s, \ \text{a.e.} \ t \in \,]0, T[$$

which is just estimate (2.18). In particular, we have proved that $u \in W^{1,\infty}(0, T; X)$. To prove that u satisfies a.e. on $]0, T[$ in Eq. (2.1) we shall proceed as in the proof of Theorem 1.5. Let $t_0 \in \,]0, T[$ be such that $\frac{\mathrm{d}u}{\mathrm{d}t}(t_0)$ exists. Then inequality (2.2) implies (see (1.41)),

$$\left\langle \frac{u(t)-u(t_0)}{t-t_0}, u(t_0)-x \right\rangle_s \leqslant \frac{1}{t-t_0}\int_{t_0}^t \langle f(\tau)+y, u(\tau)-x\rangle_s\,\mathrm{d}\tau$$

where the right-hand side is upper-semicontinuous and therefore integrable. Thus, letting $t \to t_0$ we find

$$\left\langle \frac{\mathrm{d}u}{\mathrm{d}t}(t_0), u(t_0)-x \right\rangle_s \leqslant \langle f(t_0)+y, u(t_0)-x\rangle_s. \tag{2.20}$$

To obtain (2.20) we have also assumed that t_0 is a Lebesgue point for the function $t \to \langle f(t)+y, u(t)-x\rangle_s$ but no loss of generality results because almost everywhere on $]0, T[$ this is the case. It follows by (2.20) that there exists some

$f_0 \in F(u(t_0) - x)$ such that

$$\left(\frac{du}{dt}(t_0) - y - f(t_0), f_0\right) \leqslant 0. \tag{2.21}$$

On the other hand, for every $0 \leqslant h \leqslant t_0$, $u(t_0 - h)$ can be represented as

$$u(t_0 - h) = u(t_0) - h\frac{d}{dt}u(t_0) + g(h)$$

where $\lim_{h\to 0} \|g(h)\|/h = 0$. By hypothesis there exists $[x_h, y_h] \in A$ such that $u(t_0 - h) + hf(t_0) = x_h - hy_h$ (We recall that $f(t) \in C$ on $[0, T]$.) In (2.21) we take $x = x_h$ and $y = y_h$ to obtain

$$(u(t_0) - x_h + g(h), f_0) \leqslant 0$$

or

$$\|u(t_0) - x_h\|/h \leqslant \|g(h)\|/h \to 0 \quad \text{as} \quad h \to 0.$$

Hence $\lim_{h\to 0} y_h = \frac{du}{dt}(t_0) - f(t_0)$ and the closedness of A implies $\frac{du}{dt}(t_0) - f(t_0) \in Au(t_0)$ as claimed. Thus the proof is complete.

Remarks. 1°. Theorem 2.2 and its proof remain valid if the condition that X be reflexive is weakened to the condition that every X-valued Lipschitz continuous function on $[0, T]$ is almost everywhere differentiable on $]0, T[$. This is the case for $X = l_1$.

2°. In Theorem 2.2 the assumption $f \in W^{1,1}(0, T; X)$ is unnecessarily strong to prove that the integral solutions with initial values in $D(A)$ satisfy almost everywhere on $]0, T[$ Eq. (2.1), i.e. are strong solutions. It turns out that in some cases it suffices to assume that $f(t)$ is of *bounded variation* on $[0, T]$ (see M. G. Crandall and A. Pazy [3], H. Brézis [8], for this treatement).

If X and X^* are uniformly convex, Theorem 2.2 can be strengthened to generalize Theorem 1.6.

COROLLARY 2.1. *In addition to the conditions of Theorem 2.2, let X and X^* be uniformly convex. Then the solution $u(t)$ of the initial value problem* (2.1) *is everywhere differentiable from the right and*

$$\frac{d^+}{dt}u(t) = (Au(t) + f(t))^0, \quad \textit{for } 0 \leqslant t < T. \tag{2.22}$$

Proof. Since for every t the operator $u \to Au + f(t)$ satisfies the condition (1.28), it follows by Theorem 1.6 that for every $u \in D(A)$ the set $Au + f$ has a unique element of minimum norm $(Au+f)^0$. Moreover, (2.18) clearly implies that

$$\frac{du}{dt}(t) = (Au(t)+f(t))^0 \qquad \text{a.e. } t \in]0, T[.$$

On the other hand, we have

$$\frac{d}{dh}(u(t+h)-u(t)) \in Au(t+h)+f(t+h).$$

Since A is dissipative, we see by multiplying by $F(u(t+h)-u(t))$ and integrating over $[0, h]$ that

$$\|u(t+h)-u(t)\| \leqslant h|Au(t)+f(t)| \quad + \int_0^h \|f(s+h)-f(s)\| \, ds.$$

therefore,

$$\limsup_{h\to 0} \left\| \frac{u(t+h)-u(t)}{h} \right\| \leqslant |Au(t)+f(t)|.$$

The last argument in the proof of Theorem 1.6 can now be repeated here, and the proof is complete.

Definition 2.2. The function $u \in C(0, T; X)$ is called a *weak solution* of the initial value problem (2.1) if there are sequences $\{u_n\} \subset W^{1,\infty}(0, T; X)$ and $\{f_n\} \subset L^1(0, T; X)$ such that

(a) $\dfrac{du_n}{dt}(t) \in Au_n(t)+f(t)$ a.e. $t \in]0, T[$, $n = 1, 2, \ldots$

(b) $\lim_{n\to\infty} u_n(t) = u(t)$ uniformly on $[0, T]$.

(c) $u(0) = u_0$ and $\lim_{n\to\infty} f_n = f$ in $L^1(0, T; X)$.

If A is dissipative it follows at once that every weak solution is also an integral solution.

Corollary 2.2. *Let X be reflexive and let $A \subset X \times X$ satisfy the assumptions of Theorem 2.1. Then for every $u_0 \in \overline{D(A)}$ and $f \in L^1(0, T; X)$, $f(t) \in C$ a.e. on $]0, T[$, the initial value problem (2.1) has a unique weak solution $u(t) \in \overline{D(A)}$ a.e. on $[0, T]$.*

Proof. For simplicity we shall assume that $C = X$. Let $(u_0, f) \in D(A) \times W^{1,1}(0, T; X)$ and let $u \in W^{1,\infty}(0, T; X)$ be the solution of (2.1) provided by Theorem 2.2. We set $u = K(u_0, f)$ and observe that

$$\|K(x,f)(t) - K(y,g)(t)\| \leqslant \|x-y\| + \int_0^t \|f(s)-g(s)\| \, ds, \quad 0 \leqslant t \leqslant T.$$

for every $x, y \in D(A)$ and f, g in $W^{1,1}(0, T; X)$. Let us denote again by K the extension of K on $\overline{D(A)} \times L^1(0, T; X)$. Obviously, $u(t) = K(u_0, f)$ is a weak solution of (2.1) for every $u_0 \in \overline{D(A)}$ and $f \in L^1(0, T; X)$ and this concludes the proof.

Theorem 2.2 and its consequences remain valid if the condition that A is dissipative is weakened to the condition that $A + \omega I$ is dissipative for some real ω. It would thus be possible to follow the method given above. However, we think that it is useful for the reader to be aware of another approach to the existence theory of nonlinear differential equations associated with dissipative sets, available in the case in which X^* is uniformly convex. For simplicity we shall confine ourselves to consider the case in which $C = X$.

THEOREM 2.3. *Let X^* be uniformly convex and let A be an m-dissipative subset of $X \times X$. Let $(u_0, f) \in D(A) \times W^{1,1}(0, T; X)$ and let ω be any real and fixed number. Then there exists a unique function u: $[0, T] \to X$ which satisfies*

(i) $u \in W^{1,\infty}(0, T; X)$,

(ii) $u(t) \in D(A)$ a.e. *on* $]0, T[$ *and*

$$\begin{aligned} &\frac{du}{dt}(t) \in Au(t) + \omega u(t) + f(t) \qquad \text{a.e. } t \in]0, T[\\ &u(0) = u_0. \end{aligned} \tag{2.23}$$

Moreover, if X is uniformly convex, then $u(t)$ is everywhere differentiable from the right and

$$\frac{d^+}{dt} u(t) \in (Au(t) + \omega u(t) + f(t))^0, \quad \textit{for } 0 \leqslant t < T. \tag{2.24}$$

Proof. 1°. *Uniqueness.* Let u and v be two solutions to problem (2.1), i.e.,

$$\frac{d}{dt}(u(t) - v(t)) \in Au(t) - Av(t) + \omega(u(t) - v(t))$$

and $u(0) - v(0) = 0$. Multiplying the latter by $F(u(t) - v(t))$ and using Lemma 1.2, yields

$$\frac{d}{dt}\|u(t) - v(t)\| \leqslant \omega\|u(t) - u(t)\|, \quad \text{a.e. on }]0, T[$$

which implies that $u = v$.

2°. *Existence.* Consider the approximate equations

$$\begin{aligned} &\frac{du_n}{dt} = A_n u_n + \omega u_n + f, \quad 0 \leqslant t \leqslant T, \\ &u_n(0) = u_0, \end{aligned} \tag{2.25}$$

where

$$A_n = n((1 - n^{-1}A)^{-1} - 1), \qquad n = 1, 2, \ldots$$

We recall that for every n, A_n is everywhere defined and Lipschitz on X. Thus Eq. (2.25) has a unique classical solution $u_n \in C^1(0, T; X)$ (see Proposition 1.1).

Let m, n be arbitrary positive integers. Using again Lemma 1.2 we find

$$\frac{1}{2}\frac{\mathrm{d}}{\mathrm{d}t}\|u_n(t) - u_m(t)\|^2 = \omega\|u_n(t) - u_m(t)\|^2 +$$

$$+ (A_nu_n(t) - A_mu_m(t),\ F(u_n(t) - u_m(t))), \quad \text{a.e. on }]0, T[$$

and therefore

$$\|u_n(t) - u_m(t)\|^2 \leqslant 2\int_0^t \mathrm{e}^{2\omega(t-s)}(A_nu_n(s) - A_mu_m(s);\ F(u_n(s) - u_m(s))\,\mathrm{d}s.$$

Set $J_n = (1 - n^{-1}A)^{-1}$. Since $A_nu_n \in A\,J_nu_n$ and A is dissipative, the last inequality implies that

$$\|u_n(t) - u_m(t)\|^2 \leqslant 2\int_0^t \mathrm{e}^{2\alpha(t-s)}\|A_nu_n(s) - A_mu_m(s)\|\ \|F(u_n(s) -$$

$$- u_m(s)) - F(J_nu_n(s) - J_mu_m(s))\|\,\mathrm{d}s. \tag{2.26}$$

On the other hand, differentiating Eq. (2.25) yields

$$\frac{\mathrm{d}^2u_n}{\mathrm{d}t^2} = \omega\frac{\mathrm{d}u_n}{\mathrm{d}t} + \frac{\mathrm{d}}{\mathrm{d}t}A_nu_n + \frac{\mathrm{d}f}{\mathrm{d}t}, \qquad \text{a.e. } t \in]0, T[. \tag{2.27}$$

By the dissipativeness of A_n we have

$$\left(\frac{\mathrm{d}}{\mathrm{d}t}(A_nu_n),\ F\left(\frac{\mathrm{d}u_n}{\mathrm{d}t}\right)\right) \leqslant 0, \qquad \text{a.e. } t \in]0, T[.$$

Thus, multiplying Eq. (2.27) by $F\left(\frac{\mathrm{d}u_n}{\mathrm{d}t}\right)$ and integrating this inequality over $[0, t]$ we obtain

$$\left\|\frac{\mathrm{d}u_n}{\mathrm{d}t}(t)\right\| \leqslant \mathrm{e}^{\omega t}\|A_nu_0 + \omega u_0 + f(0)\| + \int_0^t \mathrm{e}^{\omega(t-s)}\left\|\frac{\mathrm{d}f}{\mathrm{d}s}(s)\right\|\mathrm{d}s. \tag{2.28}$$

Since $\|A_n u_0\| \leqslant |Au_0|$, it follows from (2.28) that $\|A_n u_n(t)\|$ are uniformly bounded on $[0, T]$. Let $n \to \infty$ then $\|u_n(t) - J_n u_n(t)\| = n^{-1} \|A_n u_n(t)\| \to 0$ uniformly on $[0, T]$. Recall that the uniform convexity of X^* implies in particular that the duality mapping F of X is uniformly continuous on the bounded sets of X. Thus (2.26) shows that $\lim_{n\to\infty} u_n(t) = u(t)$ exists and that this limit is uniform on $[0, T]$. Also estimate (2.28) implies that $u(t)$ is absolutely continuous on $[0, T]$ and $\dfrac{du}{dt} \in L^\infty(0, T; X)$, i.e., $u \in W^{1,\infty}(0, T; X)$. We shall prove that $u(t)$ satisfies almost everywhere on $]0, T[$ Eq. (2.23). The argument we shall use is quite similar to that used in the proof of Theorem 2.2. Let $[x, y] \in A$ and let $x_n = x - n^{-1}y$. Since $y = A_n x_n$ and A_n is dissipative, it follows from Eq. (2.25) that

$$\|u_n(t) - x_n\|^2 - \|u_n(t_0) - x_n\|^2 \leqslant 2\int_{t_0}^{t} (y + \omega u_n(s) + f(s),$$
$$F(u_n(s) - x_n))\,ds$$

and letting $n \to \infty$,

$$\|u(t) - x\|^2 - \|u(t_0) - x\|^2 \leqslant 2\int_{t_0}^{t} (y + \omega u(s) + f(s),\ F(u(s) - x))\,ds$$

or

$$\left(\frac{u(t) - u(t_0)}{t - t_0},\ F(u(t_0) - x)\right) \leqslant \frac{1}{t - t_0}\int_{t_0}^{t} (y + \omega u(s) + f(s),$$
$$F(u(s) - x))\,ds.$$

Let t_0 be such that $\dfrac{du}{dt}(t_0)$ exists. Then the last inequality implies that

$$\left(\frac{du}{dt}(t_0) - y - \omega u(t_0) - f(t_0),\ F(u(t_0) - x)\right) \leqslant 0.$$

Since $[x, y] \in A$ was arbitrary and A is maximal dissipative we have $u(t_0) \in D(A)$ and $\dfrac{du}{dt}(t_0) \in Au(t_0) + \omega u(t_0) + f(t_0)$. The proof of (2.24) is quite similar to that of Corollary 2.1 and therefore we shall omit it.

Remark. Under the assumptions of Theorem 2.3, suppose that there exists a closed convex cone C of X such that

$$(1 - \lambda A)^{-1}\, C \subset C, \quad \text{for every } \lambda > 0$$

and

$$u_0 \in C,\ f(t) \in C, \quad \text{for every } t \in [0, T].$$

Then the solution $u(t)$ of problem (2.23) also satisfies

$$u(t) \in C \quad \text{for every } t \in [0, T].$$

To prove this it suffices to observe in the above proof that $u_n(t) \in C$ for every $t \in [0, T]$. This is an immediate consequence of the fact that the integral Volterra operator associated with the approximate problem (2.25) maps the closed set $\{u \in C(0, T; X);\ u(t) \in C \text{ on } [0, T]\}$ into itself.

2.2. Periodic solutions

In this section we are interested in the problem

$$\begin{aligned} &\frac{du}{dt}(t) \in Au(t) + f(t), \quad \text{a.e. on }]0, T[\\ &u(0) = u(T). \end{aligned} \tag{2.29}$$

THEOREM 2.4. *Let X be reflexive and let A be an m-dissipative set in $X \times X$. In addition, suppose that there is $\omega > 0$ such that $A + \omega I$ is dissipative. Let $f \in W^{1,1}(0, T; X)$ be such that $f(0) = f(T)$. Then problem (2.29) has a unique solution $u \in W^{1,\infty}(0, T; X)$. Moreover, if X and X^* are uniformly convex, then $u(t)$ is everywhere differentiable from the right and satisfies*

$$\frac{d^+}{dt} u(t) = (Au(t) + f(t))^0, \text{ for every } t \in [0, T[. \tag{2.30}$$

Proof. According to Corollary 2.2 for every $x \in \overline{D(A)}$, the equation

$$\begin{aligned} &\frac{du}{dt} \in Au + f \quad 0 < t < T \\ &u(0) = x \end{aligned} \tag{2.31}$$

has a unique weak solution $u \in C(0, T; X)$. Let $K_T\colon \overline{D(A)} \to \overline{D(A)}$ be the operator defined by $K_T(x) = u(T)$. Since $A + \omega I$ is dissipative it follows by Definition 2.2 that

$$\|K_T(x) - K_T(y)\| \leqslant e^{-\omega T}\|x - y\|, \text{ for every } x, y \in \overline{D(A)}.$$

Hence K_T has a unique fixed point $u_0 \in \overline{D(A)}$. Let $u(t)$ be the corresponding weak solution of (2.31). Obviously $u(0) = u(T)$. We shall prove that $u(t)$ satisfies the conditions required by our theorem. Let $\{u_n\} \subset W^{1,\infty}(0, T; X)$

and $\{f_n\} \subset W^{1,1}(0, T; X)$ be such that $u_n \to u$ in $C(0, T; X)$, $f_n \to f$ in $L^1(0, T; X)$ as $n \to \infty$ and

$$\frac{du_n}{dt} \in Au_n + f_n \quad \text{a.e. on }]0, T[.$$

For each n, we have

$$\|u_n(t+h) - u_n(t)\| \leqslant e^{-\omega t}\|u_n(h) - u_n(0)\| + \int_0^t e^{-\omega(t-s)}\|f_n(s+h) - f_n(s)\|ds$$

while since $u_n \to u$, it follows that

$$\|u(t+h) - u(t)\| \leqslant e^{-\omega t}\|u(h) - u(0)\| + \int_0^t e^{-\omega(t-s)}\|f(s+h) - f(s)\|ds \tag{2.32}$$

for every t and $t+h$ in $[0, T]$ and $h > 0$. We extend $f(t)$ and $u(t)$ as $f(t-T)$ and $u(t-T)$, respectively, for $T \leqslant t \leqslant 2T$. Thus, we may assume that $u(t)$ is an integral solution of the equation $\frac{du}{dt} \in Au + f$ on $[0, 2T]$ and therefore, estimate (2.32) holds for every t and h in $[0, T]$. Taking $t = T$ in (2.32) and recalling that $u(0) = u(T)$, we obtain

$$\|u(h) - u(0)\| \leqslant (1 - e^{-\omega T})^{-1}\int_0^T e^{-\omega(T-s)}\|f(s+h) - f(s)\|ds. \tag{2.33}$$

From estimates (2.32) and (2.33) it follows that $u(t)$ is absolutely continuous on $[0, T]$ and

$$\left\|\frac{du}{dt}(t)\right\| \leqslant e^{-\omega t}(1 - e^{-\omega T})^{-1}\int_0^T e^{-\omega(T-s)}\left\|\frac{df}{ds}(s)\right\| ds +$$
$$+\int_0^t e^{-\omega(t-s)}\left\|\frac{df}{ds}(s)\right\| ds.$$

We conclude the proof by the same argument as that used in the proof of Theorem 2.2 (see also the proof of Theorem 2.3).

2.3. Examples

In this section we give some simple examples to which the preceding theory applies.

Example 1. We are given a Hilbert space H and a reflexive Banach space V such that $V \subset H$ and the inclusion mapping of V into H is continuous. Suppose that V is dense in H and denote by V' the dual space of V. Identifying H with its own dual we can write

$$V \subset H \subset V'$$

The norms in V and H shall be denoted by $\|\cdot\|$ and $|\cdot|$, respectively. We use the notation (v_1, v_2) to indicate pairing between the element v_1 of V' and the element v_2 of V; if $v_1, v_2 \in H$ this is the ordinary inner product in H. Finally, we shall denote by $\|\cdot\|_*$ the norm (dual) in V'.

Besides these spaces we are given a single-valued, monotone operator $A: V \to V'$ which is assumed to be hemicontinuous and coercive from V to V'.

We begin with the following simple existence theorem.

THEOREM 2.5. *Let $f \in W^{1,1}(0, T; H)$ and let $u_0 \in V$ such that $Au_0 \in H$. Then there exists one and only one function $u(t)$: $[0, T] \to V$ which satisfies*

$$u \in W^{1,\infty}(0, T; H), \; Au \in L^\infty(0, T; H). \tag{2.34}$$

$$\frac{du}{dt}(t) + Au(t) = f(t), \quad \text{a.e. } t \in]0, T[. \tag{2.35}$$

$$u(0) = u_0 \tag{2.36}$$

Moreover, $u(t)$ is everywhere differentiable from the right (in the topology of H) and

$$\frac{d^+}{dt}u(t) + Au(t) = f(t), \quad \text{for every } 0 \leqslant t < T. \tag{2.37}$$

Proof. Let A_H be the operator defined by

$$A_H u = Au \text{ for } u \in D(A_H) = \{u \in V;\; Au \in H\}. \tag{2.38}$$

By hypothesis, the operator $u \to u + Au$ is monotone, hemicontinuous and coercive from V to V'. We may conclude therefore, that $R(1 + A)$ is all of V' (see Theorems 1.3 and 1.5 in Chapter II). This clearly implies that $R(1 + A_H) = H$, i.e., A_H is maximal monotone in $H \times H$. The general theory (i.e., Theorem 2.2 and Corollary 2.1) can therefore be applied to the present situation, with $X = H$ and $A = -A_H$.

From Theorem 2.4 we can obtain, in a similar way, an existence result for the periodic problem associated with the operator A. We propose that the reader formulate the corresponding theorem. We are now able to derive from Theorem 2.5 other existence theorems for some special cases.

THEOREM 2.6. *Let V, H be given satisfying the above hypotheses and let A be a hemicontinuous and monotone operator from V to V' which is assumed to satisfy*

$$(Au, u) \geqslant \omega \|u\|^p + C_1, \quad \forall\, u \in V. \tag{2.39}$$

$$\|Au\|_* \leqslant C_2(1 + \|u\|^{p-1}), \quad \forall\, u \in V. \tag{2.40}$$

where $\omega > 0$, $p \geqslant 2$ and C_1, C_2 are some constants.

Given $u_0 \in H$ *and* $f \in L^{p'}(0, T; V')$, *where* $\frac{1}{p} + \frac{1}{p'} = 1$, *then there exists a unique function* $u(t)$ *which is* V'*-valued absolutely continuous on* $[0, T]$ *and satisfies*

$$u \in C(0, T; H) \cap L^p(0, T; V), \frac{du}{dt} \in L^{p'}(0, T; V'). \tag{2.41}$$

$$\frac{du}{dt}(t) + Au(t) = f(t) \text{ a.e. } t \in]0, T[. \tag{2.42}$$

$$u(0) = u_0. \tag{2.43}$$

Proof. First we assume that $u_0 \in D(A_H)$ and $f \in W^{1,1}(0, T; H)$ and denote by $u \in W^{1,\infty}(0, T; H)$ the solution of the Cauchy problem (2.42), (2.43), provided by Theorem 2.5. Note that conditions (2.39) and (2.40) imply the existence of a constant C_1 such that

$$\frac{1}{2}\frac{d}{dt}|u(t)|^2 + \omega\|u(t)\|^p \leqslant \|f(t)\|_* \|u(t)\|^{p-1} + C_1, \text{ a.e. on }]0, T[.$$

Integrating this inequality from 0 to t, after some calculation, we obtain

$$|u(t)|^2 + \int_0^t \|u(s)\|^p ds \leqslant C(|u_0|^2 + \int_0^t \|f(s)\|_*^{p'} ds + 1), \ 0 \leqslant t \leqslant T. \tag{2.44}$$

Hence

$$\int_0^T \left\|\frac{du}{dt}(t)\right\|_*^{p'} dt \leqslant C\left(|u_0|^2 + \int_0^T \|f(t)\|_*^{p'} dt + 1\right), \quad C > 0. \tag{2.45}$$

We assert that $D(A_H)$ is a dense subset of H. Indeed let u be any arbitrary element of H and let $u_\varepsilon = (1 + \varepsilon A_H)^{-1}u$. The equality

$$u_\varepsilon + \varepsilon Au_\varepsilon = u$$

then implies, by virtue of (2.39), and (2.40), that

$$|u_\varepsilon|^2 + \omega\varepsilon\|u_\varepsilon\|^p \leqslant |u_\varepsilon|\,|u| + C\varepsilon, \ \varepsilon > 0$$

and

$$\|u_\varepsilon - u\|_* \leqslant \varepsilon\|Au_\varepsilon\|_* \leqslant C\varepsilon(\|u_\varepsilon\|^{p-1} + 1).$$

This shows that u_ε are bounded in H and $u_\varepsilon \to u$ in V' as $\varepsilon \to 0$. This clearly implies that $u_\varepsilon \rightharpoonup u$ in H as $\varepsilon \to 0$, and therefore we may conclude that $D(A_H)$ is dense in H as asserted.

Now let $u_0 \in H$ and $f \in L^{p'}(0, T; V')$. Choose $\{u_n^0\} \subset D(A_H)$ and $\{f_n\} \subset W^{1,1}(0, T; H)$ such that $u_n^0 \to u_0$ in H and $f_n \to f$ in $L^{p'}(0, T; V')$ as $n \to \infty$. We have used here the density of $W^{1,1}(0, T; H)$ in $L^{p'}(0, T; V')$ which is an immediate consequence of the fact that H is dense in V'. Let $u_n \in W^{1,\infty}(0, T; H)$ be the solution of (2.42) with the initial value u_n^0 and with f_n instead of f. Since A is monotone we have

$$\frac{1}{2}\frac{d}{dt}|u_n(t) - u_m(t)|^2 \leqslant (f_n(t) - f_m(t), u_n(t) - u_m(t)), \quad \text{a.e. on }]0, T[$$

and therefore

$$|u_n(t) - u_m(t)|^2 \leqslant |u_n^0 - u_m^0|^2 +$$
$$+ 2\left(\int_0^t \|f_n(s) - f_m(s)\|_*^{p'}\,\mathrm{d}s\right)^{1/p'} \left(\int_0^t \|u_n(s) - u_m(s)\|^p\,\mathrm{d}s\right)^{1/p}. \tag{2.46}$$

On the other hand, it follows by (2.44) and (2.45) that u_n are bounded in $L^p(0, T; V)$ and $\frac{\mathrm{d}u_n}{\mathrm{d}t}$ are bounded in $L^{p'}(0, T; V')$. Comparing this with inequality (2.46) we find that $\lim_{n\to\infty} u_n(t) = u(t)$ exists in H and the limit is uniform on $[0, T]$. Moreover, extracting a subsequence if necessary, we may assume that

$$u_n \rightharpoonup u \quad \text{in } L^p(0, T; V)$$

$$\frac{\mathrm{d}u_n}{\mathrm{d}t} \rightharpoonup \frac{\mathrm{d}u}{\mathrm{d}t} \quad \text{in} \quad L^{p'}(0, T; V')$$

where $\frac{\mathrm{d}u}{\mathrm{d}t}$ denotes the derivative of u in the sense of V'-valued distributions on $]0, T[$. In particular, we have proved that $u \in W^{1,p'}(0, T; V')$ and this implies (see Theorem 2.2, Chapter I) that $u(t) : [0, T] \to V'$ is absolutely continuous, almost everywhere differentiable on $]0, T[$ and its ordinary derivative belongs to $L^{p'}(0, T; V')$, i.e. $u \in A^{1,p'}(0, T; V')$. The argument below is thus to show that $u(t)$ satisfies a.e. on $]0, T[$ Eq. (2.42). Let $x \in V$ be fixed. Then

$$\frac{1}{2}\frac{\mathrm{d}}{\mathrm{d}t}|u_n(t) - x|^2 + (Au_n(t), u_n(t) - x) = (f_n(t), u_n(t) - x), \quad \text{a.e. on }]0, T[.$$

Integrating this equality over $]s, t[$ and using the monotonicity of A, we obtain

$$\left(\frac{u_n(t) - u_n(s)}{t - s},\ u_n(s) - x\right) \leqslant \frac{1}{t - s}\int_s^t (-Ax + f_n(\tau), u_n(\tau) - x)\,\mathrm{d}\tau$$

and hence

$$\left(\frac{u(t) - u(s)}{t - s},\ u(s) - x\right) \leqslant \frac{1}{t - s}\int_s^t (-Ax + f(\tau), u(\tau) - x)\,\mathrm{d}\tau.$$

This implies

$$\left(\frac{\mathrm{d}u}{\mathrm{d}t}(t_0) - f(t_0) + Ax, u(t_0) - x\right) \leqslant 0, \quad \forall x \in V \tag{2.47}$$

for every $t_0 \in]0, T[$ such that $\frac{\mathrm{d}}{\mathrm{d}t}u(t_0)$ exists (in V') and $\lim_{h\to 0}\frac{1}{h}\int_{t_0}^{t_0+h} f(s)\,\mathrm{d}s = f(t_0)$. In other words, inequality (2.47) holds almost everywhere on

$]0, T[$. Thus the arbitrariness of $x \in V$ as well as the fact that A is maximal monotone in $V \times V'$ imply (2.42). The proof of Theorem 2.6 is complete.

Finally, we state a regularity result for this equation.

THEOREM 2.7. *Let V be a Hilbert space and let A be a hemicontinuous operator from V to V' satisfying*

$$(Au - Av, u - v) \geqslant \omega \|u - v\|^2 \text{ for every } u, v \in V. \tag{2.48}$$

$$\|Au\|_* \leqslant C(\|u\| + 1) \quad \text{for every } u \in V \tag{2.49}$$

where $\omega > 0$.
Let $u_0 \in H$ and $f \in L^2(0, T; V')$ be given such that

$$t^\alpha \frac{\mathrm{d}f}{\mathrm{d}t} \in L^2(0, T; V') \tag{2.50}$$

for some $\alpha \geqslant 1$. Then, solution u of the initial value problem (2.42), (2.43) *satisfies*

$$t^\alpha \frac{\mathrm{d}u}{\mathrm{d}t} \in L^2(0, T; V) \cap L^\infty(0, T; H). \tag{2.51}$$

Proof. We know from Theorem 2.6 that $u \in L^2(0, T; V)$ and $\dfrac{\mathrm{d}u}{\mathrm{d}t} \in L^2(0, T; V')$. In particular, this implies that

$$\frac{1}{2}(|u(t)|^2 - |u(s)|^2) = \int_s^t \left(\frac{\mathrm{d}u}{\mathrm{d}t}(\tau), u(\tau)\right)\mathrm{d}\tau, \text{ for every } s, t \in [0, T].$$

In other words, $|u(t)|^2$ is absolutely continuous and $\dfrac{1}{2}\dfrac{\mathrm{d}}{\mathrm{d}t}|u(t)|^2 = \left(\dfrac{\mathrm{d}u}{\mathrm{d}t}(t), u(t)\right)$ a.e. on $]0, T[$. Thus from hypothesis (2.48) of our theorem, it follows that

$$\frac{1}{2}\frac{\mathrm{d}}{\mathrm{d}t}(t^{2\alpha}|u(t+h) - u(t)|^2) + \omega t^{2\alpha}\|u(t+h) - u(t)\|^2 \leqslant$$

$$\leqslant t^{2\alpha}\|f(t+h) - f(t)\|_* \|u(t+h) - u(t)\| + 2\alpha t^{2\alpha-1}|u(t+h) - u(t)|^2.$$

Integrating over $[0, t]$ and using Schwarz's inequality, after rearrangement we obtain

$$t^{2\alpha}|u(t+h) - u(t)|^2 + \int_0^t s^{2\alpha}\|u(s+h) - u(s)\|^2 \,\mathrm{d}s \leqslant$$

$$\leqslant C\left(\int_0^t s^{2\alpha}\|f(s+h) - f(s)\|_*^2 \,\mathrm{d}s + \int_0^t s^{2(\alpha-1)}\|u(s+h) - u(s)\|_*^2 \,\mathrm{d}s\right), \tag{2.52}$$

$$0 \leqslant t \leqslant T.$$

Here we have also used the obvious inequality: $|u(s+h)-u(t)|^2 \leqslant \|u(t+h)-u(t)\|_* \, \|u(t+h)-u(t)\|$. Since $t^\alpha \dfrac{df}{dt} \in L^2(0, T; V')$, Proposition 2.1 (Chapter I), together with estimate (2.52), then implies that $u(t)$ is H-absolutely continuous on every interval $[\delta, T]$, $0 < \delta < T$, and

$$t^{2\alpha}\left|\frac{du}{dt}(t)\right|^2 + \int_0^t s^{2\alpha}\left\|\frac{du}{ds}(s)\right\|^2 ds \leqslant C\left(\int_0^t s^{2\alpha}\left\|\frac{df}{ds}(s)\right\|_*^2 ds + \right.$$

$$\left. + t^{2(\alpha-1)}\int_0^t \left\|\frac{du}{ds}(s)\right\|_*^2 ds\right), \quad \text{for } 0 \leqslant t \leqslant T,$$

as claimed.

Remark. From Theorem 2.1, (see also Corollary 2.2) we know that for every $u_0 \in \overline{D(A_H)} = H$ and $f \in L^1(0, T; H)$ the initial value problems (2.42), and (2.43) have a unique generalized solution $u(t)$. It is an interesting fact that under slightly different additional assumptions on f this solution is actually a strong solution of Eq. (2.42). In particular, for $f = 0$ it follows that the semigroup $S(t)$ generated on H by $-A_H$, has *a smoothing effect on the initial data*, i.e.,

$$S(t)u_0 \in D(A_H) \text{ for every } u_0 \in H = \overline{D(A_H)},\ t > 0 \tag{2.53}$$

and

$$t\frac{d}{dt}S(t)u_0 \in L^\infty(0; T; H) \text{ for some } T > 0. \tag{2.54}$$

Other remarkable classes of nonlinear semigroups with this property will be considered in the next chapter.

The general theory given in this section can be applied to partial differential equations of the parabolic type. A typical example is the initial boundary value problem

$$\begin{aligned} &\frac{\partial u}{\partial t} + \sum_{|\alpha| \leqslant m} (-1)^\alpha D^\alpha A_\alpha(x, u, \ldots D^m u) = f \\ &D^\beta u = 0 \quad \text{on }]0, T[\times \Gamma;\ |\beta| \leqslant m - 1 \\ &u(0, x) = u_0(x) \text{ on } \Omega, \end{aligned} \tag{2.55}$$

in the cylinder $]0, T[\times\Omega$, where $A_\alpha(x, \xi)$ satisfy conditions I and II from Section 1.3, Chapter II and in addition

III′ $a(u, u) \geqslant \omega\|u\|_{m,p}^p + C$ for every $u \in W^{m,p}(\Omega)$

where $\omega >$ and

$$a(u, v) = \sum_{|\alpha| \leqslant m} \int_\Omega A_\alpha(x, u, \ldots, D^m u) D^\alpha v \, dx; \quad u, v \in W^{m,p}(\Omega).$$

We have proved that the operator $A : W_0^{m,p}(\Omega) \to (W_0^{m,p}(\Omega))' = W^{-m,p'}(\Omega)$ defined by

$$(Au, v) = a(u, v) \quad \text{for every } u, v \in W_0^{m,p}(\Omega)$$

is monotone and demicontinuous. Moreover, it follows by I and III′ that

$$\|Au\|_{-m,p'} \leqslant C(1 + \|u\|_{m,p}^{p-1}), \qquad \forall\, u \in W_0^{m,p}(\Omega)$$

and

$$(Au, u) \geqslant \omega \|u\|_{m,p}^{p} + C, \qquad \forall\, u \in W_0^{m,p}(\Omega).$$

We take $V = W_0^{m,p}(\Omega)$, $H = L^2(\Omega)$ and $V' = W^{-m,p'}(\Omega)$ where $\frac{1}{p} + \frac{1}{p'} = 1$, $2 \leqslant p < \infty$. Problem (2.55) (taken in an appropiate sense) can be written as

$$\frac{du}{dt} + Au(t) = f(t) \qquad \text{a.e. } t \in]0, T[$$

$$u(0) = u_0$$

where $u : [0, T] \to V$. Theorem 2.6 may therefore be applied to yield the following result.

Given:

$$u_0 \in L^2(\Omega),\ f \in L^{p'}(0, T; W^{-m,p'}(\Omega)), \tag{2.56}$$

there exists a unique solution $u(t.x)$ *for problem* (2.55) *satisfying*

$$u \in C(0, T; L^2(\Omega)) \cap L^p(0, T; W_0^{m,p}(\Omega)) \tag{2.57}$$

$$\frac{\partial u}{\partial t} \in L^{p'}(0, T; W^{-m,p'}(\Omega)). \tag{2.58}$$

Moreover, if $Au_0 \in L^2(\Omega)$, $f \in L^1(0, T; L^2(\Omega))$ *and* $\frac{\partial f}{\partial t} \in L^1(0, T; L^2(\Omega))$ *then*

$$\frac{\partial u}{\partial t} \in L^\infty(0, T; L^2(\Omega)). \tag{2.59}$$

Example 2. Let Ω be a bounded domain in R^n with a smooth boundary Γ and let β be a maximal monotone subset of $R^1 \times R^1$ such that $0 \in D(\beta)$. Consider the following semi-linear parabolic problem

$$\frac{\partial u}{\partial t} - \Delta u + \beta(u) \ni f, \qquad \text{on }]0, T[\times \Omega$$

$$u(t, x) = 0, \qquad \text{on }]0, T[\times \Gamma \tag{2.60}$$

$$u(0, x) = u_0(x), \qquad \text{on } \Omega$$

where $f(t, x)$ is a given L^p-function on $]0, T[\times \Omega$, $p > 1$. Set

$$\tilde{\beta}(u) = \{v \in L^p(\Omega);\ v(x) \in \beta(u(x)) \text{ a.e. on } \Omega\}, \quad u \in D(\tilde{\beta})$$

where

$$D(\tilde{\beta}) = \{u \in L^p(\Omega);\ \exists\ v \in L^p(\Omega), \text{ such that } v(x) \in \beta(u(x)), \text{ a.e. on } \Omega\}.$$

Let $A \subset L^p(\Omega) \times L^p(\Omega)$ be the operator defined by

$$Au = \Delta u - \tilde{\beta}(u), \qquad \forall\ u \in D(A)$$

where $D(A) = W_0^{1,p}(\Omega) \cap W^{2,p}(\Omega) \cap D(\tilde{\beta})$.

Then problem (2.60) may be rewritten as

$$\frac{\mathrm{d}u}{\mathrm{d}t} = Au + f, \qquad 0 < t < T$$

$$u(0) = u_0$$

where u is regarded as a function from $[0, T]$ to $L^p(\Omega)$. By Proposition 3.7 (Chapter II) the operator A defined above is m-dissipative in $L^p(\Omega) \times L^p(\Omega)$. We recall that for $1 < p < \infty$, the space $X = L^p(\Omega)$ is uniformly convex. Corollary 2.1 of Theorem 2.2 therefore yields the following result:

Let $u_0 \in W_0^{1,p}(\Omega) \cap W^{2,p}(\Omega) \cap D(\tilde{\beta})$ *and* $f \in L^1(0, T; L^p(\Omega))$ *be given such that* $\frac{\partial f}{\partial t} \in L^1(0, T; L^p(\Omega))$. *Then problem* (2.60) *has a unique solution* u *satisfying:*

$$u \in C(0, T; L^p(\Omega)),\ \frac{\partial u}{\partial t} \in L^\infty(0, T; L^p(\Omega)) \tag{2.61}$$

$$u(t, .) \in W_0^{1,p}(\Omega) \cap W^{2,p}(\Omega), \ \textit{for every } t \in [0, T[. \tag{2.62}$$

Moreover, $\dfrac{\partial^+}{\partial t} u(t, .)$ *exists everywhere on* $[0, T[$ *and*

$$\frac{\partial^+}{\partial t} u(t, x) = (\Delta u(t, x) - \beta(u(t, x)) + f(t, x))^0 \quad \text{a.e. } x \in \Omega \tag{2.63}$$

for every $0 \leqslant t < T$.

Remark. Suppose that $0 \in \beta(0)$. It turns out that if $f(t, x) \geqslant 0$ a.e. on $]0, T[\times \times \Omega$ and if $u_0(x) \geqslant 0$ a.e. on Ω, then the solution $u(t, x)$ for problem (2.60) is positive on $]0, T[\times \Omega$. To prove this it suffices to show that for every $\lambda > 0$ and any $f \in C = \{u \in L^p(\Omega);\ u(x) \geqslant 0 \text{ a.e. on } \Omega\}$ the solution $u \in W_0^{1,p}(\Omega) \cap W^{2,p}(\Omega) \cap D(\tilde{\beta})$ for the equation

$$u - \lambda \Delta u + \lambda \tilde{\beta}(u) \ni f \text{ on } \Omega$$

belongs to C (see *Remark* after the proof of Theorem 2.3).

Example 3. *Riccati equations.* Let V and H be a pair of separable Hilbert spaces such that V is dense in H and the inclusion mapping of V into H is completely continuous. Notations for inner products and norms will be the same as in *Example* 1.

Besides these spaces we are given a linear continuous operator A from V to V', which is assumed to satisfy

$$(Au, u) \geqslant \omega \|u\|^2, \text{ for every } u \in V \tag{2.64}$$

where $\omega > 0$.

We shall denote by $(H, H)_{\text{H.S.}}$ the space of Hilbert-Schmidt operators mapping H into itself. We shall use similar notations for the space of Hilbert-Schmidt operators mapping H into V, or V' into H. The definition and the main properties of Hilbert-Schmidt operators were given in Chapter II, Section 3.4, to which we shall refer in the sequel.

We are interested in the problem

$$\begin{gathered} \frac{\mathrm{d}}{\mathrm{d}t} P(t) + A^* P(t) + P(t) A + P^2(t) = F(t), \quad \text{a.e. } t \in]0, T[\\ P(0) = P_0 \end{gathered} \tag{2.65}$$

where $P_0 \in (H, H)_{\text{H.S.}}$ and $F(t)$ is a given $(H, H)_{\text{H.S.}}$-valued function on $[0, T]$. The solution $P(t)$ of the Cauchy problem (2.65) will be a function of t on $[0, T]$ with values in the space $(H, H)_{\text{H.S.}}$. We have denoted by A^* the adjoint operator of A.

Equation (2.65) arises in the *synthesis* of optimal control problems governed by linear parabolic differential equations (see J. L. Lions [2]) and in physics (see e.g. A. Bove, G. Da Prato and G. Fano [1]). In the control theory with quadratic cost criterion, the solution $P(t)$ of Eq. (2.65) represents the "feed-back" operator associated with the control problem given, and for every t it is self-adjoint continuous and positive on H.

This is the motivation for the fact that we look for solutions $P(t)$ which satisfy

$$P(t) \in C, \quad \text{for every } t \in [0, T]$$

where C is the closed convex cone of $(H, H)_{\text{H.S.}}$ defined by

$$C = \{P \in (H, H)_{\text{H.S.}};\ P = P^*, \quad P \geqslant 0\}. \tag{2.66}$$

Here $P \geqslant 0$ means that $(Px, x) \geqslant 0$ for all $x \in H$.

The main existence result is the following.

THEOREM 2.8. *Let V, H, A be given, satisfying the above conditions. Let P_0, F be given satisfying*

$$P_0 \in (H, H)_{\text{H.S.}},\ P_0 = P_0^*,\ P_0 \geqslant 0. \tag{2.67}$$

$$F \in L^2(0, T; (H, V')_{\text{H.S.}} + (V, H)_{\text{H.S.}}),$$

$$F(t) = F^*(t),\ F(t) \geqslant 0, \text{ a.e. on }]0, T[. \tag{2.68}$$

Then there exists one and only one solution $P\colon [0, T] \to (H, H)_{\text{H.S.}}$ for problem (2.65), *which satisfies*

$$P \in C(0, T; (H, H)_{\text{H.S.}}) \cap L^2(0, T; (H, V)_{\text{H.S.}} \cap (V', H)_{\text{H.S.}}). \tag{2.69}$$

$$\frac{\mathrm{d}P}{\mathrm{d}t} \in L^2(0, T; (H, V')_{\text{H.S.}} + (V, H)_{\text{H.S.}}). \tag{2.70}$$

$$P(t) = P^*(t),\ P(t) \geqslant 0 \textit{ for every } t \in [0, T]. \tag{2.71}$$

Moreover, if $F \in L^1(0, T; (H, H)_{\text{H.S.}}),\ \dfrac{\mathrm{d}F}{\mathrm{d}t} \in L^1(0, T;\ (H, H)_{\text{H.S.}})$ *and if* $A^*P_0 + P_0A \in (H, H)_{\text{H.S.}}$ *then*

$$\frac{\mathrm{d}P}{\mathrm{d}t} \in L^\infty(0, T; (H, H)_{\text{H.S.}}). \tag{2.72}$$

Proof. We set

$$\mathscr{H} = (H, H)_{\text{H.S.}} \text{ and } \mathscr{V} = (H, V)_{\text{H.S.}} \cap (V', H)_{\text{H.S.}}$$

Then the dual space of $\mathscr{V}$ is $\mathscr{V}' = (H, V')_{\text{H.S.}} + (V, H)_{\text{H.S.}}$ and the following inclusion relations hold

$$\mathscr{V} \subset \mathscr{H} \subset \mathscr{V}'.$$

The notations for inner products and norms in $\mathscr{V}$, $\mathscr{H}$ and $\mathscr{V}'$ will be the same as in Chapter II Section 3.3. Let $\mathscr{L} : \mathscr{V} \to \mathscr{V}'$ be the operator defined by

$$\mathscr{L}(P) = A^*P + PA$$

By Lemma 3.3 (Chapter II, Section 3.3) we know that $\mathscr{L}$ is continuous from $\mathscr{V}$ to $\mathscr{V}'$ and

$$[\mathscr{L}(P), P] \geqslant \alpha \|P\|^2_{\mathscr{V}}, \quad \forall\, P \in \mathscr{V}. \tag{2.73}$$

We also recall that

$$[P^2 - Q^2,\ P - Q] \geqslant 0, \text{ for all } P, Q \in C \tag{2.74}$$

and

$$\|P^2\|_{\mathscr{H}} \leqslant \|P\|^2_{\mathscr{H}}, \quad \text{for all } P \in \mathscr{H}. \tag{2.75}$$

Define

$$\mathscr{A}(P) = \mathscr{L}(P) + P^2, \quad P \in D(\mathscr{A})$$

where

$$D(\mathscr{A}) = \{P \in C;\ \mathscr{L}(P) \in \mathscr{H}\}$$

The operator $\mathscr{A}$ is obviously monotone and closed in $\mathscr{H} \times \mathscr{H}$. Moreover, in view of Theorem 3.9 (Chapter II, Section 3.3), $R(\mathscr{I} + \lambda\mathscr{A}) \supset C$ for all $\lambda > 0$. ($\mathscr{I}$ denotes the identity operator in $\mathscr{H}$.)

First suppose that $F \in W^{1,1}(0, T; \mathscr{H})$ and $A^*P_0 + P_0A \in \mathscr{H}$, i.e., $P_0 \in D(\mathscr{A})$. Now we refer to Corollary 2.1, where $A = \mathscr{A}$ and $X = \mathscr{H}$, to obtain a solution $P \in W^{1,\infty}(0, T; \mathscr{H})$ for the equation

$$\begin{aligned} &\frac{\mathrm{d}P}{\mathrm{d}t} + \mathscr{A}(P) = F, \quad \text{a.e. on }]0, T[\\ &P(0) = P_0 \end{aligned} \tag{2.76}$$

satisfying

$$P = P^*,\ P \geqslant 0 \quad \text{on } [0, T].$$

This proves the last part of Theorem 2.8.

We now turn to P_0 and F satisfying conditions (2.67) and (2.68). Then there exist two sequences $\{P_0^n\} \subset D(\mathscr{A})$ and $\{F_n\} \subset W^{1,1}(0, T; \mathscr{H})$ such that

$$P_0^n \in C, \;\; F_n(t) \in C, \;\; \text{a.e.} \;\; t \in]0, T[, \;\; n = 1, 2, \ldots$$

and

$$\lim_{n \to \infty} P_0^n = P_0 \text{ in } \mathscr{H}$$

respectively

$$\lim_{n \to \infty} F_n = F \text{ in } L^2(0, T; \mathscr{H}).$$

Let $P_n \in W^{1,\infty}(0, T; \mathscr{H})$ be the corresponding solution of (2.65) (equivalently (2.76)).

Inasmuch as the operator P^2 is monotone in $\mathscr{H} \times \mathscr{H}$ (see (2.74)), we find that

$$\frac{1}{2} \frac{\mathrm{d}}{\mathrm{d}t} \|P_n(t) - P_m(t)\|_{\mathscr{H}}^2 + [\mathscr{L}(P_n(t) - P_m(t)), P_n(t) - P_m(t)] \leqslant$$
$$\leqslant [F_n(t) - F_m(t), P_n(t) - P_m(t)] \text{ a.e. on }]0, T[.$$

This implies in view of (2.73) that

$$\|P_n(t) - P_m(t)\|_{\mathscr{H}}^2 + \alpha \int_0^t \|P_n(s) - P_m(s)\|_{\mathscr{V}}^2 \,\mathrm{d}s \leqslant \|P_n^0 - P_m^0\|_{\mathscr{H}}^2 +$$
$$+ C \int_0^t \|F_n(s) - F_m(s)\|_{\mathscr{V}'}^2 \,\mathrm{d}s, \quad 0 \leqslant t \leqslant T$$

and we may therefore conclude as an intermediate step that there exists $P \in C(0, T; \mathscr{H}) \cap L^2(0, T; \mathscr{V})$ such that

$$\lim_{n \to \infty} P_n = P \text{ in } C(0, T; \mathscr{H}) \cap L^2(0, T; \mathscr{V}).$$

Thus from (2.75) we infer that

$$\lim_{n - \infty} P_n^2(t) = P^2(t) \text{ in } \mathscr{H}$$

and the limit is uniform in $t \in [0, T]$. Moreover, since $\mathscr{A}$ is continuous and linear from $\mathscr{V}$ to $\mathscr{V}'$, it follows that

$$\mathscr{A} P_n \to \mathscr{A} P \quad \text{in } L^2(0, T; \mathscr{V}')$$

and therefore

$$\frac{dP_n}{dt} \to \frac{dP}{dt} \quad \text{in } L^2(0, T; \mathscr{V}')$$

This shows that P is a solution of Eq. (2.65). The uniqueness of P follows from the monotonicity of $\mathscr{L}$ and P^2 by using a standard device. The proof is complete.

COROLLARY 2.3. *Let P_0 and F be given satisfying conditions* (2.67) *and* (2.68). *In addition, suppose that*

$$t\frac{d}{dt} F \in L^2(0, T; (H, V')_{\text{H.S.}} + (V, H)_{\text{H.S.}}) \tag{2.77}$$

Then solution $P(t)$ for the initial value problem (2.65) *satisfies*

$$t\frac{d}{dt} P \in L^2(0, T; (H, V)_{\text{H.S.}} \cap (V', H)_{\text{H.S.}}) \cap L^\infty(0, T; (H, H)_{\text{H.S.}}). \tag{2.78}$$

The proof of this corollary is the same as that of Theorem 2.7 and therefore we shall omit it.

Remark. It is clear that the proof of Theorem 2.8 goes through if one considers the Riccati equations of the following form

$$\frac{d}{dt} P + A^* P + PA + P^n = F$$

where n is a positive integer. In this connection, we could also consider much more general nonlinearities, but we simply refer to related problems in G. Da Prato [1], [3] and L. Tartar [1].

We conclude this section with a simple example to which Theorem 2.8 can be applied.

Let Ω be a bounded domain in R^n with a sufficiently smooth boundary Γ and let $H^m(\Omega)$, $H_0^m(\Omega)$ be the usual Sobolev spaces. We shall apply Theorem 2.8 in the case $V = H_0^1(\Omega)$ and $H = L^2(\Omega)$. For A we take

$$Au = -\Delta u, \quad u \in H_0^1(\Omega).$$

Identifying the operator $P \in (H, H)_{\text{H.S.}}$ with its kernel $P(x, \xi) \in L^2(\Omega_x \times \Omega_\xi)$, Theorem 2.8 yields the following existence result (see also Chapter II, Section 3.4).

COROLLARY 2.4. *Let* $P_0 \in L^2(\Omega_x \times \Omega_\xi)$ *and* $F \in L^2(]0, T[\times \Omega_x \times \Omega_\xi)$ *be given, satisfying*

$$P_0(x, \xi) = P_0(\xi, x), \ P_0(x, \xi) \geqslant 0, \text{ a.e. } on \ \Omega_x \times \Omega_\xi. \tag{2.79}$$

$$F(t, x, \xi) = F(t, \xi, x), \ F(t, x, \xi) \geqslant 0, \text{ a.e. } on \ ]0, T[\times\Omega_x\times\Omega_\xi. \tag{2.80}$$

Then there exists a unique function $P(t, x, \xi)$ *satisfying:*

$$\frac{\partial P}{\partial t}(t, x, \xi) - \Delta_x P(t, x, \xi) - \Delta_\xi P(t, x, \xi) +$$

$$+ \int_\Omega P(t, x, \zeta)\, P(t, \zeta, \xi) \mathrm{d}\zeta = F(t, x, \xi). \tag{2.81}$$

$$P(0, x, \xi) = P_0(x, \xi), \ \text{ a.e. } on \ \Omega_x \times \Omega_\xi. \tag{2.82}$$

$$P(t, x, \xi) = 0 \text{ a.e. } on \ ]0, T[\times \Gamma_x \times \Gamma_\xi. \tag{2.83}$$

$$P(t, x, \xi) = P(t, \xi, x), \ P(t, x, \xi) \geqslant 0 \text{ a.e. } on \ ]0, T[\times \Omega_x \times \Omega_\xi. \tag{2.84}$$

$$P \in C(0, T; L^2(\Omega_x \times \Omega_\xi)) \cap L^2(0, T; \ H_0^1(\Omega_x \times \Omega_\xi)). \tag{2.85}$$

$$\frac{\partial P}{\partial t} \in L^2(0, T; H^{-1}(\Omega_x \times \Omega_\xi)). \tag{2.86}$$

§3. Differential equations associated with continuous dissipative operators

This paragraph is concerned with the initial value problem

$$\frac{\mathrm{d}u}{\mathrm{d}t} \in Au + Bu \quad 0 < t < \infty \tag{3.1}$$

$$u(0) = u_0$$

in a general Banach space X, where A is an m-dissipative set of $X \times X$ and B is continuous and dissipative from X to itself.

3.1. A general existence result

The main result is the following theorem.

THEOREM 3.1. *Let A be an m-dissipative subset of $X \times X$. Let B be a continuous, everywhere defined, nonlinear dissipative operator from X into itself.*

Then for each $u_0 \in \overline{D(A)}$ there exists a unique continuous function $u : [0, \infty[\to X$ such that $u(0) = u_0$ and the following inequality holds

$$\|u(t) - x\|^2 \leqslant \|u(s) - x\|^2 + 2\int_s^t \langle y + Bu(\tau), u(\tau) - x\rangle_s \, d\tau, \qquad (3.2)$$

for every $[x, y] \in A$ and $0 \leqslant s \leqslant t \leqslant T$.

We shall divide the proof of this theorem into several steps.

LEMMA 3.1. *Suppose that the assumptions of Theorem 3.1 are satisfied. Then for each u_0 in $\overline{D(A)}$ there are an interval $[0, T_{u_0}]$ and a function $u(t)$ which is defined on $[0, T_{u_0}]$ and satisfies (3.2) for every $[x, y]$ in A and all $0 \leqslant s \leqslant t \leqslant T_{u_0}$.*

Proof. Let $u_0 \in \overline{D(A)}$ and let V be a neighbourhood of u_0 for which there is a positive M such that $\|Bu\| \leqslant M$ for every $u \in V$. Suppose for simplicity that V is the ball having the center u_0 and of radius ρ, and choose $T > 0$ such that

$$2MT + \sup\{\|S(t)u_0 - u_0\|;\ 0 \leqslant t \leqslant T\} \leqslant M, \qquad (3.3)$$

where $S(t)$ is the semigroup of nonlinear contractions generated by A on $\overline{D(A)}$, i.e.,

$$S(t)x = \lim_{n\to\infty}\left(1 - \frac{t}{n} A\right)^{-n} x, \quad x \in \overline{D(A)}, \quad t \geqslant 0.$$

For every positive integer n consider a partition $\{t_i^n\}_{i=0}^{N_n}$ of $[0, T]$ defined by

$$0 = t_0^n < t_1^n \ldots < t_{N_n}^n = T; \ t_i^n = t_{i-1}^n + \delta_i^n; \quad i = 1, \ldots N_n$$

and let $u_n(t) : [0, T] \to X$ be the continuous function, inductively defined on $[0, T]$ by

$$u_n(t) = v_n^i(t - t_{i-1}^n), \text{ for } t_{i-1}^n \leqslant t \leqslant t_i^n,$$

where $v_n^i(t)$ is the integral solution of the equation

$$\frac{dv_n}{dt} - Av_n \ni Bu_n(t_{i-1}^n),$$

$$v_n(0) = u_n(t_{i-1}^n).$$

We take $u_n(0) = u_0$ for all n. Clearly, v_n^i exists and is unique by Theorem 2.1. Recalling that $v(t) = S(t)u_n(t_{i-1}^n)$ is the integral solution of the equation

$$\frac{dv}{dt} \in Av(t)$$

$$v(0) = u_n(t_{i-1}^n),$$

inequality (2.4) in Theorem 2.1, yields

$$\|u_n(t) - S(t - t_{i-1}^n)u_n(t_{i-1}^n)\| \leqslant (t - t_{i-1}^n)\|Bu_n(t_{i-1}^n)\|,$$

and therefore

$$\|u_n(t) - u_n(t_{i-1}^n)\| \leqslant \|S(t - t_{i-1}^n)u_n(t_{i-1}^n) - u_n(t_{i-1}^n)\| + \\ + \delta_i^n\|Bu_n(t_{i-1}^n)\|, \quad t_{i-1}^n \leqslant t \leqslant t_i^n. \tag{3.4}$$

Let g_n be the step function defined by

$$g_n(t) = Bu_n(t_{i-1}^n), \quad \text{for } t_{i-1}^n \leqslant t < t_i^n$$

and note that $u_n(t)$ is the integral solution on $[0, T]$ of the equation

$$\frac{du_n}{dt} \in Au_n + g_n \tag{3.5}$$

Again using inequality (2.4), we get

$$\|u_n(t) - S(t)u_0\| \leqslant \int_0^t \|g_n(\tau)\| \, d\tau, \quad 0 \leqslant t \leqslant T. \tag{3.6}$$

From relations (3.3) and (3.6) one easily follows by induction that $u_n(t_i^n) \in V$ for all i and n.

Next, we choose inductively the partition $\{t_i^n\}_{i=1}^N$ such that

$$\|Bz - Bu_n(t_{i-1}^n)\| \leqslant 1/n \tag{3.7}$$

for all $z \in X$ satisfying

$$\|z - u_n(t_{i-1}^n)\| \leqslant M\delta_i^n + \sup\{\|S(t)u_n(t_{i-1}^n) - u_n(t_{i-1}^n)\|; \; 0 \leqslant t \leqslant \delta_i^n\}. \tag{3.8}$$

Without any loss of generality we may assume that δ_i^n is the largest number for which (3.7) holds.

We shall prove that there is an integer N such that $t_N^n = T$. Suppose, for contradiction that this is not the case. Therefore, there is $s_0 \in]0, T[$ such that $\lim_{i\to\infty} t_i^n = s_0$. First we shall prove that $\lim_{i\to\infty} u_n(t_i^n)$ exists. For $j \geqslant i$ we set $\delta_{ij}^n = t_j^n - t_i^n$ and use inequality (2.4) with $u(t) = u_n(t)$ and $v(t) = u_n(t + \delta_{ij}^n)$, to obtain

$$\|u_n(t_j^n) - u_n(t_i^n)\| \leqslant \|u_n(\delta_{ij}^n) - u_n(0)\|^2 + \\ + 2\int_0^{t_i^n} \|g_n(\tau + \delta_{ij}^n) - g_n(\tau)\| \, \|u_n(\tau + \delta_{ij}^n) - u_n(\tau)\| \, d\tau. \tag{3.9}$$

Let $\varepsilon > 0$ and let h be fixed but sufficiently large such that

$$0 < s_0 - t_i^n \leqslant \varepsilon^2/16\, M\rho, \quad \text{for all } i \geqslant h. \tag{3.10}$$

Inasmuch as $u_n(t)$ is uniformly continuous on $[0, t_h^n]$ we have

$$\|u_n(\tau + \delta_{ij}^n) - u_n(\tau)\| \leqslant \varepsilon/4MT \text{ for } 0 \leqslant \tau \leqslant \tau + \delta_{ij}^n \leqslant t_h^n$$

and all sufficiently large $i \leqslant j$. Then we obtain from (3.9) and (3.10) that

$$\|u_n(t_j^n) - u_n(t_i^n)\|^2 \leqslant \varepsilon^2/2 + \varepsilon$$

for all sufficiently large $i \leqslant j$. Hence $\lim\limits_{i\to\infty} u_n(t_i^n) = z_0$ exists. Let $\alpha > 0$ be such that $\|Bz - Bz_0\| \leqslant 1/2n$ whenever $\|z - z_0\| \leqslant \alpha$ and let $\beta > 0$ be such that

$$\|S(t)u_n(t_j^n) - u_n(t_j^n)\| \leqslant \alpha/4, \quad \forall\, t \in [0, \beta],\ j = 0, 1, \ldots$$

Choose i provided that

$$\delta_i^n < \delta^n = \max(\beta, \alpha/2M), \quad \|u_n(t_{i-1}^n) - z_0\| \leqslant \alpha/4.$$

If

$$\|z - u_n(t_{i-1}^n)\| \leqslant M\delta^n + \sup\{\|S(t)u_n(t_{i-1}^n) - u_n(t_{i-1}^n)\|;\ 0 \leqslant t \leqslant \delta^n\}$$

then $\|z - u_n(t_{i-1}^n)\| \leqslant 3\alpha/4$, and this implies that $\|z - z_0\| \leqslant \alpha$. Consequently

$$\|Bz - Bu_n(t_{i-1}^n)\| \leqslant \|Bz - Bz_0\| + \|Bz_0 - Bu_n(t_{i-1}^n)\| \leqslant 1/n.$$

and this contradicts our assumption that δ_i^n is the largest number for which inequality (3.7) holds. The contradiction shows that there exists a finite partition $\{t_i^n\}$ with the required properties.

We next show that the sequence $\{u_n\}$ converges uniformly to a function u satisfying (3.2). Let m, n be positive integers. Once more using Theorem 2.1 (inequality (2.4)) for $u = u_n$ and $v = u_m$ we get

$$\|u_n(t) - u_m(t)\|^2 \leqslant \|u_n(0) - u_m(0)\|^2 +$$
$$+ 2\int_0^t \langle g_n(\tau) - g_m(\tau),\ u_n(\tau) - u_m(\tau)\rangle_s \,d\tau \quad \text{for } 0 \leqslant s \leqslant t \leqslant T. \tag{3.11}$$

Let $t \in [0, T]$ and let i and j be some positive integers such that $t \in [t_{j-1}^m, t_j^m] \cap [t_{i-1}^n, t_i^n]$. From inequalities (3.4) and (3.7) we have

$$\|Bu_n(t) - Bu_n(t_{i-1}^n)\| \leqslant 1/n,\ \|Bu_m(t) - Bu_m(t_{j-1}^m)\| \leqslant 1/m. \tag{3.12}$$

and since B is dissipative, (3.11) and (3.12) imply that

$$\|u_n(t) - u_m(t)\|^2 \leqslant 2(1/n + 1/m) \int_0^t \|u_n(\tau) - u_m(\tau)\| \, d\tau, \ 0 \leqslant t < T.$$

Consequently

$$\|u_n(t) - u_m(t)\| \leqslant T(1/n + 1/m), \ 0 \leqslant t < T.$$

We set $u(t) = \lim_{n \to \infty} u_n(t)$ and note that the limit is uniform on $[0, T]$. Since the function $\langle , \rangle_s$ is upper-semicontinuous, we can pass to the limit as $n \to \infty$ in (3.5), i.e.,

$$\frac{1}{2} \|u_n(t) - x\|^2 \leqslant \frac{1}{2} \|u_n(s) - x\|^2 + \int_s^t < g_n(\tau) + y, \ u_n(\tau) - x >_s d\tau,$$

to conclude that $u(t)$ satisfies (3.2) on $[0, T]$ for every $[x, y] \in A$. Therefore, Lemma 3.1 has been proved. We can now derive an important existence result due to R. H. Martin Jr. [1].

Corollary 3.1. *Let X be a real Banach space and let B be a continuous, everywhere defined, dissipative operator from X to itself. Then for each $u_0 \in X$ there exists a unique function $u \in C^1(0, \infty; X)$ satisfying*

$$\frac{du}{dt}(t) = Bu(t), \quad \textit{for } t \geqslant 0 \tag{3.13}$$

$$u(0) = u_0.$$

Proof. Let u_0 be any fixed element of X. In view of Lemma 3.1 there is a function $u(t)$ defined on an interval $[0, T]$ such that $u(0) = u_0$, and satisfying the inequality

$$\|u(t) - x\|^2 \leqslant \|u(s) - x\|^2 + 2 \int_s^t \langle Bu(\tau), \ u(\tau) - x \rangle_s d\tau, \ 0 \leqslant s \leqslant t \leqslant T \tag{3.14}$$

for every $x \in X$.
Recalling that

$$\langle u(t) - u(s), \ u(s) - x \rangle_s \leqslant \frac{1}{2} (\|u(t) - x\|^2 - \|u(s) - x\|^2)$$

and using the continuity of B, we get

$$\limsup_{s \to t} \left(\frac{u(t) - u(s)}{t - s} - Bu(t), f_x \right) \leqslant 0, \ \ 0 \leqslant t \leqslant T \tag{3.15}$$

where f_x is some element of $F(u(t) - x)$ and x is arbitrary in X. In particular we may take x in X such that $u(t) - x = \dfrac{u(t) - u(s)}{t - s} - Bu(t)$. Then, (3.15) implies that

$$\lim_{s \to t} \left\| \frac{u(t) - u(s)}{t - s} - Bu(t) \right\| = 0$$

and this shows that $u(t)$ is differentiable everywhere on $[0, T]$ and satisfies Eq. (3.13) on this interval. Since B is dissipative, $u(t)$ is the unique solution of (3.13) on this interval, and it can be extended outside $[0, T]$. Let $[0, T_{\max}[$ be the largest interval of existence of $u(t)$. We shall prove that $T_{\max} = \infty$. Suppose, for contradiction that this is not the case, i.e., $T_{\max} < \infty$. Then, from Proposition 1.2, $\lim_{t \to T_{\max}} u(t) = u(T_{\max})$ exists. Since B is continuous, this implies that $\lim_{t \to T_{\max}} \dfrac{\mathrm{d}u}{\mathrm{d}t} = Bu(T_{\max})$ and, by the first part of the proof, it follows that $u(t)$ can be defined as a solution of (3.13) in a neighbourhood of $T_{\max}$. But this contradicts the hypothesis, and the proof of Corollary 3.1 is complete.

A lemma of interest is the following.

LEMMA 3.2. *Let B be continuous, dissipative and everywhere defined in X. Let x and y be given in X. Then*

$$(Bx - By, f) \leqslant 0 \quad \textit{for every } f \in F(x - y). \tag{3.16}$$

Proof. Corollary 3.1 implies that B is the generator of a semigroup $S(t)$ of contractions defined on X, i.e.,

$$Bx = \lim_{t \to 0} (S(t)x - x)/t, \qquad \forall x \in X.$$

Thus assertion of the lemma follows easily from the definition of F.

Proof of Theorem 3.1. From Lemma 3.1, for every $u_0 \in \overline{D(A)}$ there exists at least a function $u(t, u_0)$ which is defined on $[0, T_{u_0}]$ and satisfies (3.2) for every $[x, y] \in A$. From Lemma 3.2 it follows that $u(t, u_0)$ is an integral solution on $[0, T_{u_0}]$ of the equation $\dfrac{\mathrm{d}u}{\mathrm{d}t} \in Au + Bu$. Thus, from Theorem 2.1 we infer that

$$\|u(t, u_0) - u(t, v_0)\| \leqslant \|u_0 - v_0\|, \quad \text{for } 0 \leqslant t \leqslant \inf\,(T_{u_0}, T_{v_0}) \tag{3.17}$$

where u_0 and v_0 are arbitrary in $\overline{D(A)}$. This implies, in particular, the uniqueness of $u(t, u_0)$ and the semigroup property

$$u(t + t', u_0) = u(t, u(t', u_0)), \quad \text{for} \quad 0 \leqslant t + t' \leqslant T_{u_0}. \tag{3.18}$$

Then by using a standard device (see the proof of Corollary 3.1), one concludes that u can be defined on $[0, \infty[$. The uniqueness follows as mentioned above, from Theorem 2.1.

REMARK. These are examples due to J. Dieudonne and J. York which show that the classical existence theorem of Peano fails in infinite dimensional Banach spaces. More recently A. Cellina [1] has proved that in any non-reflexive Banach space Peano's theorem is not necessarily true.

3.2. Continuous perturbations of m-dissipative operators

We are able now to prove the following perturbation theorem

THEOREM 3.2. *Let X be a real Banach space and let A be an m-dissipative subset of $X \times X$. Let B be a continuous, everywhere defined, dissipative operator from X to itself. Then $A + B$ is m-dissipative in $X \times X$.*

Proof. It follows from Lemma 3.2 that $A+B$ is a dissipative subset of $X\times X$. To prove that $A + B$ is m-dissipative we must show that for any y_0 in X the equation $x - Ax - Bx \ni y_0$ has a solution $x \in D(A)$. Without any loss of generality we may assume that $y_0 = 0$. This can be achieved by shifting the range of B.

Let u_0 be arbitrary but fixed in $D(A)$. From Lemma 3.1 there is a unique continuous function $u\colon [0, \infty[\to X$ such that $u(0) = u_0$ and satisfy

$$\|u(t) - x\|^2 \leqslant \|u(s) - x\|^2 + 2\int_s^t \langle y + Bu(\tau), u(\tau) - x\rangle_s \, d\tau \tag{3.19}$$

for every $[x, y] \in A - I$ and $0 \leqslant s \leqslant t$.

Applying inequality (2.16) with $\omega = 1$ and $v(t) = u(t + h)$, we get

$$\|u(t + h) - u(t)\| \leqslant \exp(-t)\|u(h) - u_0\|, \quad \forall\, t, h \geqslant 0 \tag{3.20}$$

because B is dissipative.

Next, we take $v = u_0$ and $s = 0$ in (2.16). After some calculation involving Gronwal's lemma we obtain

$$\|u(h) - u_0\| \leqslant (1 - \exp(-h))\,(|Au_0| + \|Bu_0\| + \|u_0\|), \quad \forall\, h > 0 \tag{3.21}$$

where $|Au_0| = \inf\{\|y\|; y \in Au_0\}$. This shows that $\lim_{t\to\infty} u(t) = x_0$ exists. Moreover, we have

$$\lim_{t\to\infty} \sup\,(\|u(t + h) - u(t)\|/h; \quad 0 < h < \infty) = 0$$

so that (3.19) and

$$2\langle u(t + h) - u(t), u(t) - x\rangle_s \leqslant \|u(t + h) - x\|^2 - \|u(t) - x\|^2, \quad \forall\, x \in X.$$

yields

$$\lim_{t\to\infty}\sup\left(\sup_{h>0}\frac{1}{h}\int_t^{t+h}\langle y+Bu(\tau),u(\tau)-x\rangle_s\mathrm{d}\tau\right)\geqslant 0$$

for every $x \in D(A)$ and $y \in Ax - x$. Since $\langle,\rangle_s$ is upper-semicontinuous this clearly implies that

$$\langle y + Bx_0, x_0 - x\rangle_s \geqslant 0, \qquad \forall\,[x, y] \in A - I$$

and hence $x_0 - Ax_0 - Bx_0 \ni 0$, because A is m-dissipative. Therefore, Theorem 3.2 has been proved. In particular, we have the following.

COROLLARY 3.2. *Let B be everywhere defined, continuous and dissipative from a Banach space X to itself. Then B is m-dissipative in $X \times X$.*

As a consequence of Theorem 3.2 proved above we shall derive some existence results for time-dependent differential equations associated to dissipative and continuous operators. Other applications will be given in the next section.

COROLLARY 3.3. *Let X be a real Banach space and let $A(t)u$: $[0, T] \times X \to X$ be a nonlinear mapping satisfying:*

(i) *For every $t \in [0, T]$, the operator $u \to A(t, u)$ is everywhere defined and dissipative in $X \times X$.*

(ii) *The mapping $(t, u) \to A(t)u$ is continuous from $[0, T] \times X$ to X.*

Then for any $u_0 \in X$, the initial value problem

$$\frac{\mathrm{d}}{\mathrm{d}t}u(t) = A(t)\,u(t) \qquad 0 \leqslant t \leqslant T$$
$$u(0) = u_0 \tag{3.22}$$

has a unique solution $u \in C^1(0, T; X)$.

Proof. Let $\tilde{A}$ be the realization of $A(t)$ on $C(0, T; X)$, i.e.,

$$(\tilde{A}u)(t) = A(t)u(t) \quad \text{for} \quad u \in C(0, T; X), \quad 0 \leqslant t \leqslant T$$

where $C(0, T; X)$ denotes, as usual, the space of all X-valued continuous functions on $[0, T]$ endowed with the norm

$$|||u||| = \sup\{\|u(t)\|; \qquad 0 \leqslant t \leqslant T\}.$$

Obviously, the operator $\tilde{A}$ is everywhere defined and continuous on $Y = C(0, T; X)$. Moreover, for every $\lambda > 0$, and u, v in Y we have

$$|||u - v||| \leqslant |||u - v - \lambda(\tilde{A}u - \tilde{A}v)|||$$

because $A(t)$ is dissipative (see Chapter II, Proposition 3.1). Thus, $\tilde{A}$ is also dissipative in $Y \times Y$. Let L be the operator defined by

$$Lu = -\frac{\mathrm{d}u}{\mathrm{d}t}, \qquad \text{for} \quad u \in D(L)$$

where

$$D(L) = \{u \in C^1(0, T; X); u(0) = u_0\}.$$

Clearly, for every $f \in C(0, T; X)$ the equation $u - \lambda Lu = f$ has a unique solution $u(t) \in C(0, T; X)$ given by

$$u(t) = \exp\left(-\frac{t}{\lambda}\right) u_0 + \frac{1}{\lambda}\int_0^t \exp\left(\frac{s-t}{\lambda}\right) f(s)\, ds, \quad 0 \leqslant t \leqslant T. \tag{3.23}$$

Moreover, the mapping $f \to u$ is non-expansive on Y. Thus, we have shown that L is m-dissipative in $Y \times Y$, so that Theorem 3.2 implies that $A + L$ also is m-dissipative in $Y \times Y$. In other words, for every $\lambda > 0$ the initial value problem

$$\begin{aligned} &\lambda u(t) + \frac{du}{dt}(t) = A(t)\, u(t), \quad 0 \leqslant t \leqslant T \\ &u(0) = u_0 \end{aligned} \tag{3.24}$$

has a unique solution $u \in C^1(0, T; X)$.

Obviously, we can apply the above argument to $A^\lambda(t)u = \exp(-\lambda t)A(t)$ $(\exp(\lambda t)u)$ to deduce that $v(t) = \exp(\lambda t)u(t)$ is a solution of Eq. (3.22). This concludes the proof.

COROLLARY 3.4. *In addition to the hypotheses of Corollary* 3.3, *we assume that*

(iii) $A(0)u = A(T)u$ *for every* $u \in X$.

Then for every $\lambda > 0$, *the periodic problem*

$$\begin{aligned} &\lambda u(t) + \frac{du}{dt}(t) = A(t)u(t), \quad 0 \leqslant t \leqslant T \\ &u(0) = u(T) \end{aligned} \tag{3.25}$$

has a unique solution $u \in C^1(0, T; X)$.

Proof. This existence result can be established in the same way as Corollary 3.3, with the difference that operator L must be replaced by L_π defined by

$$L_\pi u = -\frac{du}{dt} \quad \text{for } u \in D(L_\pi) = \{u \in C^1(0, T; X); u(0) = u(T)\}.$$

We shall omit the details.

3.3. Semi-linear second order elliptic equations in L^1

Let β be a maximal monotone single-valued operator from R^1 into itself such that $0 \in D(\beta)$. Let A be an unbounded linear operator on $L^1(\Omega)$. Here Ω is a measurable subset on R^n which for simplicity will be assumed to be bounded.

We shall study in this section the equation

$$Au(x) - \beta(u(x)) = f(x), \quad \text{a.e. on } \Omega \tag{3.26}$$

where $f \in L^1(\Omega)$ is given.

Let us now specify the conditions assuring the existence of a solution for this equation.

(i) *Operator A is linear, densely defined and m-dissipative in $L^1(\Omega)$. For any $\lambda > 0$ and $f \in L^1(\Omega)$,*

$$\operatorname*{ess\,sup}_{x \in \Omega} (1 - \lambda A)^{-1} f(x) \leqslant \max \{0, \operatorname*{ess\,sup}_{x \in \Omega} f(x)\}. \tag{3.27}$$

(ii) *There exists $\omega > 0$ such that*

$$\omega \| u \|_{L^1(\Omega)} \leqslant \| Au \|_{L^1(\Omega)}, \text{ for all } u \in D(A). \tag{3.28}$$

The main result is the following

THEOREM 3.3. *If conditions* (i) *and* (ii) *are satisfied, then for every $f \in L^1(\Omega)$, Eq.* (3.26) *has a unique solution $u \in D(A)$.*

Proof. Let $B \subset L^1(\Omega) \times L^1(\Omega)$ be defined by

$$Bu = \{f \in L^1(\Omega);\ f(x) = -\beta(u(x)), \text{ a.e. } x \in \Omega\}. \tag{3.29}$$

In other words, B is the realization of $-\beta$ on $L^1(\Omega)$. It is immediately clear that B is m-dissipative in $L^1(\Omega) \times L^1(\Omega)$. We must prove that the range of $A + B$ is all of $L^1(\Omega)$. The proof of this fact is more complicated and therefore we shall divide it into several steps.

a) *$A + B$ has closed range.* Let f and f^* in $R(A + B)$ and let u, u^* be the corresponding solutions of (3.26). Let $g = f - Au = Bu$ and $g^* = f^* - Au^* = Bu^*$. Define

$$\operatorname{sign}[u] = \begin{cases} 1 & \text{for } u > 0 \\ [-1,1] & \text{for } u = 0 \\ -1 & \text{for } u < 0 \end{cases}$$

and note that the duality mapping F of $L^1(\Omega)$ to its dual space $L^\infty(\Omega)$ is given by

$$(F(u))(x) = \|u\|_{L^1(\Omega)} \operatorname{sign}[u(x)],\ u \in L^1(\Omega). \tag{3.30}$$

From the Hille-Yosida theorem, condition (i) implies that A is the infinitesimal generator of a linear contraction semigroup on $L^1(\Omega)$. This implies in particular that

$$(Au - Au^*, h) \leqslant 0, \quad \text{for all } h \in F(u - u^*) \tag{3.31}$$

where, as usual, (,) denotes the pairing between $L^\infty(\Omega)$ and $L^1(\Omega)$. We multiply the equation

$$A(u - u^*) + g - g^* = f - f^*$$

by

$$h(x) = \begin{cases} 1 & \text{on } [u > u^*] \cup [g > g^*] \\ 0 & \text{on } [u = u^*] \cap [g = g^*] \\ -1 & \text{on } [u < u^*] \cup [g < g^*] \end{cases}$$

We note that $h \in L^\infty(\Omega)$ and $h(x) \in \operatorname{sign}\,[u(x) - u^*(x)] \cap \operatorname{sign}\,[g(x) - g^*(x)]$ a.e. on Ω. By (3.30) and (3.31) we infer that

$$\|g - g^*\|_{L^1(\Omega)} \leqslant \|f - f^*\|_{L^1(\Omega)}. \tag{3.32}$$

Let $u_n \in D(A) \cap D(B)$, $Au_n + Bu_n = f_n$ and $f_n \to f$ in $L^1(\Omega)$ as $n \to \infty$. By (3.32) we have

$$\|A(u_n - u_m)\|_{L^1(\Omega)} \leqslant 2\|f_n - f_m\|_{L^1(\Omega)}.$$

Thus by (3.28) and the closedness of A (this is again a consequence of the m-dissipativeness), $u_n \to u$ and $Au_n \to Au$ in $L^1(\Omega)$ as $n \to \infty$. Since B is m-dissipative, this clearly implies that $f \in Au + Bu$. Therefore, we have proved that $A + B$ has closed range.

b) *$A + B$ has dense range*. First we shall prove that $L^\infty(\Omega) \subset R(1 - A - B)$. Let $f \in L^\infty(\Omega)$ be arbitrary but fixed. Let β_f be the maximal monotone graph in $R^1 \times R^1$ defined by

$$\beta_f(r) = \begin{cases} \beta_1(\|f\|_\infty) & \text{if } r > (1 + \beta)^{-1}(\|f\|_\infty) \\ \beta(r) & \text{if } r \in D(\beta) \text{ and } |r + \beta(r)| \leqslant \|f\|_\infty \\ \beta_1(-\|f\|_\infty) & \text{if } r < (1+\beta)^{-1}(-\|f\|_\infty) \end{cases} \tag{3.33}$$

where $\beta_1 = \beta(1 + \beta)^{-1} = 1 - (1 + \beta)^{-1}$ and $\|f\|_\infty = \operatorname{ess\,sup}_{x \in \Omega} |f(x)|$. Note that β_f is everywhere defined and maximal monotone in $R^1 \times R^1$. Since it is single-valued, this implies that β_f is a continuous monotone function from R^1 into itself. We set

$$B_f(u)(x) = -\beta_f(u(x)), \quad \text{for } u \in L^1(\Omega), \text{ a.e. on } \Omega$$

and observe that the operator B_f is everywhere defined, continuous and dissipative on $L^1(\Omega)$. Theorem 3.2 may therefore be applied to conclude that equation

$$u - Au - B_f u = f \tag{3.34}$$

has a unique solution $u \in D(A)$. Since A is m-dissipative, we have

$$u(x) = (1 - A)^{-1}f(x) + (1 - A)^{-1}B_f u(x), \text{ a.e. } x \in \Omega$$

and hypothesis (3.27) implies that

$$|u(x) - B_f u(x)| \leqslant \|f\|_\infty \quad \text{a.e. } x \in \Omega.$$

Comparing this with (3.33) we see that $u \in D(\beta)$ and $\beta_f(u) = \beta(u)$ a.e. on Ω. Thus, from Eq. (3.34), we infer that

$$u - Au - Bu = f$$

Clearly, A may be replaced by $\frac{1}{\lambda} A$ in the above argument, to deduce that for every $f \in L^\infty(\Omega)$ and $\lambda > 0$ the equation $\lambda u - Au - Bu = f$ has a solution $u_\lambda \in D(A) \cap D(B)$. Using once again hypothesis (ii) and estimate (3.32) we obtain

$$\omega \|u_\lambda\|_{L^1(\Omega)} \leqslant \|Au_\lambda\|_{L^1(\Omega)} \leqslant 2\|f\|_{L^1(\Omega)}.$$

Hence $\lambda u_\lambda \to 0$ in $L^1(\Omega)$ as $\lambda \to 0$. This implies that f belongs to the closure of the range of $A + B$, i.e. $f \in \overline{R(A + B)}$.

We conclude the proof of Theorem 3.3 by observing that (a) and (b) imply $R(A + B) = L^1(\Omega)$ as claimed. The uniqueness is implied by (3.32).

Remarks 1°. Theorem 3.3 can be established without requiring that β be single-valued. We refer to H. Brézis and W. A. Strauss [1] for the proof in this general case.

2°. It is clear that the proof of Theorem 3.3 works if the condition that A be linear and m-dissipative is weakened to the condition that A be a general m-dissipative subset of $X \times X$. In this case condition (ii) has to be replaced by

$$\|A^{-1}u - A^{-1}v\|_{L^1(\Omega)} \leqslant \omega \|u - v\|_{L^1(\Omega)} \text{ for all } u, v \in L^1(\Omega).$$

We conclude this section with the following example. Let us assume that Ω is an open bounded subset of R^n with smooth boundary. Define

$$A = \Delta, \quad D(A) = \{u \in W_0^{1,1}(\Omega);\ \Delta u \in L^1(\Omega)\}$$

where, as usual, Δ denotes the Laplacian. (Actually, in what follows, Δ can be replaced by a second order elliptic differential operator with smooth coefficients.) It turns out that operator A satisfies hypotheses (i) and (ii) stated above (see Theorem 8 in H. Brézis and W.A. Strauss [1]).

Theorem 3.3 therefore yields the following result (to be compared with Proposition 3.7 in Chapter II).

COROLLARY 3.5. *Let β be a maximal monotone, single-valued operator from R^1 to itself such that $0 \in D(\beta)$. Let $f \in L^1(\Omega)$. Then there exists a unique solution $u \in W_0^{1,1}(\Omega)$ of the equation*

$$\Delta u(x) - \beta(u(x)) = f(x), \quad \text{a.e. on } \Omega.$$

§ 4. Time-dependent nonlinear differential equations

In this paragraph we shall briefly consider first-order evolution equations of the type

$$\frac{du}{dt} \in A(t)u + f(t), \quad 0 < t < T$$
$$u(0) = u_0 \tag{4.1}$$

where $A(t)$ is a family of nonlinear m-dissipative operators in a Banach space X. In order to avoid making the treatement too ponderous, we have not put the results of this last paragraph into a general framework.

4.1. Evolution equations associated with dissipative sets

The main result is the following theorem.

THEOREM 4.1. *Let X^* be uniformly convex and let C be a closed convex cone of X. Let $\{A(t); 0 \leqslant t \leqslant T\}$ be a family of closed and dissipative sets in $X \times X$ satisfying*

(i) *$C \subset R(1 - \lambda A(t))$ for any $\lambda > 0$ and $t \in [0, T]$. Moreover,*

$$(1 - \lambda A(t))^{-1}C \subset C, \quad \forall \lambda > 0, \quad \forall t \in [0, T]. \tag{4.2}$$

(ii) *$D(A(t)) = D$ is independent of t. There is $M > 0$ such that*

$$\|(1 - \lambda A(t))^{-1}x - (1 - \lambda A(s))^{-1}x\| \leqslant M\lambda|t - s|\,(1 + \|x\| + \|A_\lambda(t)x\|) \tag{4.3}$$

for every $x \in C$ and $0 \leqslant t,\ s \leqslant T$.
Let $f \in W^{1,1}(0, T; X)$ and $u_0 \in D$ be given satisfying

$$u_0 \in C, \quad f(t) \in C, \quad \text{for every } t \in [0, T]. \tag{4.4}$$

Then the initial value problem (4.1) has a unique solution $u \in W^{1,\infty}(0, T; X)$ which satisfies

$$u(t) \in C, \quad \text{for } 0 \leqslant t \leqslant T. \tag{4.5}$$

Proof. Consider the approximate equations

$$\frac{du_\lambda}{dt}(t) = A_\lambda(t)u_\lambda(t) + f(t) \qquad 0 \leqslant t \leqslant T$$
$$u_\lambda(0) = u_0 \tag{4.6}$$

which are clearly equivalent to

$$u_\lambda(t) = \exp(-t/\lambda)\, u_0 + \frac{1}{\lambda}\int_0^t \exp(-(t-s/\lambda)\,[\lambda f(s) + (1-\lambda A(s))^{-1} u_\lambda(s)\, ds$$

By Picard fixed-point theorem, it follows then that Eq. (4.6) has a unique solution $u \in C^1(0, T; X)$ such that $u_\lambda(t) \in C$ on $[0, T]$. Note that assumption (4.3) implies

$$\frac{d}{dt}\|A_\lambda(t)x\| \leqslant M(1 + \|x\| + \|A_\lambda(t)x\|) \text{ a.e. on }]0, T[. \tag{4.7}$$

Using Eqs. (4.6) and inequality (4.7) we deduce

$$\frac{1}{2}\frac{d}{dt}\left\|\frac{du_\lambda}{dt}(t)\right\|^2 \leqslant \left\|\frac{df}{dt}(t)\right\|\left\|\frac{du_\lambda}{dt}(t)\right\| + M(1 + \|u_\lambda(t)\| +$$
$$+ \|A_\lambda(t)\, u_\lambda(t)\|)\left\|\frac{du_\lambda}{dt}\right\|, \text{ a.e. on }]0, T[$$

because

$$\left(\frac{d}{dt}(A_\lambda(s)\, u_\lambda(t)),\ F\left(\frac{du_\lambda}{dt}(t)\right)\right) \leqslant 0, \qquad \text{a.e } t \in]0, T[.$$

The last estimate, combined with

$$\left\|\frac{du_\lambda}{dt}(0)\right\| = \|A_\lambda(0)\, u_0 + f(0)\| \leqslant |A(0)\, u_0| + \|f(0)\|$$

yields

$$\left\|\frac{du_\lambda}{dt}(t)\right\| \leqslant |A(0)\, u_0| + \|f(0)\| + M\int_0^t \left\|\frac{df}{ds}(s)\right\| ds +$$
$$+ M\int_0^t (1 + \|u_\lambda(s)\| + \|A_\lambda u_\lambda(s)\|)\, ds, \quad 0 \leqslant t \leqslant T$$

and by a simple calculation involving Eq. (4.6) and Gronwal's lemma, one finds that

$$\left\| \frac{du_\lambda}{dt}(t) \right\| + \|A_\lambda(t)u_\lambda(t)\| \leqslant C \quad \text{on } [0, T], \tag{4.8}$$

where C is independent of λ. The argument used in the proof of Theorem 2.3 can now be repeated here to deduce that $u_\lambda(t) \to u(t)$ uniformly on $[0, T]$ as $\lambda \to 0$. Thus estimate (4.8) implies that $\frac{du}{dt} \in L^\infty(0, T; X)$, i.e. $u \in W^{1,\infty}(0, T; X)$. To prove that $u(t)$ satisfies a.e. on $]0, T[$ Eq. (4.1), one uses the last argument in the proof of Theorem 2.2.

Remark. Similar results for the corresponding periodic problem can be obtained by a straightforward extension of the ideas in Section 2.2.

4.2. Evolution equations associated with nonlinear monotone hemicontinuous operators

We are given a real Hilbert space H and a real reflexive Banach space V with the following properties: V is dense in H and the inclusion mapping of V into H is continuous. We identify H with its own dual and denote V' as the dual space of V. Notations for norms and inner products will be the same as in Section 2.3, Example 1.

Secondly, we have a family of nonlinear operators $A(t): V \to V'$ defined on V for almost every $t \in]0, T[$, which are assumed to satisfy:

(i) *$A(t)$ is monotone and hemicontinuous from V to V' for almost every $t \in]0, T[$.*

(ii) *Function $A(t)u(t): [0, T] \to V'$ is measurable for every $u \in L^p(0, T;V)$.*

(iii) *There is a constant C such that*

$$\|A(t)u\|_* \leqslant C(\|u\|^{p-1} + 1) \quad \textit{for } u \in V,\ t \in]0, T[. \tag{4.9}$$

(iv) *There are constants $\alpha, \omega(\omega > 0)$ such that*

$$(A(t)\, u, u) \geqslant \omega \|u\|^p + \alpha \quad \textit{for } u \in V,\ t \in]0, T[\tag{4.10}$$

where $2 \leqslant p < \infty$.

The following result is an extension of Theorem 2.6 to the time-dependent case. However, the approach that will be used to prove it is entirely different.

THEOREM 4.2. *Given $u_0 \in H$ and $f \in L^{p'}(0, T; V')$; $1/p + 1/p' = 1$. Under the above hypotheses, there exists a unique function $u(t)$ which is V'-valued absolutely continuous on $[0, T]$ and satisfies*

$$u \in L^p(0, T; V) \cap C(0, T; H). \tag{4.11}$$

$$\frac{du}{dt} \in L^{p'}(0, T; V'). \tag{4.12}$$

$$\frac{du}{dt}(t) + A(t)u(t) = f(t), \quad \text{a.e. on }]0, T[. \tag{4.13}$$

$$u(0) = u_0. \tag{4.14}$$

Proof. We shall use the following notations

$$\mathscr{V} = L^p(0, T; V), \quad \mathscr{H} = L^2(0, T; H), \quad \mathscr{V}' = L^{p'}(0, T; V').$$

Clearly $\mathscr{V}$ and $\mathscr{V}'$ are dual pairs and the following inclusion relations hold

$$\mathscr{V} \subset \mathscr{H} \subset \mathscr{V}'.$$

Let u_0 be arbitrary but fixed in H. Let $B: \mathscr{V} \to \mathscr{V}'$ be defined by

$$Bu = \frac{du}{dt}, \quad \text{for } u \in D(B)$$

where

$$D(B) = \left\{u \in \mathscr{V} \text{ such that } \frac{du}{dt} \in \mathscr{V}' \text{ and } u(0) = u_0\right\}.$$

$\dfrac{d}{dt}$ is taken in the sense of vectorial distributions on $]0, T[$. We note that $u \in D(B)$ implies that $u \in W^{1,p'}(0, T; V')$ so that $u(0)$ makes sense.

We shall prove Theorem 4.2 by using a perturbation result for maximal monotone operators in $\mathscr{V} \times \mathscr{V}'$. The following lemma precedes the main argument.

LEMMA 4.1. *Operator B is maximal monotone in $\mathscr{V} \times \mathscr{V}'$.*

Proof. It is easy to see that B is monotone in $\mathscr{V} \times \mathscr{V}'$. To prove that it is maximal monotone it suffices to show that $R(B + \Phi) = \mathscr{V}'$ where Φ is defined by

$$\Phi(u)(t) = F(u(t)) \|u(t)\|^{p-2}, \quad u \in \mathscr{V}, \quad t \in]0, T[$$

and $F\colon V \to V'$ is the duality mapping of V. This may be seen from Theorem 1.2 in Chapter II (or rather its proof) observing that $u \to \Phi(u)/\|u\|^{p-2}$ is the duality mapping of $\mathscr{V}$. Let f be any arbitrary element of $\mathscr{V}'$. The equation

$$Bu + \Phi(u) = f$$

can be written as

$$\frac{du}{dt} + F(u)\|u\|^{p-2} = f, \qquad \text{a.e. } t \in]0, T[$$

$$u(0) = u_0 \tag{4.15}$$

In view of the renorming theorem stated in Chapter I, we may assume once again that V and V' are strictly convex. Then F is demicontinuous and therefore, referring to Theorem 2.6, problem (4.15) has a unique solution which satisfies

$$u \in C(0, T; H) \cap L^p(0, T; V), \quad \frac{du}{dt} \in L^{p'}(0, T; V').$$

This shows that B is maximal monotone. Incidentally, we have also proved that $D(B) \subset C(0, T; H)$.

We continue the proof of Theorem 4.2 by defining

$$(Au)(t) = A(t)u(t), \quad \text{for} \quad u \in \mathscr{V} \text{ and a.e. on }]0, T[.$$

Note that hypotheses (i) (ii) and (iii) imply that A is monotone, everywhere defined and hemicontinuous from $\mathscr{V}$ to $\mathscr{V}'$. Moreover, it follows by hypothesis (iv) that

$$\langle Au, u\rangle/\|u\|_{\mathscr{V}} \to +\infty \text{ as } \|u\|_{\mathscr{V}} \to +\infty \tag{4.16}$$

where $\langle\,,\rangle$ is the natural pairing between $\mathscr{V}$ and $\mathscr{V}'$. Invoking Lemma 4.1 proved before and Corollary 1.1 given in Chapter II, we deduce that $A + B$ is maximal monotone in $\mathscr{V} \times \mathscr{V}'$.

Since by relation (4.16), $A + B$ is also coercive on $\mathscr{V}$, it follows by Theorem 1.6 (Chapter II) that $R(A + B) = \mathscr{V}'$. This completes the proof of Theorem 4.2.

Remark. Theorem 4.2 remains valid if one only requires condition $1 < p < \infty$ rather than condition $2 \leqslant p < \infty$ of the theorem. To see this it suffices to take in the above proof $\mathscr{V} = L^p(0, T; V) \cap L^2(0, T; H)$ and $\mathscr{V}' = L^{p'}(0, T; V') + L^2(0, T; H)$.

Bibliographical notes

§1. The study of nonlinear semigroups of contractions emerges in connection with the existence of strong solutions for differential equations associated with nonlinear dissipative operators and was approached for the first time in

Hilbert spaces by Y. Komura [1] (We must also mention the pioneering paper of I. Segal [1].) His results were later improved and extended to a general framework by T. Kato [2], [3], M. G. Crandall and A. Pazy [1], [2], [3], H. Brézis and A. Pazy [1], M. G. Crandall and T. Liggett [1], M. G. Crandall [2]. Theorems 1.2 and 1.3 in the form in which they have been presented in Section 1.1 are attributed to M. G. Crandall and A. Pazy [1]. Theorems 1.3 and 1.4 are due to M. G. Crandall and T. Ligget [1], extending previous results of Y. Komura [1], T. Kato [1], M. G. Crandall and A. Pazy [1], H. Brézis and A. Pazy [1]. Theorem 1.4 part i, ii) and Theorem 1.6 are due to H. Brézis and A. Pazy [1]. Theorem 1.7 was originally proved by M. G. Crandall and A. Pazy [2], in uniformly convex Banach spaces and it was extended by H. Brézis [4] to the case of a general Banach space X with X^* uniformly convex. Proposition 1.1 is due to I. Miyadera. and S. Oharu [1], extending a previous result for linear contractions by P Chernoff [1].

Among other recent results in this field we must mention those obtained by M. G. Crandall [2] regarding the generalized domain for semigroup generators in nonreflexive Banach spaces. Related to this problem are the works of Y. Konishi [3], [4]. M. G. Crandall [4] applies the theory of nonlinear semigroups to obtain existence results for first-order quasilinear partial differential equations. Asymptotic properties of discrete semigroups of contractions are discussed in Pazy's paper [2] (see also A. Corduneanu [2]). Nonlinear semigroups in Banach lattices are studied in the works of K. Sato [1], [2], B. Calvert [1] and Y. Konishi [2].

§2. The integral solutions of the quasi-autonomous differential equations were introduced and studied by Ph. Bénilan and H. Brézis [1]. In general Banach spaces, Theorem 2.1 was formulated and proved by Ph. Bénilan [1], [2]. More general results covering the time-dependent case can be found in the work of M. G. Crandall and A. Pazy [3]. We mention here the author's paper [1] and that of H. Attouch-A. Damlamian [1] regarding the differential equations of the form $\frac{du}{dt} = Au + B(t)u + f$, $u(0) = u_0$ (see also the recent paper by R. V. Bressan [1]). Results similar to those in Section 2.2. have been obtained by O. Vejvoda [1], F. Browder [8] and more recently by I. Straskraba and O. Vejvoda [1]. Theorem 2.4 was established by H. Brézis [3] and we refer the reader to Ph. Bénilan and H. Brézis [1] and H. Brézis [8] for other results regarding periodic problems. Results similar to Theorems 2.5 and 2.6 were obtained by many authors (see e.g. F. Browder [5], [7], J. L. Lions and W. Strauss [1], J. L. Lions [1] and W. A. Strauss [1], [2]). The existence of almost periodic solutions of certain nonlinear differential equations having a form as Eq. (2.1) have been investigated in a series of papers by M. Biroli [1], [2]. A typical result (M. Biroli [2]) can be formulated as follows: Consider Eq. (2.1) under the following conditions: (a) A is maximal monotone in $H\times H$ and $(Au - Av, u - v) \geqslant \omega \|u - v\|^2$ (b) H is a real Hilbert space and V is a uniformly convex Banach space with its norm densely and continously imbedded in H. (c) The injection map of V into H is compact. Let $f \in C(R^1, H)$ be such that $\frac{df}{dt} \in L^\infty(R^1, H)$. If $f(t)\colon R^1 \to H$ is almost periodic, then Eq. (2.1) has a unique solution $u(t)$ which is almost periodic in V. Numerous results concerning the almost periodicity of the solutions of certain partial differential equations having the special form encountered in this chapter may be found in the monograph by L. Amerio and G. Prouse [1]. Further references can be found in this book and also in that by C. Corduneanu [2]. The first result being of the same type as Theorem 2.7 was established by the author [5], but in the form in which it has been presented belongs to K. Gröger [1]. Theorem 2.8 was originally established by R. Temam [1]. The simple proof that we have presented here uses an argument due to H. Brézis [3].

Using different methods, other existence results for Riccati differential equations in Banach spaces have been obtained by G. Da Prato [1], [2], [3], J. L. Lions [2], L. Tartar [1], A.J. Pritchard [1], A.J. Pritchard and R. Curtain [1].

In the work of M. Slemrod [1] is given another interesting application of the nonlinear semigroup theory in the synthesis problem for certain classes of distributed boundary control systems.

§ 3. The results of Sections 3.1 and 3.2 are due to the author [6], extending some results of G.F. Webb [1] for the case in which A is linear. Recently, M. Pierre [1] obtained essentially the same conclusions merely assuming that B is continnuous on $D(A)$.

Under further assumptions on B or X the case A-nonlinear has been treated by several authors. We have mentioned the works of F. Browder [7], G. Da Prato [2], M. Iannelli [1] and A. Corduneanu [1]. In nonreflexive Banach spaces, the existence result stated in Corollary 3.1 was first established by R. H. Martin, Jr. [1]. The same author (R. H. Martin, Jr. [2], [3]) obtained other remarkable results regarding the existence of solutions on closed subsets of a Banach space. For other papers which deal with the existence theory of differential equations associated with continuous dissipative operators we must mention those of D. L. Lovelady [1], [2], D. L. Lovelady and R. H. Martin [1], N. Pavel [1], [2]. Further references can be found in the survey of N. Pavel [3].

Other perturbation results together with several applications to partial differential equations and functional differential equations can be found in the works by G. F. Webb [3], [4], [5], [6]. Theorem 3.3 was established independently by Y. Konishi [7] and H. Brézis and W. Strauss [1].

§ 4. The first theorem of the same type as Theorem 4.1 was established by T. Kato [2] assuming that $A(t)$ is single-valued. The paper [3] by M.G. Crandall and A. Pazy allows a more general t-dependence of A and their results are available in general Banach spaces. In this work they give general conditions on $A(t)$ guaranteeing the existence of an associated evolution operator. When $A(t)$ is linear, a discussion on the results of this type may be found in a book by K. Yosida [1]. In concluding these notes, we must mention that the semigroup approach was recently used by C.V. Pao [1] and A. Belleni-Morante [1] in the existence theory of nonlinear transport equation. In this respect the work of A. Belleni-Morante and L. Barone [1] can also be cited.

Chapter IV

Nonlinear differential equations in Hilbert spaces

In this chapter, we shall consider the sharpening of some of the results found in Chapter III in the case in which the basic space X is Hilbert. In particular, one obtains a complete generation theory for nonlinear semigroups of contractions in a Hilbert space.

The smoothness of the solutions to nonlinear differential equations of the subgradient type, as well as its applications to the existence theory of evolution variational inequations, constitutes the core of the present chapter. This discussion gives us an opportunity to develop that part of the machinery which allows us to treat other important classes of nonlinear functional-differential equations.

§1. Nonlinear semigroups in Hilbert spaces

1.1. Nonlinear version of the Hille-Yosida theorem

As seen in Chapter I, a densely defined linear operator A is a generator of a semigroup of linear contractions on a Banach space X if and only if A is m-dissipative. We shall see that if X is a Hilbert space then essentially the same relationship persists between the nonlinear semigroups of contractions on closed convex subsets of H and (multivalued) m-dissipative operators in H.

We shall start with a density theorem extending a previous result from the linear theory of semigroups (see Section 3 in Chapter I).

THEOREM 1.1. *Let $S(t)$ be a semigroup of nonlinear contractions defined on a closed convex subset C of a real Hilbert space H. Then the generator A_0 of $S(t)$ is densely defined on C.*

We shall prove Theorem 1.1 by following a finite sequence of lemmas. Let $A_h: C \to H$ be the operator defined by

$$A_h x = h^{-1}(S(h)x - x), \quad h > 0$$

Since A_h is dissipative, we know that $(1 - \lambda A_h)^{-1}$ is non-expansive on $D((1 - \lambda A_h)^{-1})$ for each λ and h positive. Let us verify

LEMMA 1.1. (a) $C \subset \bigcap_{\lambda, h>0} D((1 - \lambda A_h)^{-1})$

(b) *For every* $x \in C$, *function* $(h, \lambda) \to (1 - \lambda A_h)^{-1}x$ *is separately continuous*

Proof. (a) Let x be arbitrary but fixed in C. Equation

$$(1 - \lambda A_h)y = x$$

is equivalent to

$$y = \frac{h}{h + \lambda} x + \frac{\lambda}{h + \lambda} S(h)y. \tag{1.2}$$

It is easy to see that the mapping defined by the right side of Eq. (1.2) maps C into itself and is lipschitzian with Lipschitz constant $\lambda(\lambda + h)^{-1}$. We conclude that Eq. (1.2) has a unique solution

$$y_{\lambda,h} = (1 - \lambda A_h)^{-1}x.$$

(b) It is an immediate consequence of the fact that $S(h)x$ is continuous in h and nonexpansive in x.

We also note, for later use, some relations involving $y_{\lambda,h}$

$$S(h)y_{\lambda,h} - x = \left(1 + \frac{h}{\lambda}\right)(y_{\lambda,h} - x) \tag{1.3}$$

$$A_h y_{\lambda,h} = \lambda^{-1}(y_{\lambda,h} - x) \tag{1.4}$$

$$|y_{\lambda,h} - x| \leqslant \frac{\lambda}{h} |x - S(h)x|. \tag{1.5}$$

Here and throughout this Section $|\cdot|$ will denote the norm in H.

LEMMA 1.2. *Let* $\varepsilon > 0$ *and* $0 < \delta < 1$ *be such that* $h \in]0, \delta[$, *implies* $|S(h)x - x| \leqslant \varepsilon$. *If* $nh = t \in]0, \delta[$, *n being a positive integer, then*

$$|y_{\lambda,h} - y_{\lambda,t}|^2 \leqslant 2\varepsilon |y_{\lambda,h} - x|. \tag{1.6}$$

Proof. Without loss of generality, we may assume that $x = 0$. Then, according to relation (1.3), we have

$$S(h)y_{\lambda,h} = \left(1 + \frac{h}{\lambda}\right) y_{\lambda,h}.$$

Consequently,

$$|y_{\lambda,h} - S((j-1)h)\, y_{\lambda,t}|^2 \geqslant |S(h)\, y_{\lambda,h} - S(jh)\, y_{\lambda,t}|^2 =$$

$$= \left|\left(1 + \frac{h}{\lambda}\right) y_{\lambda,h} - S(jh) y_{\lambda,t}\right|^2 \geqslant$$

$$\geqslant |y_{\lambda,h} - S(jh) y_{\lambda,t}|^2 + \frac{2h}{\lambda}(y_{\lambda,h}, y_{\lambda,h} - S(jh) y_{\lambda,t}) \qquad j = 1, 2, \ldots, n$$

and also

$$|y_{\lambda,h} - S(t) y_{\lambda,t}|^2 = \left| y_{\lambda,h} - \left(1 + \frac{t}{\lambda}\right) y_{\lambda,t}\right|^2 \geqslant$$

$$\geqslant |y_{\lambda,h} - y_{\lambda,t}|^2 + \frac{2t}{\lambda}(y_{\lambda,t}, y_{\lambda,t} - y_{\lambda,h}).$$

By summing these $n + 1$ inequalities, one finally obtains,

$$|y_{\lambda,h}|^2 + |y_{\lambda,t}|^2 \leqslant \frac{1}{n} \sum_{j=1}^{n} (y_{\lambda,h}, S(jh)\, y_{\lambda,t}) + (y_{\lambda,h}, y_{\lambda,t}).$$

Since $jh \in]0, \delta[$, this implies that

$$|S(jh)\, y_{\lambda,t}| \leqslant |y_{\lambda,t}| + |S(jh)\, 0| \leqslant |y_{\lambda,t}| + \varepsilon.$$

Thus,

$$|y_{\lambda,h}|^2 + |y_{\lambda,t}|^2 \leqslant |y_{\lambda,h}|\, (\varepsilon + |y_{\lambda,t}|) + (y_{\lambda,h}, y_{\lambda,t}).$$

Finally,

$$|y_{\lambda,h} - y_{\lambda,t}|^2 \leqslant 2\varepsilon\, |y_{\lambda,h}|$$

which is the desired result.

LEMMA 1.3. *Let $x \in C$ and ε, δ as in Lemma 1.2. Then, for all $\lambda > 0$ and $h \in]0, \delta[$, the following inequality holds*

$$|y_{\lambda,h} - x| \leqslant 2\varepsilon\left(1 + \frac{4\lambda}{\delta}\right). \tag{1.7}$$

In particular, for every $\lambda > 0$ the function $h \to y_{\lambda,h}$ is bounded on $]0, \delta[$.

Proof. Again without loss of generality we shall assume that $x = 0$. If $h \in]\delta/2, \delta]$, then by (1.5) it follows that

$$|y_{\lambda,h}| \leqslant 2\lambda\, \varepsilon/\delta. \tag{1.8}$$

If $h \in]0, \delta/2[$, there exists a positive integer n such that $nh = t \in [\delta/2, \delta]$. Thus from Lemma 1.2 it follows that

$$|y_{\lambda,h} - y_{\lambda,t}|^2 \leqslant 2\varepsilon|y_{\lambda,h}| \leqslant 2\varepsilon|y_{\lambda,h} - y_{\lambda,t}| + 2\varepsilon\,|y_{\lambda,t}|$$

and by (1.8)

$$|y_{\lambda,h} - y_{\lambda,t}| \leqslant \varepsilon\left(1 + \left(1 + \frac{4\lambda}{\delta}\right)^{1/2}\right).$$

Together with (1.8) this last inequality implies the desired estimate (1.7).

LEMMA 1.4. *For all $\lambda > 0$, $\lim\limits_{h\to 0} y_{\lambda,h} = y_\lambda$ exists and belongs to $D(A_0)$.*

Proof. According to Lemma 1.3, there exists a positive constant M such that $|y_{\lambda,h} - x| \leqslant M$ for all $h, \lambda \in]0, 1]$. Then inequality (1.6) implies that

$$|y_{\lambda,h} - y_{\lambda,t}| \leqslant (2M\varepsilon)^{1/2} \qquad t = nh \in]0, \delta[.$$

Consequently,

$$|y_{\lambda,h} - y_{\lambda,t}| \leqslant 2(2M\varepsilon)^{1/2} \tag{1.9}$$

for all $h, t \in]0, \delta[$ such that t/h is a rational number. Since for every $\lambda > 0$ the function $h \to y_{\lambda,h}$ is continuous on $]0, \infty[$, inequality (1.9) clearly extends to all $h, t > 0$. In particular, one concludes that $\lim\limits_{h\to 0} y_{\lambda,h} = y_\lambda$ exists for all $\lambda > 0$.

In order to prove that $y_\lambda \in D(A_0)$, it suffices to show that

$$|S(t)y_\lambda - y_\lambda| \leqslant Mt \text{ for all } t \in [0, 1]$$

where M is independent of t.

Indeed, for $t = nh$ we have

$$|S(t)y_{\lambda,h} - y_{\lambda,h}| \leqslant \sum_{j=1}^{n} |S(j-1)h)\, y_{\lambda,h} - S(jh)\, y_{\lambda,h}| \leqslant$$

$$\leqslant n\,|y_{\lambda,h} - S(h)y_{\lambda,h}| = \frac{t}{\lambda}\,|x - y_{\lambda,h}|.$$

If we consider t fixed and $n \to \infty$, then $y_{\lambda,h} \to y_\lambda$.

Thus, the above inequality implies that

$$|S(t)y_\lambda - y_\lambda| \leqslant t/\lambda\,|x - y_\lambda| \text{ for all } \lambda > 0$$

which concludes the proof of Lemma 1.4.

Proof of Theorem 1.1. Let x be arbitrary but fixed in C. We shall prove that y_λ, given by Lemma 1.4, converges strongly to x as $\lambda \to 0$. In fact, letting $h \to 0$ in estimate (1.7) one obtains

$$|y_\lambda - x| \leqslant 2\varepsilon\,(1 + 4\lambda/\delta) \text{ for all } \lambda > 0.$$

As ε is arbitrary, we deduce that $\lim\limits_{\lambda \to 0} y_\lambda = x$, i.e. $x \in \overline{D(A_0)}$. This completes the proof of Theorem 1.1.

In a Hilbert space H there exists a one to one correspondence between semigroups of nonlinear contractions on closed convex subsets of H and maximal dissipative subsets of $H \times H$.

This is one of the principle motivations for developing the existence theory of *multivalued* differential equations.

THEOREM 1.2. *Let C be a nonempty closed convex subset of H and let S be a semigroup of nonlinear contractions defined on C. Then there exists a unique maximal dissipative set $A \subset H \times H$ such that A^0 is the generator of S.*

Let A be a maximal dissipative set in $H \times H$. Then there is a unique semigroup S on $\overline{D(A)}$ such that A^0 is the generator of S.

Proof. Let $S(t)$ be a semigroup of nonlinear contractions defined on the closed convex subset C of H. Let A_0 denote the infinitesimal generator of $S(t)$. According to Theorem 1.1, $\overline{D(A_0)} = C$ so that there exists a maximal dissipative extension A of A_0 such that $\overline{D(A)} = C$ (see Theorem 3.2 in Chapter II). We shall prove that $S(t)$ is generated by A (see Definition 1.3, Chapter III). By Theorem 1.6 in Chapter III there exists a semigroup of nonlinear contractions $T(t)$ defined on $\overline{D(A)} = C$ such that

$$\frac{\mathrm{d}}{\mathrm{d}t} T(t)x = A^0 T(t)x \in AT(t)x, \qquad \text{a.e. on }]0, \infty[,\ x \in D(A).$$

On the other hand, we have

$$\frac{\mathrm{d}}{\mathrm{d}t} S(t)x = A_0 S(t)\, x \in A\; S(t)x \quad , \text{ a.e. } t > 0$$

for all $x \in D(A_0)$. Consequently

$$\frac{\mathrm{d}}{\mathrm{d}t} |S(t)x - T(t)x|^2 \leqslant 0 \quad , \text{ a.e. } t > 0$$

for all $x \in D(A_0)$. In other words, $S(t)x = T(t)x$ for all $t \geqslant 0$ and $x \in \overline{D(A_0)} = \overline{D(A)} = C$. By virtue of Theorems 1.2 and 1.6 from Chapter III, one deduces also that $A_0 = A^0$. The uniqueness of A is a direct consequence of Theorem 1.7 proved in the last chapter. We shall conclude the proof by observing that the last part of Theorem 1.2 was proved under more general assumptions in Chapter III (Corollary 1.1).

Theorem 1.2. is apparently incomplete because the nonlinear semigroups of contractions defined on arbitrary closed subsets of H have not been considered. This situation is clarified by the following remarkable result due to Y. Komura [2].

THEOREM 1.3. *Let $S(t)$ be a semigroup of nonlinear contractions defined on a closed subset C of a Hilbert space H. Then, there exists a semigroup of nonlinear contractions $\tilde{S}(t)$ defined on $\overline{\text{conv } C}$, which extends $S(t)$.*

The proof of this theorem may be found in Komura's paper [2]. Another proof, based on a selection theorem for multivalued semigroups of contractions, was given by H. Brézis and A. Pazy [2]. Since we shall not use the theorem in the sequel, and its proof is complicated, we shall omit it.

Remark. Theorems 1.1, 1.2 and 1.3 in this section constitute a complete generation theory for nonlinear semigroups of contractions in a Hilbert space. The extension of these results to *semigroups of type ω* (see Section 1.1, Chapter III) is straightforward. As seen in Chapter III, Section 1.3, Theorems 1.1 and 1.2 fail in general Banach spaces X. Using an argument similar to that used in the proof of Theorem 1.1 it follows that every contraction semigroup S on a finite dimensional space can be obtained from a dissipative set A, via the exponential formula

$$S(t) = \lim_{n \to \infty} \left(1 - \frac{t}{n} A\right)^{-n}, \qquad t \geqslant 0.$$

However, even if X is two-dimensional, it turns out that A is not unique if X fails to be strictly convex (see M. G. Crandall and T. Liggett [1]).

1.2. Exponential formulae

This section is concerned with representation formulae of semigroups, in terms of their generators. We shall start with the following general result.

THEOREM 1.4. *Let C be a closed convex subset of a Hilbert space H and let A be a maximal dissipative subset of H × H. Let* $\{T(t);\ t \geqslant 0\}$ *be a family of nonlinear contractions from C into itself satisfying*

$$\lim_{h\to 0}\left(1 - \lambda\frac{T(h) - 1}{h}\right)^{-1} x = (1 - \lambda A)^{-1}x, \tag{1.10}$$

for all $\lambda > 0$ *and* $x \in \overline{D(A)} \cap C$.

Then

$$\lim_{n\to\infty}\left[T\left(\frac{t}{n}\right)\right]^n x = S(t)x, \quad \forall x \in \overline{D(A)} \cap C, \tag{1.11}$$

uniformly in t on every compact subset of $[0, \infty[$. *(Here* $S(t)$ *is the semigroup generated by A on* $\overline{D(A)}$.)

Proof. Let $A^h\colon C \to H$ denote the operator $(T(h) - 1)/h$ and $S^h(t)$ the semigroup generated by A^h on C (see Proposition 1.1, Chapter III). We shall denote also $A^h_\lambda = \lambda^{-1}((1 - \lambda A^h)^{-1} - 1)$ and by $S^h_\lambda(t)$ the nonlinear semigroup associated with A^h_λ. According to Theorem 1.4 in Chapter III, $\lim_{\lambda\to 0} S^h_\lambda(t)x = S^h(t)x$ for every $x \in C = D(A^h)$. More precisely, the following inequality holds

$$|S^h(t)x - S^h_\lambda(t)x| \leqslant 3(\lambda + \sqrt{t\lambda})\,|A^h x|, \qquad x \in C. \tag{1.12}$$

Let $S_\lambda(t)$ be the semigroup generated by A_λ on H. From Eq. (1.10) one gets

$$\lim_{h\to 0} A^h_\lambda x = A_\lambda x, \qquad \forall \lambda > 0;\ x \in \overline{D(A)} \cap C, \tag{1.13}$$

uniformly in x on every bounded subset of $\overline{D(A)} \cap C$. Let x be arbitrary but fixed in $D(A) \cap C$. We have

$$\frac{d}{dt}(S_\lambda(t)x - S^h_\lambda(t)x) = A_\lambda S_\lambda(t)x - A^h_\lambda S^h_\lambda(t)\,x.$$

Since A^h_λ is dissipative, one obtains

$$|S_\lambda(t)\,x - S^h_\lambda(t)\,x| \leqslant \int_0^t |A_\lambda S_\lambda(s)\,x - A^h_\lambda S_\lambda(s)\,x|\,ds, \qquad t \geqslant 0$$

and by (1.13)

$$|S_\lambda(t)\, x - S_\lambda^h(t)\, x| \leqslant \omega_\lambda(h) \tag{1.14}$$

where $\lim_{h\to 0} \omega_\lambda(h) = 0$.

Once again using inequality (1.34) from Chapter III, one obtains

$$|S(t)x - S_\lambda(t)x| \leqslant 3(\lambda + \sqrt{t\lambda})|\, A^0x|, \quad \forall t, \lambda > 0. \tag{1.15}$$

Combining this estimate with (1.12) and (1.14) one finally gets,

$$|S(t) - S^h(t)x| \leqslant 3(\lambda + \sqrt{t\lambda})\, (|A^0x| + |A^hx|) + \omega_\lambda(h), \qquad \forall \lambda > 0. \tag{1.16}$$

for all λ, h and $t > 0$.

On the other hand we have

$$|S^h(nh)x - T^n(h)x| \leqslant \sqrt{n}|x - T(h)x|, \quad n = 1, 2, \ldots$$

in view of Proposition 1.2 from Chapter III. Consequently

$$\left|S^{\frac{t}{n}}(t)x - T^n\left(\frac{t}{n}\right)x\right| \leqslant \frac{t}{\sqrt{n}}\left|A^{\frac{t}{n}}x - A_\lambda^{\frac{t}{n}}x\right| + \frac{t}{\sqrt{n}}\left|A_\lambda^{\frac{t}{n}}x\right|, \quad \forall \lambda > 0.$$

Together with (1.16) this inequality gives

$$\left|S(t)x - T^n\left(\frac{t}{n}\right)x\right| \leqslant \frac{t}{\sqrt{n}}\, (|A_\lambda^{\frac{t}{n}}x| + |A^{\frac{t}{n}}x - A_\lambda^{\frac{t}{n}}x|) +$$

$$+ 3(\lambda + \sqrt{t\lambda})\, (|A^0x| + |A^{\frac{t}{n}}x - A_\lambda^{\frac{t}{n}}x| + |A_\lambda^{\frac{t}{n}}x|) + \omega_\lambda\left(\frac{t}{n}\right) \tag{1.17}$$

for all $\lambda, t > 0$ and $n = 1, 2, \ldots$ Since $A_\lambda x \to A^0x$ as $\lambda \to 0$, from (1.13) and (1.17) one deduces (1.11) for every $x \in D(A) \cap C$. Since $S(t)$ and $T(t)$ are contractions the result follows for $x \in \overline{D(A)} \cap C$. This completes the proof.

Remark. In particular, taking $T(t) = (1 - tA)^{-1}$ and $C = \overline{D(A)}$, in Theorem 1.4, one obtains the exponential formula

$$S(t)x = \lim_{n\to\infty} \left(1 - \frac{t}{n}A\right)^{-n} x, \qquad \forall\, x \in \overline{D(A)}, \quad t \geqslant 0.$$

The result we shall prove now is well known in the linear case as the *Trotter product formula*.

THEOREM 1.5. *Let A and B be two maximal dissipative subsets of $H \times H$. Assume that the closure $\overline{A+B}$ of the subset $A+B$ is maximal dissipative and let $S(t)$ be the semigroup generated by $\overline{A+B}$ on $\overline{D(A) \cap D(B)}$. Then,*

$$S(t)x = \lim_{n\to\infty} \left[\left(1 - \frac{t}{n} A\right)^{-1} \left(1 - \frac{t}{n} B\right)^{-1} \right]^n x, \tag{1.18}$$

for every $x \in \overline{D(A) \cap D(B)}$, uniformly in t on every compact of positive half axis.

Proof. We apply Theorem 1.4 where $C = H$ and $T(t) = (1 - tA)^{-1}(1 - tB)^{-1}$. For concluding the proof it suffices to show that for every $x \in H$ the following equality holds

$$\lim_{h\to 0} \left(1 - \lambda \frac{T(h) - 1}{h}\right)^{-1} x = (1 - \lambda\overline{A+B})^{-1}x, \quad \forall \lambda > 0. \tag{1.19}$$

We set $y_h = \left(1 - \frac{\lambda}{h}(T(h) - 1)\right)^{-1} x$ and $z_h = (1 - hB)^{-1} y_h$.

By an elementary calculation one gets

$$\left(1 + \frac{h}{\lambda}\right) y_h - \frac{h}{\lambda} x = (1 - hA)^{-1} z_h$$

and

$$w_h + \frac{1}{\lambda}(y_h - x) \in A\left(\left(1 + \frac{h}{\lambda}\right) y_h - \frac{h}{\lambda} x\right),$$

where

$$w_h = \frac{1}{h}(y_h - z_h) \in -Bz_h.$$

Let v be arbitrary in $D(A) \cap D(B)$ and let $(u_1, u_2) \in H \times H$ be such that $u_1 \in Av$ and $u_2 \in Bv$.

We have

$$\left(u_1 - w_h - \frac{1}{\lambda}(y_h - x),\ v - \left(1 + \frac{h}{\lambda}\right) y_h + \frac{h}{\lambda} x\right) \leqslant 0$$

and

$$(u_2 + w_h,\ v - z_h) \leqslant 0.$$

because A and B are dissipative.

Thus; we obtain after some rearrangements

$$\frac{1}{\lambda}\left(1+\frac{h}{\lambda}\right)|y_h|^2 \leqslant \left(w_h - u_1 - \frac{1}{\lambda}x,\ v - \left(1+\frac{h}{\lambda}\right)y_h + \frac{h}{\lambda}x\right) +$$

$$+\frac{1}{\lambda}\left(y_h,\ v + \frac{h}{\lambda}x\right)$$

respectively

$$h|w_h|^2 \leqslant -(u_2 + w_h,\ v - y_h) - h(u_2, w_h),$$

which imply

$$\frac{1}{\lambda}\left(1+\frac{h}{\lambda}\right)|y_h|^2 + h|w_h|^2 \leqslant -\left(u_1 + u_2 + \frac{1}{\lambda}x,\ v - y_h\right) -$$

$$-\frac{h}{\lambda}(u_1 - w_h,\ x - y_h) + \frac{1}{\lambda}\left(y_h,\ v + \frac{h}{\lambda}x\right) - h(u_2, w_h) \tag{1.20}$$

From this it easily follows that $|y_h|$ and $h|w_h|^2$ are bounded as $h \to 0$. Without loss of generality we shall assume that y_h converges weakly to y as $h \to 0$. From estimate (1.20) we obtain

$$\frac{1}{\lambda}|y|^2 \leqslant -\left(u_1 + u_2 + \frac{1}{\lambda}x,\ v - y\right) + \frac{1}{\lambda}(y, v).$$

Therefore

$$\left(u_1 + u_2 + \frac{1}{\lambda}x - \frac{1}{\lambda}y,\ v - y\right) \leqslant 0, \quad [v, u_1] \in A; \quad [v, u_2] \in B. \tag{1.21}$$

Since $\overline{A+B}$ is maximal dissipative we deduce that $\frac{1}{\lambda}(y - x) \in \overline{A+B}\,y$.

Consequently

$$y = (1 - \lambda\overline{A+B})^{-1}x.$$

Coming back to estimate (1.20) one finds

$$\limsup_{h\to 0} \frac{1}{\lambda}|y_h|^2 \leqslant -\left(u_1 + u_2 + \frac{1}{\lambda}x,\ v - y\right) + \frac{1}{\lambda}(y, v).$$

Therefore

$$\limsup_{h\to 0} \frac{1}{\lambda} |y_h|^2 \leqslant -\left(\xi + \frac{1}{\lambda} x, v - y\right) + \frac{1}{\lambda} (y, v)$$

for every $[v, \xi] \in \overline{A + B}$.

Taking $v = y$ we obtain

$$\limsup_{h\to 0} |y_h|^2 \leqslant |y|^2.$$

which implies that y_h converges strongly to y.

THEOREM 1.6. *Let A and B be single-valued maximal dissipative operators such that $A + B$ is maximal dissipative. Then for every $x \in \overline{D(A) \cap D(B)}$ holds*

$$\lim_{n\to\infty} \left[S_A\left(\frac{t}{n}\right) S_B\left(\frac{t}{n}\right)\right]^n x = S_{A+B}(t)x \tag{1.22}$$

and the limit is uniform in t on every bounded interval.

Proof. Clearly $T(h) = S_A(h)\, S_B(h)$ is non-expansive on $C = \overline{D(A) \cap D(B)}$. We shall prove that $\lim_{h\to 0} \frac{T(h)x - x}{h} = Ax + Bx$ for every $x \in D(A) \cap \cap D(B)$. We have

$$\frac{T(h)x - x}{h} = \frac{S_A(h)x - x}{h} + \frac{S_A(h)S_B(h)x - S_A(h)x}{h}.$$

Since $\lim_{h\to 0} \frac{S_A(h)x - x}{h} = Ax$, it suffices to show that

$$y_h = \frac{S_A(h)S_B(h)x - S_A(h)x}{h} \to Bx \quad \text{as} \quad h \to 0.$$

As $S_A(h)$ is nonexpansive, we have

$$|y_h| \leqslant \left|\frac{S_B(h)x - x}{h}\right| \leqslant |Bx|, \quad \forall h > 0. \tag{1.23}$$

On the other hand,

$$\left(\frac{S_A(h)u - u}{h} + \frac{S_B(h)x - x}{h} - \frac{S_A(h)x - x}{h} - y_h, u - S_B(h)x\right) \geqslant 0, \tag{1.24}$$

because $S_A(h) - 1$ is dissipative. We choose a subsequence $\{y_{h_n}\}$ weakly convergent to y as $h_n \to 0$. By (1.24) it follows that

$$(Au + Bx - \mathrm{A}x - y, u - x) \geqslant 0, \quad \forall u \in D(A).$$

Since A is maximal dissipative, this implies that $y = Bx$. Then, from condition (1.23) one deduces that y_h converges strongly to Bx as $h \to 0$. Now, we shall prove that

$$z_h = \left(1 - \frac{\lambda}{h}(T(h) - 1)\right)^{-1} x \to (1 - \lambda(A + B))^{-1}x \quad \text{as } h \to 0$$

which implies formula (1.22) by Theorem 1.4.

By the dissipativity of $T(h) - 1$ we obtain,

$$\left(\frac{z_h - x}{\lambda} - \frac{T(h)u - u}{h}, \ z_h - u\right) \leqslant 0,$$

which implies that $|z_h|$ is bounded as $h \to 0$. Extracting further sequences if necessary, we may assume that $\{z_h\}$ converges weakly to z as $h \to 0$. Passing to the limit through this sequence, yields

$$\left(\frac{z - x}{\lambda} - Au - Bu, \ z - u\right) \leqslant 0 \text{ for every } u \in D(A) \cap D(B)$$

and therefore

$$z = (1 - \lambda(A + B))^{-1}x$$

because $A + B$ is maximal dissipative. On the other hand, the inequality

$$\limsup_{h\to 0} \frac{1}{\lambda}|z_h|^2 \leqslant \left((A + B)u + \frac{x}{\lambda}, \ z - u\right) + \left(\frac{z}{\lambda}, \ u\right)$$

implies that $\limsup_{h\to 0} |z_h|^2 \leqslant |z|^2$. Hence $\{z_h\}$ is strongly convergent to z as $h \to 0$. This completes the proof of Theorem 1.6.

1.3. Invariant sets with respect to nonlinear semigroups

Let C be a closed convex subset of a Hilbert space H and let A be a maximal monotone subset of $H \times H$. We denote the semigroup of nonlinear contractions generated by $-A$ on $\overline{D(A)}$ by $S(t)$.

The problem we shall consider here is to find sufficient conditions such that $S(t)(\overline{D(A)} \cap C) \subset C$ for all $t \geqslant 0$.

Let $P_{\overline{D(A)}}$ denote the projection operator of H into $\overline{D(A)}$.

THEOREM 1.7. *Let C be a closed convex subset of H and A maximal monotone in $H \times H$ such that $P_{\overline{D(A)}} C \subset C$. Then the following four conditions are equivalent:*

(i) $(1 + \lambda A)^{-1}C \subset C$ *for all* $\lambda > 0$.

(ii) $(A^0x, x - P_Cx) \geqslant 0$ *for every* $x \in D(A)$.

(iii) *dist* $(S(t)x, C) \leqslant$ *dist* (x, C) *for all* $x \in \overline{D(A)}$.

(iv) $S(t)(\overline{D(A)} \cap C) \subset C$ *for all* $t \geqslant 0$.

In particular, (ii) $\Leftrightarrow$ (iv) leads to the following result.

COROLLARY 1.1. *Let A be a maximal monotone subset of $H \times H$ and let C be a closed convex subset of H such that $C \subset \overline{D(A)}$, $P_C(D(A)) \subset D(A)$ and*

$$-A^0x \in \overline{\bigcup_{\substack{\lambda>0\\ u\in C}} \lambda(u - x)} \quad \textit{for all } x \in D(A) \cap C.$$

Then

$$S(t)C \subset C \textit{ for every } t \geqslant 0.$$

We shall derive Theorem 1.7 as a consequence of a general result concerning convex functions. The notations are those used in Chapter II, § 2.

PROPOSITION 1.1. *Let A be a maximal monotone subset of $H \times H$ and let $\varphi : H \to]-\infty, +\infty]$ be a lower-semicontinuous convex function nonidentically $+\infty$. Assume that*

$$\varphi(P_{\overline{D(A)}}x) \leqslant \varphi(x) \quad \textit{for all } x \in H. \tag{1.25}$$

Then the following conditions are equivalent:

(j) $\varphi((1 + \lambda A)^{-1}x) \leqslant \varphi(x)$ *for all* $x \in H$ *and* $\lambda > 0$.

(jj) $(A_\lambda x, y) \geqslant 0$ *for every* $[x, y] \in \partial\varphi$ *and* $\lambda > 0$.

(jjj) $(A_\lambda x, \partial\varphi_\mu(x)) \geqslant 0$ *for every* $x \in H$ *and* $\lambda, \mu > 0$.

(jv) $(A^0x, \partial\varphi_\mu(x)) \geqslant 0$ *for every* $x \in D(A)$ *and* $\mu > 0$.

(v) $\varphi(S(t)x) \leqslant \varphi(x)$, *for every* $x \in \overline{D(A)}$, *and* $t > 0$

where $A_\lambda = \lambda^{-1}(1 - (1 + \lambda A)^{-1})$.

Proof. (j) $\Rightarrow$ (jj). Let $[x, y]$ be arbitrary in $\partial\varphi$. By definition of $\partial\varphi$ (see Section 2, Chapter II) we have

$$0 \leqslant \varphi(x) - \varphi((1 + \lambda A)^{-1}x) \leqslant (y, x - (1 + \lambda A)^{-1}x) = \lambda(y, A_\lambda x).$$

(jj) $\Rightarrow$ (jjj). We recall (see Theorem 2.2 in Chapter II) that

$$\partial\varphi_\mu = (\partial\varphi)_\mu = \mu^{-1}(1 - (1 + \mu\partial\varphi)^{-1}x) \text{ for all } \mu > 0.$$

Hence

$$(A_\lambda(1 + \mu\partial\varphi)^{-1}x, \partial\varphi_\mu(x)) \geqslant 0$$

because $\partial\varphi_\mu(x) \in \partial\varphi(1 + \mu\partial\varphi)^{-1}x)$. Since A_λ is monotone, one finds that

$$(A_\lambda x - A_\lambda(1 + \mu\partial\varphi)^{-1}x,\ \partial\varphi_\mu(x)) \geqslant 0$$

and therefore

$$(A_\lambda x, \partial\varphi_\mu(x)) \geqslant 0.$$

(jjj) $\Rightarrow$ (jv). It suffices to use the fact that $\lim_{\lambda\to 0} A_\lambda x = A^0 x$ for every $x \in D(A)$.

(jv) $\Rightarrow$ (v). If $x \in D(A)$ then the function $t \to S(t)x$ is locally Lipschitzian on $[0, \infty[$ and therefore so is the function $t \to \varphi_\mu(S(t)x)$. We note that

$$\frac{\mathrm{d}}{\mathrm{d}t}\varphi_\mu(S(t)x) = \left(\partial\varphi_\mu(S(t)x),\ \frac{\mathrm{d}}{\mathrm{d}t}\ S(t)x\right), \text{ a.e. } t > 0.$$

Since $\frac{\mathrm{d}}{\mathrm{d}t} S(t)x = -A^0 S(t)x$, we deduce that $\frac{\mathrm{d}}{\mathrm{d}t}\ \varphi_\mu(S(t)x) \leqslant 0$, a.e. $t > 0$. In other words the function $t \to \varphi_\mu(S(t)x)$ is monotonically nonincreasing on $[0, \infty[$.

Then, one deduces (v) by noting that $\lim_{\mu\to 0} \varphi_\mu(x) = \varphi(x)$ for every $x \in H$ (see Theorem 2.2, Chapter II).

(v) $\Rightarrow$ (j). We set $T(t) = S(t)\, P_{\overline{D(A)}}$ and $x_t = \left(1 - \frac{\lambda}{t}(T(t) - 1)\right)^{-1} x.$

We have

$$x_t = \frac{t}{t+\lambda}\, x + \frac{\lambda}{t+\lambda}\, S(t)P_{\overline{D(A)}}x_t$$

Hence

$$\varphi(x_t) \leqslant \frac{t}{t+\lambda}\ \varphi(x) + \frac{\lambda}{t+\lambda}\,\varphi(x_t),$$

and therefore

$$\varphi(x_t) \leqslant \varphi(x), \text{ for every } \ t \geqslant 0. \tag{1.26}$$

In addition we have

$$(x_t - S(t)\ \mathrm{Proj}_{\overline{D(A)}}x_t - (u - S(t)u),\ x_t - u) \geqslant 0, \quad \forall\, u \in H,$$

because the operator $S(t)P_{\overline{D(A)}}$ is nonexpansive. Consequently

$$\left(\frac{x - x_t}{\lambda} - \frac{u - S(t)u}{t},\quad x_t - u\right) \geqslant 0, \quad \forall t > 0.$$

We take $u \in D(A)$ and we pass $t \to 0$ in this inequality. One obtains

$$\left(A^0 u + \frac{x_0 - x}{\lambda},\quad u - x_0\right) \geqslant 0, \quad \forall u \in D(A)$$

where x_0 is a weak-limit point of x_t. According to Theorem 3.4 from Chapter II one concludes that $x_0 \in D(A)$ and $\dfrac{x - x_0}{\lambda} \in Ax_0$. Hence we have proved that $\lim\limits_{t \to 0} x_t = (1 + \lambda A)^{-1} x_0$ in the weak topology of H. As every lower-semicontinuous convex function on H is also weakly lower-semicontinuous one obtains the desired result, by passing to the limit in (1.26).

Proof of Theorem 1.7. We apply Proposition 1.1 where $\varphi = I_C$ is the indicator function of the convex C, i.e.,

$$I_C(x) = \begin{cases} 0, & x \in C, \\ \infty, & x \in H \setminus C. \end{cases}$$

Obviously condition (i) is equivalent to (j) and (iv) is equivalent to (v). First note that

$$(1 + \mu \partial I_C)^{-1} x = P_C x \text{ for all } x \in H, \text{ and } \mu > 0. \tag{1.27}$$

Indeed, from the definition of ∂I_C we have

$$((1 + \mu \partial I_C)^{-1} x,\ (1 + \mu \partial I_C)^{-1} x - u) \leqslant (x,\ (1 + \mu \partial I_C)^{-1} x - u)$$

for every $u \in C$.

This implies that

$$|(1 + \mu \partial I_C)^{-1} x - x| \leqslant |x - u| \text{ for every } u \in C,$$

as claimed. In particular, the equality (1.27) shows that (ii) and (jv) are equivalent. On the other hand,

$$(I_C)_\mu x = \frac{1}{2\mu} \inf \{|x - u|^2;\ u \in C\} = \frac{1}{2\mu} |P_C x - x|^2.$$

Therefore, condition (iii) is equivalent to

$$(I_C)_\mu(S(t)x) \leqslant (I_C)_\mu(x) \quad \text{for all } t \geqslant 0. \tag{1.28}$$

However, as we have observed in the Proof of Proposition 1.1, (1.28) is equivalent to (v), i.e.

$$I_C(S(t)x) \leqslant I_C(x) \quad \text{for all } x \in D(A).$$

Thus, the proof of Theorem 1.7 is complete. The next theorem precises Theorem 1.6 proved in Chapter III in the case of Hilbert spaces.

THEOREM 1.8. *Let A be a closed monotone set in $H \times H$ satisfying*

$$\overline{\text{conv}\, D(A)} \subset R(1 + \lambda A) \quad \textit{for every } \lambda > 0 \tag{1.29}$$

and let C be a closed convex subset of $\bigcap_{\lambda > 0} R(1 + \lambda A)$.

If $S(t)$ is the semigroup generated by $-A$ then the following conditions are equivalent:

(a) $(1 + \lambda A)^{-1} C \subset C$ *for every* $\lambda > 0$.

(b) $S(t)(\overline{D(A)} \cap C) \subset \overline{D(A)} \cap C$ *for every* $t \geqslant 0$ *and* $P_{\overline{D(A)}}\, C \subset C$.

(c) $D(A) \cap C$ *is dense in* $\overline{D(A)} \cap C$, $A + \partial I_C$ *is maximal monotone,* $(A + \partial I_C)^0 x = A^0 x$ *for every* $x \in D(A) \cap C$ *and* $P_{\overline{D(A)}}\, C \subset C$.

Proof. First we shall prove that $\overline{D(A)}$ is convex. To this purpose consider a maximal monotone extension $\tilde{A}$ of A satisfying $D(\tilde{A}) \subset \overline{\text{conv}\, D(A)}$ (see Chapter II, Theorem 3.2). Obviously $D(A) \subset D(\tilde{A}) \subset \overline{\text{conv}\, D(A)}$. Let $x \in D(\tilde{A})$ and x_λ be the solution of equation

$$x_\lambda + \lambda y_\lambda = x, \; [x_\lambda, y_\lambda] \in A \tag{1.30}$$

which exists from assumption (1.29). Since $A \subset \tilde{A}$ we also have $x_\lambda = (1 + \lambda \tilde{A})^{-1} x$, $y_\lambda = \tilde{A}_\lambda x$. By Proposition 3.6 given in Chapter II, $x_\lambda \to x$ and $y_\lambda \to \tilde{A}^0 x = y$ as $\lambda \to 0$. Inasmuch as A is closed, we infer that $[x, y] \in A$. Thus $D(A) = D(\tilde{A})$ and $\tilde{A}^0 x = A^0 x$. In particular this shows that $\overline{D(A)}$ is convex. We already know that A^0 is the infinitesimal generator of a semigroup S defined on $\overline{D(A)}$ (see Theorem 1.6 in Chapter III). By Theorem 1.2, $S(t)$ coincides with the semigroup generated by $\tilde{A}$. Moreover, we observe that $(1 + \lambda \tilde{A})^{-1} x = (1 + \lambda A)^{-1} x$ for every $x \in C$ and $\lambda > 0$. Thus we may assume in Theorem 1.8 that A is maximal monotone in $H \times H$ and we apply Theorem 1.7 to conclude that conditions (a) and (b) are equivalent. Next we prove that (a) $\Rightarrow$ (c). To show this it suffices to observe that A and ∂I_C satisfy the assumptions of Theorem 3.8 given in Chapter II. This implies that $A + \partial I_C$ is maximal monotone and $\overline{D(A) \cap C} = \overline{D(A)} \cap C = \overline{D(A + \partial I_C)}$. Moreover, in view of Proposition 1.1, part (jj), we have

$$(A_\lambda x, (A + \partial I_C)^0 x) \geqslant (A_\lambda x, y) \geqslant |A_\lambda x|^2, \quad \forall\, \lambda > 0$$

where y is some element of Ax. Remembering that $A_\lambda x \to A^0 x$ this yields $|A^0 x| \leqslant |(A + \partial I_C)^0 x|$. Hence $A^0 x = (A + \partial I_C)^0 x$.

Finally we prove that (c) $\Rightarrow$ (b). From Theorem 1.6 cited above the semigroup $S(t)$ generated by A on $\overline{D(A)}$ coincides on $\overline{D(A)} \cap C$ with the semigroup generated by $A + \partial I_C$. Thus $S(t)x \in \overline{D(A) \cap C} = \overline{D(A)} \cap C$ for every $t \geqslant 0$ and $x \in \overline{D(A)} \cap C$ and the proof is complete.

We conclude this paragraph with some facts concerning $\partial\varphi$ — *monotone operators.*

Definition 1.1. Let A be a maximal monotone subset in $H \times H$ and let $\varphi: H \to]-\infty, +\infty]$ be a lower-semicontinuous, convex and proper function. A is called $\partial\varphi$ — *monotone* if

$$\varphi((1 + \lambda A)^{-1}x - (1 + \lambda A)^{-1}y) \leqslant \varphi(x - y) \tag{1.31}$$

for all x, y in H and for every $\lambda > 0$.

PROPOSITION 1.2. *Let $\varphi: H \to]-\infty, +\infty]$ be a lower-semicontinuous, proper convex function such that*

$$\varphi(P_{\overline{D(A)}}x - P_{\overline{D(A)}}y) \leqslant \varphi(x - y) \quad \textit{for all } x, y \textit{ in } H. \tag{1.32}$$

Then the following conditions are equivalent

(a) *A is $\partial\varphi$ — monotone.*

(b) $(A_\lambda x - A_\lambda y, z) \geqslant 0$ *for every $z \in \partial\varphi(x - y)$ and $\lambda > 0$.*

(c) $(A_\lambda x - A_\lambda y, \partial\varphi_\mu(x - y)) \geqslant 0$, *for all $x, y \in H$, $\lambda, \mu > 0$.*

(d) $(A^0 x - A^0 y, \partial\varphi_\mu(x - y)) \geqslant 0$ *for all $x, y \in D(A)$ and $\mu > 0$.*

Proof. Denote by $\mathscr{H}$ the product space $H \times H$ with the natural inner product. Let $\mathscr{A}$ be the maximal monotone set in $\mathscr{H} \times \mathscr{H}$ defined by

$$\mathscr{A}[x, y] = [Ax, Ay], \; D(\mathscr{A}) = D(A) \times D(A).$$

Clearly $\mathscr{A}_\lambda[x, y] = [A_\lambda x, A_\lambda y]$ and $\mathscr{A}^0[x, y] = [A^0 x, A^0 y]$. Finally, let $\Phi: \mathscr{H} \to]-\infty, \infty]$ be defined by

$$\Phi[x, y] = \varphi(x - y).$$

It is easy to verify that

$$\partial\Phi\,[x, y] = \{[z, -z];\; z \in \partial\varphi(x - y)\}$$

$$\partial\Phi_\mu[x, y] = \{\partial\varphi_{2\mu}(x - y), -\partial\varphi_{2\mu}(x - y)\}$$

$$\Phi_\mu\,[x, y] = \varphi_{2\mu}(x - y).$$

Now we refer to Proposition 1.1 to complete the proof.

§ 2. Smoothing effect on initial data

As seen in Chapter III, if A is a maximal monotone set of $H \times H$, then for every $u_0 \in \overline{D(A)}$ and $f \in L^1(0, T; H)$ the initial value problem

$$\frac{du}{dt} + Au \ni f \quad \text{a.e. } t \in]0, T[\tag{2.1}$$

$$u(0) = u_0$$

has a unique weak solution $u \in C(0, T; H)$. Moreover, if $u_0 \in D(A)$ and $f \in W^{1,1}(0, T; H)$ then $u(t)$ is a strong solution of Eq. (2.1) and it belongs to $W^{1,\infty}(0, T; H)$. The problem which arises naturally is to find sufficient conditions on A such that $u(t)$ be a strong solution of (2.1) even in the case in which $u_0 \in \overline{D(A)} \setminus D(A)$ and $f \bar{\in} W^{1,1}(0, T; H)$. Results of this kind will allow us to weaken the regularity hypotheses on the initial data as well as on the right-hand side of the equation.

Finally, we are interested in finding sufficient conditions on A such that

$$S(t)\overline{D(A)} \subset D(A) \text{ for all } t > 0$$
$$t \frac{d}{dt} S(t)x \in L^\infty(0, \infty; H) \text{ for every } x \in \overline{D(A)} \tag{2.2}$$

where $S(t)$ is the semigroup generated by $-A$ on $\overline{D(A)}$. As mentioned in Chapter I, Section 3, in the linear case, this property is characteristic of analytic semigroups of class C_0. In Chapter III (Theorem 2.7) we have determined an important class of nonlinear monotone operators in a Hilbert space with this property. This paragraph presents other results in this direction. However there are no results of any generality concerning the case in which the space is not Hilbert.

2.1. The case in which $A = \partial\varphi$

Throughout this section we shall assume familiarity with the concepts and definitions given in Chapter II concerning convex functions and subgradients. However, we shall repeat for easy reference the basic notations.

Let $\varphi : H \to]-\infty, +\infty]$ be a lower-semicontinuous, proper convex function on a real Hilbert space H. Denote by $\partial\varphi \subset H \times H$ the subdifferential of φ, i.e.

$$\partial\varphi(x) = \{y \in H;\ \varphi(x) - \varphi(u) \leqslant (y, x - u), \quad \forall\ u \in H\}$$

and denote by $D(\varphi)$ the effective domain of φ, i.e.,

$$D(\varphi) = \{x \in H;\ \varphi(x) < +\infty\}.$$

Recall that $A = \partial\varphi$ is maximal monotone in $H \times H$ and $\overline{D(A)} = \overline{D(\varphi)}$. Consider the Cauchy problem

$$\begin{aligned} &\frac{du}{dt} + \partial\varphi(u) \ni f,\ 0 < t < T \\ &u(0) = u_0. \end{aligned} \tag{2.3}$$

The main result is

THEOREM 2.1. *Let f be given in $L^2(0, T; H)$ and $u_0 \in \overline{D(A)}$. Then problem* (2.3) *has a unique solution $u \in C(0, T; H)$ which satisfies*

$$u \in W^{1,2}(\delta, T; H) \text{ for every } 0 < \delta < T. \tag{2.4}$$

$$u(t) \in D(A), \quad \text{a.e.} \quad t \in]0, T[. \tag{2.5}$$

$$\sqrt{t}\,\frac{du}{dt} \in L^2(0, T; H),\ \varphi(u) \in L^1(0, T). \tag{2.6}$$

Moreover, if $u_0 \in D(\varphi)$ then

$$\frac{du}{dt} \in L^2(0, T; H),\ \varphi(u) \in L^\infty(0, T). \tag{2.7}$$

In order to prove Theorem 2.1 we need the following lemma which also is of interest for other problems.

LEMMA 2.1. *Let $u \in W^{1,2}(0, T; H)$ be such that $u(t) \in D(\partial\varphi)$ a.e. on $]0, T[$ and there exists $g \in L^2(0, T; H)$ such that*

$$g(t) \in \partial\varphi u(t)), \quad \text{a.e.} \quad t \in]0, T[.$$

Then, function $t \to \varphi(u(t))$ is absolutely continuous on $[0, T]$ and the following equality holds

$$\frac{d}{dt}\varphi(u(t)) = \left(h(t),\ \frac{du}{dt}(t)\right), \quad \text{a.e.} \quad t \in]0, T[,$$

for all $h(t) \in \partial\varphi(u(t))$.

Proof. As usual we denote by $(\partial\varphi)_\lambda$ the operator $\lambda^{-1}(1 - (1 + \lambda\partial\varphi)^{-1})$. From Theorem 2.2, in Chapter II, we have

$$(\partial\varphi)_\lambda = \partial\varphi_\lambda \text{ for all } \lambda > 0.$$

Clearly, the function $t \to \varphi_\lambda(u(t))$ is a.e. differentiable on $]0, T[$ and by definition,

$$\frac{d}{dt}\varphi_\lambda(u(t)) = \left((\partial\varphi)_\lambda u(t),\ \frac{du}{dt}(t)\right), \quad \text{a.e.} \quad t \in]0, T[.$$

Consequently,

$$\varphi_\lambda(u(t)) - \varphi_\lambda(u(s)) = \int_s^t \left((\partial\varphi)_\lambda u(\tau), \ \frac{\mathrm{d}u}{\mathrm{d}\tau}(\tau) \right) \mathrm{d}\tau\,; \ s, t \in \,]0, T[.$$

Passing to the limit $\lambda \to 0$, yields

$$\varphi(u(t)) - \varphi(u(s)) = \int_s^t \left((\partial\varphi)^0 u(\tau), \ \frac{\mathrm{d}u}{\mathrm{d}\tau}(\tau) \right) \mathrm{d}\tau$$

which implies that the function $t \to \varphi(u(t))$ is absolutely continuous on $[0, T]$. Let $t_0 \in \,]0, T[$ be such that $u(t)$ and $\varphi(u(t))$ are differentiable in $t = t_0$ and $u(t_0) \in D(\partial\varphi)$. Let h be arbitrary in $\partial\varphi(u(t_0))$. We have

$$\varphi(v) - \varphi(u(t_0)) \geqslant (h, v - u(t_0)), \text{ for all } v \in H.$$

By taking $v = u(t_0 \pm \varepsilon)$ one finally obtains

$$\frac{\mathrm{d}}{\mathrm{d}t}\varphi(u(t_0)) = \left(h, \ \frac{\mathrm{d}u}{\mathrm{d}t}(t_0) \right)$$

which concludes the proof.

Proof of Theorem 2.1. Let x_0 be any element in $D(\partial\varphi)$ and let y_0 be such that $y_0 \in \partial\varphi(x_0)$. If we introduce the new function $\tilde{\varphi}(u) = \varphi(u) - \varphi(x_0) - (y_0, u - x_0)$, then Eq. (2.3) is equivalent to the equation

$$\frac{\mathrm{d}u}{\mathrm{d}t} + \partial\tilde{\varphi}(u) \ni f(t) - y_0.$$

Hence, we may assume without loss of generality that

$$\min\{\varphi(u);\ u \in H\} = \varphi(x_0) = 0.$$

First we shall assume that $u_0 \in D(\partial\varphi)$ and $f \in W^{1,2}(0, T; H)$, i.e., f and $\frac{\mathrm{d}f}{\mathrm{d}t}$ belong to $L^2(0, T; H)$. From Theorem 2.2 in Chapter III, problem (2.3) has a unique solution $u \in C(0, T; H)$ such that $\frac{\mathrm{d}u}{\mathrm{d}t} \in L^\infty(0, T; H)$.

We multiply Eq. 2.3 by $t\,\frac{\mathrm{d}u}{\mathrm{d}t}\cdot$ By Lemma 2.1 we have

$$t\left|\frac{\mathrm{d}u}{\mathrm{d}t}\right|^2 + t\,\frac{\mathrm{d}}{\mathrm{d}t}\varphi(u) = t\left(f, \ \frac{\mathrm{d}u}{\mathrm{d}t}\right), \quad \text{a.e.} \quad t \in \,]0, T[.$$

Hence

$$\int_0^T t\left|\frac{du}{dt}\right|^2 dt + T\varphi(u(T)) = \int_0^T t\left(f(t), \frac{du}{dt}\right) dt + \int_0^T \varphi(u(t))\, dt$$

therefore, also

$$\int_0^T t\left|\frac{du}{dt}\right|^2 dt \leqslant \int_0^T t\,|f(t)|^2\, dt + 2\int_0^T \varphi(u(t))\, dt \tag{2.8}$$

because $\varphi(u)$ is nonnegative on H.

Next, we have

$$\varphi(u(t)) \leqslant (y(t), u(t) - x_0) \text{ for all } y(t) \in \partial\varphi(u(t)).$$

In other words

$$\varphi(u(t)) \leqslant \left(f(t) - \frac{du}{dt}(t),\ u(t) - x_0\right), \quad \text{a.e.} \quad t \in]0, T[.$$

Finally

$$\int_0^T \varphi(u(t))\, dt \leqslant \frac{1}{2}\,|u(0) - x_0|^2 + \int_0^T |f(t)|\,|u(t) - x_0|\, dt.$$

On the other hand, by multiplying Eq. (2.3) with $u(t) - x_0$ and integrating over $]0, t[$ one gets

$$|u(t) - x_0| \leqslant |u(0) - x_0| + \int_0^T |f(t)|\, dt, \quad 0 \leqslant t \leqslant T$$

and therefore

$$\int_0^T \varphi(u(t))\, dt \leqslant \frac{1}{2}\left(|u(0) - x_0| + \int_0^T |f(t)|\, dt\right)^2 \cdot \tag{2.9}$$

Combining estimates (2.8) and (2.9) we obtain

$$\int_0^T t\left|\frac{du}{dt}(t)\right|^2 dt \leqslant \int_0^T t\,|f(t)|^2\, dt + \left(|u_0 - x_0| + \int_0^T |f(t)|\, dt\right)^2 \cdot \tag{2.10}$$

Now we assume that $u_0 \in D(\varphi)$ and $f \in L^2(0, T; H)$. Since $D(\partial\varphi)$ is a dense subset of $D(\varphi)$, there exists $\{x_n\} \subset D(\partial\varphi)$ such that $\lim_{n\to\infty} x_n = u_0$. Let $\{f_n\} \subset W^{1,2}(0, T; H)$ be such that $\lim_{n\to\infty} f_n = f$ in $L^2(0, T; H)$, and denote by u_n the corresponding solutions of Eq. (2.3). By the monotonicity of $\partial\varphi$ one obtains

$$|u_n(t) - u_m(t)| \leqslant |x_n - x_m| + \int_0^t |f_n(s) - f_m(s)|\, ds, \quad 0 \leqslant t \leqslant T.$$

Hence $\lim_{n\to\infty} u_n(t) = u(t)$ exists uniformly on $[0, T]$. Since $\lim_{n\to\infty} \frac{du_n}{dt} = \frac{du}{dt}$ in the sense of H-valued vectorial distributions on $]0, T[$, estimate (2.10) implies that $t \frac{du}{dt} \in L^2(0, T; H)$.

In particular, it follows that $u(t)$ is absolutely continuous on every compact of $]0, T[$. By using a standard argument (as in the proof of Theorem 2.2 Chapter III), one deduces that u satisfies Eq. (2.3) on $]0, T[$.

Now, we assume that $u_0 \in D(\varphi)$. By using again Eq. (2.3) we get

$$\left| \frac{du}{dt} \right|^2 + \frac{d}{dt} \varphi(u) \leqslant |f| \left| \frac{du}{dt} \right|, \quad \text{a.e.} \quad t \in]0, T[.$$

Hence

$$\frac{1}{2} \left| \frac{du}{dt}(t) \right|^2 + \frac{d}{dt} \varphi(u(t)) \leqslant \frac{1}{2} |f(t)|^2, \quad \text{a.e.} \quad t \in]0, T[,$$

which implies that the function $t \to \varphi(u(t)) - \frac{1}{2} \int_0^t |f(s)|^2 \, ds$ is monotone nonincreasing over $[0, T]$. Since $u_0 \in D(\varphi)$ one deduces that

$$\varphi(u(t)) \leqslant \varphi(u_0) + \frac{1}{2} \int_0^t |f(t)|^2 \, dt, \quad 0 \leqslant t \leqslant T. \tag{2.11}$$

Let ε be positive and sufficiently small. From relation (2.11) we have

$$\frac{1}{2} \int_\varepsilon^T \left| \frac{du}{dt}(t) \right|^2 dt \leqslant \varphi(u_0) + \frac{1}{2} \int_0^T |f|^2 \, dt.$$

Hence $\frac{du}{dt} \in L^2(0, T; H)$ and $\varphi(u) \in L^\infty(0, T)$. This completes the proof.

THEOREM 2.2. *We are given*

$$u_0 \in \overline{D(A)}, \; f \in W^{1,1}(0, T; H).$$

Then the solution $u(t)$ for problem (2.3) *satisfies*

$$u(t) \in D(\partial\varphi) \text{ for every } t \in]0, T[. \tag{2.12}$$

$$\frac{d^+}{dt} u(t) + (\partial\varphi(u(t)) - f(t))^0 = 0, \quad \forall\, t \in]0, T[, \tag{2.13}$$

$$t \frac{du}{dt} \in L^\infty(0, T; H). \tag{2.14}$$

Proof. Using once again Eq. (2.3), we obtain

$$\frac{\mathrm{d}}{\mathrm{d}t}|u(t+h)-u(t)| \leqslant |f(t+h)-f(t)|, \quad \text{a.e.} \quad t, t+h \in]0, T[.$$

Hence

$$\left|\frac{\mathrm{d}u}{\mathrm{d}t}(t)\right| \leqslant \left|\frac{\mathrm{d}u}{\mathrm{d}t}(s)\right| + \int_s^t \left|\frac{\mathrm{d}f}{\mathrm{d}\tau}(\tau)\right| \mathrm{d}\tau, \quad \text{a.e.} \quad 0 < s < t < T.$$

This shows that the function $t \to \left|\frac{\mathrm{d}u}{\mathrm{d}t}(t)\right| - \int_0^t \left|\frac{\mathrm{d}f}{\mathrm{d}\tau}(\tau)\right| \mathrm{d}\tau$ is nonincreasing on $]0, T[$. We then have

$$\limsup_{\substack{h\to 0\\ h>0}} \left|\frac{u(t+h)-u(t)}{h}\right| < \infty, \quad t \in]0, T[.$$

Making use of the inequality

$$\left(\frac{u(t+h)-u(t)}{h}, \ u(t)-u\right) \leqslant \frac{1}{h}\int_t^{t+h} (f(\tau)-v, \ u(\tau)-u)\, \mathrm{d}\tau$$

which holds for every $[u, v] \in \partial\varphi$ and $0 \leqslant t \leqslant t+h < T$, one deduces that the weak closure of the set $\left\{\frac{u(t+h)-u(t)}{h}, \ h > 0\right\}$ belongs to $f(t) - \partial\varphi(u(t))$. In particular, it follows that $u(t) \in D(\partial\varphi)$ for every $t \in]0, T[$.

On the other hand, the inequality

$$\frac{\mathrm{d}}{\mathrm{d}h}(u(t+h)-u(t), \ u(t+h)-u(t)) \leqslant (f(t+h)-v,$$

$$u(t+h)-u(t)), \quad \forall\, v \in \partial\varphi(u(t))$$

implies that

$$|u(t+h)-u(t) \leqslant \int_0^h |f(t+s)-v|\, \mathrm{d}s, \quad \forall t \in]0, T[; \ h > 0.$$

Hence

$$\limsup_{h\to 0} \left|\frac{u(t+h)-u(t)}{h}\right| \leqslant |(\partial\varphi(u(t)) - f(t))^0|,$$

for all $t \in]0, T[$. This shows that $\lim_{h\to 0} \frac{u(t+h)-u(t)}{h} = \frac{\mathrm{d}^+}{\mathrm{d}t} u(t)$ exists for every $t \in]0, T[$ and coincides with $-(A(u(t)) - f(t))^0$. In order to prove formula (2.13) we start with the following inequality

$$\int_0^t s\left|\frac{\mathrm{d}^+}{\mathrm{d}t}u(t)\right| \mathrm{d}s \leqslant \int_0^t s\left|\frac{\mathrm{d}^+}{\mathrm{d}t}u(s)\right| \mathrm{d}s + \int_0^t s\, \mathrm{d}s \int_s^t \left|\frac{\mathrm{d}f}{\mathrm{d}\tau}(\tau)\right| \mathrm{d}\tau.$$

Hence

$$\frac{t^2}{2}\left|\frac{d^+}{dt}u(t)\right| \leqslant \frac{1}{2}\int_0^t s^2\left|\frac{df}{ds}(s)\right| ds + \frac{t}{\sqrt{2}}\left(\int_0^t s\left|\frac{d^+}{ds}u(s)\right|^2 ds\right)^{1/2}.$$

Making use of estimate (2.10), we then have

$$t\left|\frac{d^+}{dt}u(t)\right| \leqslant \frac{1}{t}\int_0^t s^2\left|\frac{df}{ds}(s)\right| ds + \sqrt{2}\left(\int_0^t s|f(s)|^2 ds + \right. \tag{2.15}$$

$$\left. + \left(|u_0 - x_0| + \int_0^t |f(s)|\, ds\right)^2\right)^{1/2}, \quad 0 < t < T$$

which completes the proof of Theorem 2.2.

THEOREM 2.3. *Let $S(t)$ be the semigroup generated by $A = -\partial\varphi$ on $\overline{D(A)}$. Then $S(t)\,\overline{D(A)} \subset D(A)$ for all $t > 0$ and the following inequality holds*

$$\left|\frac{d^+}{dt}S(t)u_0\right| = |A^0S(t)u_0| \leqslant |A^0x_0| + \frac{1}{t}|x_0 - u_0|, \quad t > 0 \tag{2.16}$$

for every $u_0 \in \overline{D(A)}$ and $x_0 \in D(A)$.

Proof. Let $[x_0, y_0]$ be arbitrary but fixed in A. As is easily seen, the function $\tilde{\varphi}(u) = \varphi(u) - \varphi(x_0) + (y_0, x_0 - u)$ is nonnegative on H and $\tilde{\varphi}(x_0) = 0$. Then Eq. (2.3) with $f \equiv 0$ can be written as

$$\begin{aligned} &\frac{du}{dt} + \partial\tilde{\varphi}(u) \ni -y_0, \quad 0 < t < \infty \\ &u(0) = u_0. \end{aligned} \tag{2.17}$$

It follows from Theorem 2.2 applied to $\tilde{\varphi}$ that $S(t)\,\overline{D(A)} \subset D(A)$ for every $t > 0$ and $t\dfrac{d^+}{dt}S(t)u_0 \in L^\infty(0, T; H)$ for every $T > 0$ and $u_0 \in \overline{D(A)}$. In order to obtain estimate (2.16) we shall use the same argument as in the proof of Theorem 2.2.

By using Eq. (2.17) we obtain

$$\int_0^t s\left|\frac{du}{ds}\right|^2 ds \leqslant \int_0^t \tilde{\varphi}(u(s))\, ds - \int_0^t s\left(y_0, \frac{du}{ds}\right) ds, \quad 0 \leqslant t \leqslant T.$$

Hence

$$\int_0^t s\left|\frac{\mathrm{d}u}{\mathrm{d}s}\right|^2 \mathrm{d}s \leqslant \int_0^t \tilde{\varphi}(u(s))\,\mathrm{d}s + \int_0^t (y_0, u(s) - u_0)\,\mathrm{d}s - t(y_0, u(t) - u_0).$$

On the other hand

$$\int_0^t \tilde{\varphi}(u(s))\,\mathrm{d}s \leqslant \frac{1}{2}(|u_0 - x_0|^2 - |u(t) - x_0|^2) - \int_0^t (y_0, u(s) - x_0)\,\mathrm{d}s.$$

Finally

$$\int_0^t s\left|\frac{\mathrm{d}u}{\mathrm{d}s}\right|^2 \mathrm{d}s \leqslant \frac{1}{2}(|u_0 - x_0|^2 + t^2|y_0|^2), \quad \text{for} \quad 0 < t < \infty.$$

We recall that function $t \to \left|\dfrac{\mathrm{d}^+}{\mathrm{d}t}u(t)\right|$ is monotone nonincreasing on $]0, T[$. This shows that

$$\left|\frac{\mathrm{d}^+}{\mathrm{d}t}u(t)\right| \leqslant |y_0| + \frac{1}{t}|x_0 - u_0| \quad \text{for} \quad t \geqslant 0, \quad [x_0, y_0] \in A$$

which is the desired estimate.

We conclude this section by a stability result which turns out to be useful in applications.

THEOREM 2.4. *Let $\varphi : H \to]-\infty, +\infty]$ be a lower-semicontinuous convex function nonidentically $+\infty$ such that every level subset $\{u \in H;\ \varphi(u) + |u|^2 \leqslant M\}$ is compact. Let $f : [0, \infty[\to H$ be absolutely continuous on every compact of $[0, \infty[$ such that $\dfrac{\mathrm{d}f}{\mathrm{d}t} \in L^1(0, \infty; H)$. In addition, we assume that*

$$\lim_{t\to\infty} f(t) = f_\infty \in R(\partial\varphi), \tag{2.18}$$

and

$$f - f_\infty \in L^1(0, \infty; H). \tag{2.19}$$

Let $u_0 \in \overline{D(\varphi)}$ and let $u(t)$ be the solution of problem (2.3). *Then*

$$\lim_{t\to\infty} \frac{\mathrm{d}^+u}{\mathrm{d}t}(t) = 0 \text{ and } \lim_{t\to\infty} u(t) = u_\infty \in (\partial\varphi)^{-1}(f_\infty). \tag{2.20}$$

Proof. Let $\tilde{\varphi}(u) = \varphi(u) - (f_\infty, u) - \inf\{\varphi(v) - (f_\infty, v);\ v \in H\}$. By assumption, there exists $v_\infty \in D(\partial\varphi)$ such that $f_\infty \in \partial\varphi(v_\infty)$. Then the inequality

$$\varphi(v_\infty) - \varphi(v) \leqslant (f_\infty, v_\infty - v) \quad \text{for every} \quad v \in D(\varphi),$$

shows that $\inf\{\varphi(v) - (f_\infty, v)\} \neq -\infty$. Noting that $\inf\{\tilde{\varphi}(u);\ u \in H\} = 0$, we observe that Eq. (2.3) becomes

$$\frac{\mathrm{d}u}{\mathrm{d}t} + \partial\tilde{\varphi}(u) \ni f - f_\infty, \quad \text{a.e.} \quad t > 0. \tag{2.21}$$

Thus estimate (2.15) implies that

$$\left|\frac{\mathrm{d}^+}{\mathrm{d}t}u(t)\right| \leqslant \int_0^t \frac{s^2}{t^2}\left|\frac{\mathrm{d}f}{\mathrm{d}s}(s)\right| \mathrm{d}s + \frac{\sqrt{2}}{t}\left(\int_0^t s\,|f(s) - f_\infty|^2\,\mathrm{d}s\right)^{1/2} +$$

$$+\frac{1}{t}\int_0^t |f(s) - f_\infty|\,\mathrm{d}s + \frac{1}{t}|u_0 - x_0|, \quad t > 0,$$

where x_0 is arbitrary in $(\partial\tilde{\varphi})^{-1}(0) = (\partial\varphi)^{-1}(f_\infty)$. Let $\varepsilon > 0$ and $0 < T < \infty$ be such that

$$\int_T^t \left|\frac{\mathrm{d}f}{\mathrm{d}s}(s)\right| \mathrm{d}s \leqslant \varepsilon \quad \text{for all} \quad t \geqslant T$$

and

$$|f(t) - f_\infty| \leqslant \varepsilon \quad \text{for} \quad t \geqslant T.$$

Thus we have

$$\left|\frac{\mathrm{d}^+}{\mathrm{d}t}u(t)\right| \leqslant \frac{1}{t^2}\int_0^T s^2\left|\frac{\mathrm{d}f}{\mathrm{d}s}(s)\right| \mathrm{d}s + \frac{\sqrt{2}}{t}\left(\int_0^T s|f(s) - f_\infty|^2\,\mathrm{d}s + t^2\varepsilon^2\right)^{1/2} +$$

$$+\frac{1}{t}\int_0^t |f(s) - f_\infty|\mathrm{d}s + \frac{1}{t}|u_0 - x_0| + \varepsilon, \quad T \leqslant t < \infty$$

which clearly implies that $\lim\limits_{t\to\infty} \frac{\mathrm{d}^+}{\mathrm{d}t}u(t) = 0$.

Now we shall prove that $\lim\limits_{t\to\infty} u(t) = u_\infty$ exists. Let x_0 be arbitrary in $(\partial\varphi)^{-1}(f_\infty)$. Noting that $\tilde{\varphi}(x_0) = \inf \tilde{\varphi}(u) = 0$, one obtains

$$\tilde{\varphi}(u(t)) \leqslant (\partial\tilde{\varphi}(u(t)),\ u(t) - x_0) \quad \text{for all} \quad t > 0.$$

Consequently,

$$0 \leqslant \tilde{\varphi}(u(t)) \leqslant \left(f(t) - f_\infty - \frac{\mathrm{d}^+}{\mathrm{d}t}u(t),\ u(t) - x_0\right), \tag{2.22}$$

which implies that $\lim\limits_{t\to\infty} \tilde{\varphi}(u(t)) = 0$. Moreover, it follows by estimate (2.22) that $\{\varphi(u(t)) + |u(t)|^2\}$ is bounded on $[0, \infty[$. Let $\{t_n\}$ and u_∞ be such that

$\lim_{n\to\infty} t_n = +\infty$ and $\lim_{n\to\infty} u(t_n) = u_\infty$. Obviously $\tilde{\varphi}(u_\infty) = 0$, i.e., $u_\infty \in (\partial\tilde{\varphi})^{-1}(0) = (\partial\varphi)^{-1}(f_\infty)$. On the other hand, making use of the equality

$$\frac{d}{dt}(u(t) - u_\infty) + \partial\tilde{\varphi}(u(t)) \ni f(t) - f_\infty, \quad \text{a.e.} \quad t > 0$$

we deduce

$$|u_\infty - u(t)| \leqslant |u_\infty - u(t_n)| + \int_{t_n}^{t} |f_\infty - f(s)|\, ds, \quad t > t_n.$$

Since $f(t) - f_\infty \in L^1(0, \infty; H)$ and $u(t_n) \to u_\infty$, we may conclude that $u(t)$ is convergent as $t \to \infty$ and its limit is just u_∞. Thus Theorem 2.4 has been proven.

Remark. Let $\varphi : H \to]-\infty, +\infty]$ be a lower-semicontinuous convex function satisfying the assumptions of Theorem 2.4. Assuming that $0 \in R(\partial\varphi)$; in other words the *infimum* of φ on H is attained, it follows by the preceding theorem that every solution $u(t)$ of the equation

$$\frac{du}{dt} + \partial\varphi(u) \ni 0, \quad \text{a.e.} \quad t > 0 \tag{2.23}$$

approaches $u_\infty \in (\partial\varphi)^{-1}(0)$ as $t \to \infty$. In other words, there exist minimum points u_∞ of φ which can be approached by trajectories $u(t)$ of the equation Eq. (2.23). This fact is a motivation of the well-known method of "steepest descent" in its continuous form.

2.2. The case in which int $D(A) \neq \emptyset$

The basic result which we shall prove in this section is the following

THEOREM 2.5. *Let H be a real Hilbert space and let A be a maximal monotone set in $H \times H$. Suppose that* int $D(A) \neq \emptyset$. *Then for every $u_0 \in \overline{D(A)}$ and any $f \in W^{1,1}(0, T; H)$ there exists a unique function $u \in C(0, T; H)$ which satisfies*

$$u(t) \in D(A) \quad \text{a.e.} \quad \textit{on }]0, T[, \quad u(0) = u_0. \tag{2.24}$$

$$u(t) \textit{ is absolutely continuous on } [0, T]. \tag{2.25}$$

$$\frac{du}{dt}(t) + Au(t) \ni f(t), \quad \text{a.e.} \quad \textit{on }]0, T[. \tag{2.26}$$

$$t\frac{du}{dt} \in L^\infty(0, T; H), \quad \frac{du}{dt} \in L^1(0, T; H). \tag{2.27}$$

Proof. Without any loss of generality we may assume that $0 \in \operatorname{int} D(A)$. Suppose firstly that $u_0 \in D(A)$. Then there exists a unique function $u \in W^{1,\infty}(0, T; H)$ satisfying

$$\frac{\mathrm{d}u}{\mathrm{d}t} + Au(t) \ni f(t), \quad \text{a.e.} \quad \text{on }]0, T[$$
$$u(0) = u_0. \tag{2.28}$$

Since $0 \in \operatorname{int} D(A)$ it follows from Theorem 1.5 proved in Chapter II that A is locally bounded at $u = 0$. In other words there exist $\rho > 0$ and $M > 0$ such that

$$|y| \leqslant M \text{ for all } y \in A(\rho w) \text{ and } |w| \leqslant 1. \tag{2.29}$$

If we multiply equation (2.28) by $u(t) - \rho w$, use the monotonicity of A and estimate (2.29), we obtain

$$\left(\frac{\mathrm{d}u}{\mathrm{d}t}(t) - f(t),\ u(t) - \rho w\right) \leqslant M|u(t) - \rho w|, \quad \text{a.e.} \quad \text{on }]0, T[. \tag{2.30}$$

In the preceding inequality we take $w = -\dfrac{\mathrm{d}u}{\mathrm{d}t}(t) \Big/ \left|\dfrac{\mathrm{d}u}{\mathrm{d}t}(t)\right|$ and integrate over $]0, t[$. This yields

$$\rho \int_0^t \left|\frac{\mathrm{d}u}{\mathrm{d}s}(s)\right| \mathrm{d}s + \frac{1}{2}|u(t)|^2 \leqslant \frac{1}{2}|u_0|^2 +$$
$$+ \int_0^t (|f(s)| + M)(|u(s)| + \rho)\, \mathrm{d}s, \quad 0 \leqslant t \leqslant T. \tag{2.31}$$

Next we take $w=0$ in inequality (2.30) and integrate over $]0, t[$ to obtain

$$|u(t)| \leqslant |u_0| + \int_0^t (|f(s)| + M)\, \mathrm{d}s, \quad 0 \leqslant t \leqslant T.$$

Comparing this with relation (2.31) yields

$$\rho \int_0^t \left|\frac{\mathrm{d}u}{\mathrm{d}s}(s)\right| \mathrm{d}s \leqslant \rho \int_0^t (|f(s)| + C)\, \mathrm{d}s +$$
$$+ \frac{1}{2}\left(|u_0| + \int_0^t (|f(s)| + C)\, \mathrm{d}s\right)^2 \tag{2.32}$$

where C is some positive constant independent of u_0 and f. On the other hand we have for almost all $0 < s < t < T$

$$\left|\frac{du}{dt}(t)\right| \leqslant \left|\frac{du}{ds}(s)\right| + \int_s^t \left|\frac{df}{d\tau}(\tau)\right| d\tau. \tag{2.33}$$

This formula follows by inequality (2.18) of Theorem 2.2 proved in Chapter III. Combining formula (2.32) with (2.33) we obtain after some rearrangements that

$$t\left|\frac{du}{dt}(t)\right| \leqslant \int_0^t (|f(s)| + C)\, ds + \frac{1}{2\rho}\left(|u_0| + \int_0^t (|f(s)|+C)\, ds\right)^2 + \int_0^t s\left|\frac{df}{ds}(s)\right| ds, \quad \text{for} \quad 0<t\leqslant T. \tag{2.34}$$

Next, suppose that $u_0 \in \overline{D(A)}$ and again denote $u(t)$ the unique weak solution of problem (2.28) (see Corollary 2.2 in Chapter III). Since estimates (2.32) and (2.34) can be extended to all $u_0 \in \overline{D(A)}$ we may conclude that $t\,\dfrac{du}{dt} \in L^\infty(0, T; H)$ and $\dfrac{du}{dt} \in L^1(0, T; H)$. Using a standard device one finally proves that $u(t)$ satisfies almost everywhere on $]0, T[$ Eq. (2.26). Thus the proof of Theorem 2.5 is complete.

Combining the techniques developed in Section 2.1 with the above arguments we can derive similar results for perturbed differential equations of the type

$$\begin{aligned} &\frac{du}{dt} + \partial\varphi(u) + Bu \ni f(t), \quad \text{a.e. on }]0, T[\\ &u(0) = u_0. \end{aligned} \tag{2.35}$$

A typical one is stated in the following theorem.

THEOREM 2.6. *Let* $\varphi : H \to]-\infty, +\infty]$ *be a lower-semicontinuous proper convex function and let* B *be a maximal monotone set in* $H \times H$. *Suppose that* $(\operatorname{int} D(B)) \cap D(\partial\varphi) \neq \emptyset$.

Let $u_0 \in \overline{D(\varphi)} \cap \overline{D(B)}$ *and* $f \in W^{1,1}(0, T; H)$ *be given. Then Eq.* (2.35) *has a unique solution* $u \in C(0, T; H)$ *which satisfies*

$$t\,\frac{du}{dt} \in L^\infty(0, T; H), \quad \sqrt{t}\,\frac{du}{dt} \in L^2(0, T; H).$$

The proof of Theorem 2.6 depends upon *a priori* estimates of the approximating equations

$$\frac{\mathrm{d}u_\lambda}{\mathrm{d}t} + \partial\varphi(u_\lambda) + B_\lambda u_\lambda \ni f(t), \quad u_\lambda(0) = u_0. \tag{2.36}$$

To this purpose, the method used in this section is applicable with minor changes. Brézis' book [8] contains the complete proof of Theorem 2.6 as well as many other results of this type.

2.3. Applications

Throughout this section, Ω will denote an open and bounded subset of R^n with sufficiently smooth boundary Γ. By $H^k(\Omega)$ and $H_0^k(\Omega)$ will denote usual Sobolev spaces on Ω.

Example 1. Let β be a maximal monotone graph in $R^1 \times R^1$ such that $D(\beta)$ is a dense subset of R^1. We recall (see Section 2.3, Chapter II) that there exists a lower-semicontinuous convex function $j\colon R^1 \to\,]-\infty, +\infty]$ such that $j \not\equiv \infty$ and $\partial j = \beta$.

PROPOSITION 2.1. *We are given $u_0 \in L^2(\Omega)$ and $f \in L^2(0, T; L^2(\Omega))$. Then the boundary value problem*

$$\frac{\partial u}{\partial t} - \Delta u = f \ \ on\ ]0, T[\times \Omega$$

$$-\frac{\partial u}{\partial n} \in \beta(u) \quad on\ ]0, T[\times \Gamma \tag{2.37}$$

$$u(0, x) = u_0(x) \ \ on\ \Omega$$

has a unique solution $u \in C(0, T; L^2(\Omega))$ satisfying

$$\sqrt{t}\, u \in L^2(0, T; H^2(\Omega)), \ \sqrt{t}\, \partial u/\partial t \in L^2(0, T; L^2(\Omega)). \tag{2.38}$$

If, in addition, $u_0 \in H^1(\Omega)$ and $j(u_0) \in L^1(\Omega)$, then

$$u \in L^2(0, T; H^2(\Omega)), \partial u/\partial t \in L^2(0, T; L^2(\Omega)). \tag{2.39}$$

If $u_0 \in L^2(\Omega)$ and $\partial f/\partial t \in L^1(0, T; L^2(\Omega))$, then $u(t) \in H^2(\Omega)$ on $]0, T[$ and

$$tu \in L^\infty(0, T; H^2(\Omega)), \ t\partial u/\partial t \in L^\infty(0, T; L^2(\Omega)). \tag{2.40}$$

Finally, if $u_0 \in H^2(\Omega)$ *and* $\partial u_0/\partial n \in -\beta(u_0)$, a.e. *on* Γ *then*

$$u \in L^\infty(0, T; H^2(\Omega)), \ \partial u/\partial t \in L^\infty(0, T; L^2(\Omega)). \tag{2.41}$$

Proof. We apply Theorem 2.1 and Theorem 2.2 where $H = L^2(\Omega)$ and $\varphi\colon L^2(\Omega) \to\ ]-\infty, +\infty]$ is defined by

$$\varphi(u) = \begin{cases} \dfrac{1}{2}\displaystyle\int_\Omega |\text{grad } u|^2 \mathrm{d}x + \int_\Gamma j(u)\, \mathrm{d}\sigma, & \text{if } u \in H^1(\Omega),\ j(u) \in L^1(\Gamma) \\ +\infty & \text{otherwise.} \end{cases}$$

We recall (see Proposition 2.9 in Chapter II) that the subdifferential of this function is given by

$$\partial\varphi(u) = -\Delta u, \ u \in D(\partial\varphi)$$

where

$$D(\partial\varphi) = \left\{ u \in H^2(\Omega), \ -\frac{\partial u}{\partial n} \in \beta(u), \ \text{a.e. on } \Gamma \right\},$$

and

$$\|u\|_{H^2(\Omega)} \leqslant C_1 \|u - \Delta u\|_{L^2(\Omega)} + C_2 \ \text{for every } u \in D(\partial\varphi).$$

Since $\overline{D(j)} = \overline{D(\beta)} = R^1$, it follows that $\overline{D(\partial\varphi)} = L^2(\Omega)$. Thus, for concluding the proof it suffices to observe that problem (2.37) can be written as

$$\frac{\mathrm{d}u}{\mathrm{d}t} + \partial\varphi(u) = f, \quad \text{a.e.} \quad t \in\ ]0, T[$$

$$u(0) = u_0.$$

Now, we shall assume that $\partial f/\partial t \in L^1(0, \infty; L^2(\Omega))$ and that $\lim_{t\to\infty} f(t) = f_\infty$ exists in $L^2(\Omega)$. In addition, we suppose that the boundary value problem

$$\begin{aligned} &\Delta u = -f \quad \text{on } \Omega \\ &\partial u/\partial n \in -\beta(u) \quad \text{on } \Gamma \end{aligned} \tag{2.42}$$

has at least a solution $u \in H^2(\Omega)$.

PROPOSITION 2.2. *Let u be a solution of problem* (2.37) *where* $u_0 \in L^2(\Omega)$ *and f is assumed to satisfy*

$$f - f_\infty \in L^1(0, \infty; L^2(\Omega)).$$

Then $u(t)$ converges in $L^2(\Omega)$ as $t \to \infty$, to a solution u_∞ of problem (2.42). *Moreover,*

$$\lim_{t\to\infty} \partial u/\partial t = 0 \text{ in } L^2(\Omega). \tag{2.43}$$

Proof. We apply Theorem 2.3 to φ defined before. Since other hypotheses are obviously satisfied it remains only to prove that every level set $\{u \in L^2(\Omega);\ \varphi(u) + \|u\|^2_{L^2(\Omega)} \leqslant M\}$ is compact in $L^2(\Omega)$. Since the function j is bounded from below by an affine function and the "trace operator" $u \to u|_\Gamma$ is continuous from $H^1(\Omega)$ into $L^2(\Gamma)$, it follows that the above set is bounded in $H^1(\Omega)$. As the space $H^1(\Omega)$ is completely continuous imbedded in $L^2(\Omega)$, the assertion follows.

Remark. Problems of this type are of physical interest in the theory of heat transfer between solids and gases (see W. R. Mann and F. Wolf [1] and A. Friedman [1]).

Example 2. Let β be a maximal monotone subset of $R^1 \times R^1$ such that $0 \in D(\beta)$. We consider the following problem

$$\begin{aligned} &\partial u/\partial t - \Delta u + \beta(u) \ni f && \text{on }]0, T[\times \Omega. \\ &u(t, x) = 0 && \text{on }]0, T[\times \Gamma. \\ &u(0, x) = u_0(x) && \text{on } \Omega. \end{aligned} \tag{2.44}$$

The next Proposition is a sharpening of a result proved in Chapter III (see *Example* 2, Section 2.3).

PROPOSITION 2.3 *We are given* $f \in L^2(0, T; L^2(\Omega))$ *and* $u_0 \in L^2(\Omega)$ *satisfying*

$$u_0(x) \in \overline{D(\beta)}, \quad \text{a.e. } on\ \Omega. \tag{2.45}$$

Then problem (2.44) *has a unique solution* $u \in C(0, T; L^2(\Omega))$ *satisfying*

$$\sqrt{t}\, u \in L^2(0, T; H_0^1(\Omega) \cap H^2(\Omega)), \quad \sqrt{t}\, \partial u/\partial t \in L^2(0, T; L^2(\Omega)).$$

If in addition $\partial f/\partial t \in L^1(0, T; L^2(\Omega))$, *then* $u(t) \in H_0^1(\Omega) \cap H^2(\Omega)$ *for every* $t \in]0, T[$ *and*

$$t\partial u/\partial t \in L^\infty(0, T; L^2(\Omega)).$$

Proof. Let $j: R \to]-\infty, +\infty]$ be a lower-semicontinuous convex function such that $\partial j = \beta$. We define

$$\varphi(u) = \begin{cases} \dfrac{1}{2}\displaystyle\int_\Omega |\operatorname{grad} u|^2 dx + \int_\Omega j(u(x))\, dx & \text{if } u \in H_0^1(\Omega) \text{ and } j(u) \in L^1(\Omega) \\ +\infty & \text{otherwise.} \end{cases}$$

We recall (see Proposition 3.8 in Chapter II) that φ is lower-semicontinuous on $L^2(\Omega)$ and

$$\partial\varphi(u) = \{w \in L^2(\Omega);\quad w(x) \in \beta(u(x)) - \Delta u(x), \text{ a.e. on } \Omega\}.$$

Furthermore, we have

$$\overline{D(\partial\varphi)} = \{u \in L^2(\Omega);\, u(x) \in \overline{D(\beta)}, \quad \text{a.e. on } \Omega\}.$$

Clearly, no loss of generality results in assuming that $0 \in \beta(0)$. Let L denote the operator $-\Delta$ with the domain $D(L) = H_0^1(\Omega) \cap H^2(\Omega)$. Noting that

$$\int_\Omega \beta_\lambda(u)\, Lu\, dx \geqslant 0 \text{ for all } u \in D(L)$$

it follows by Proposition 1.1 that

$$\int_\Omega L_\lambda u\, w dx \geqslant 0, \text{ for all } w \in \beta(u), \quad \text{a.e. on } \Omega.$$

This implies that

$$(w, L_\lambda u) \geqslant (L_\lambda u, Lu) \geqslant |L_\lambda u|^2 \quad \text{for all } [u, w] \in \partial\varphi$$

where $(\, , \,)$ is the usual inner product in $L^2(\Omega)$. Since $\lim_{\lambda\to 0} L_\lambda u = Lu$, this shows that

$$\|Lu\|_{L^2(\Omega)} \leqslant \|(\partial\varphi)^0 u\|_{L^2(\Omega)} \text{ for every } u \in D(\partial\varphi)$$

and therefore

$$\|u\|_{H^2(\Omega)\cap H_0^1(\Omega)} \leqslant C\|(\partial\varphi)^0 u\|_{L^2(\Omega)} \text{ for every } u \in D(\partial\varphi).$$

We may therefore conclude the proof of Proposition 2.3 by applying Theorems 2.1 and 2.2.

Example 3. We consider the boundary value problem

$$
\begin{aligned}
&\partial u/\partial t - \sum_{i=1}^{n} \frac{\partial}{\partial x_i}\left(\left|\frac{\partial u}{\partial x_i}\right|^{p-2} \frac{\partial u}{\partial x_i}\right) = f && \text{on }]0, T[\times \Omega \\
&u(t, x) = 0 && \text{on }]0, T[\times \Gamma \qquad (2.46)\\
&u(0, x) = u_0(x) && \text{on } \Omega
\end{aligned}
$$

where $p \in [2, +\infty[$.

Let A be the operator from $W_0^{1,p}(\Omega)$ into $W^{-1,p'}(\Omega)$ $(1/p + 1/p' = 1)$ defined by

$$(Au, v) = \sum_{i=1}^{n} \int_\Omega \left|\frac{\partial u}{\partial x_i}\right|^{p-2} \frac{\partial u}{\partial x_i} \frac{\partial v}{\partial x_i} \,\mathrm{d}x, \quad u, v \in W_0^{1,p}(\Omega).$$

It is easily seen that A coincides with the subdifferential (actually the Gâteaux differential) of the function $\varphi: W_0^{1,p}(\Omega) \to]-\infty, +\infty[$ given by

$$\varphi(u) = \frac{1}{p} \sum_{i=1}^{n} \int_\Omega \left|\frac{\partial u}{\partial x_i}\right|^{p} \mathrm{d}x.$$

Let $D = \{u \in W_0^{1,p}(\Omega);\ Au \in L^2(\Omega)\}$. Then the restriction of A to D is maximal monotone in $L^2(\Omega) \times L^2(\Omega)$ and it is just the subdifferential of the function $\psi: L^2(\Omega) \to]-\infty, +\infty]$ defined by

$$\psi(u) = \begin{cases} \varphi(u) & \text{if } u \in W_0^{1,p}(\Omega) \\ +\infty & \text{otherwise} \end{cases}.$$

In other words, $D(\psi) = W_0^{1,p}(\Omega)$. Applying once again Theorems 2.1 and 2.2, to $H = L^2(\Omega)$ and ψ defined as above, we get

PROPOSITION 2.4. *We are given* $u_0 \in L^2(\Omega)$ *and* $f \in L^2(0, T; L^2(\Omega))$. *Then problem* (2.46) *has a unique solution* $u \in C(0, T; L^2(\Omega))$ *which satisfies*

$$\sqrt{t}\,\partial u/\partial t \in L^2(0, T; L^2(\Omega)), \quad u \in L^p(0, T; W_0^{1,p}(\Omega)). \tag{2.47}$$

If $u_0 \in W_0^{1,p}(\Omega)$ *then*

$$\partial u/\partial t \in L^2(0, T; L^2(\Omega)), \quad u \in L^\infty(0, T; W_0^{1,p}(\Omega)). \tag{2.48}$$

Moreover, if $u_0 \in L^2(\Omega)$ *and* $\partial f/\partial t \in L^1(0, T; L^2(\Omega))$ *then*

$$t\partial u/\partial t \in L^\infty(0, T; L^2(\Omega)). \tag{2.49}$$

Example 4. Let β be a maximal monotone graph in $R^1 \times R^1$ such that $R(\beta) = R^1$, or equivalently

$$\lim_{|r| \to +\infty} j(r)/|r| = \infty$$

where j is such that $\beta = \partial j$. For simplicity we shall also suppose that $D(\beta) = R^1$. Hence β is locally bounded and j is continuous on R^1.

We shall consider here the following problem

$$\begin{aligned} &\frac{\partial u}{\partial t} - \Delta\beta(u) \ni f \quad \text{on }]0, T[\times \Omega \\ &u(0, x) = u_0(x) \qquad \text{on } \Omega \\ &\beta(u(t, x)) \ni 0 \qquad \text{on }]0, T[\times \Gamma \end{aligned} \tag{2.50}$$

where $f(t, x)$ and $u_0(x)$ are given functions.

Various examples of equations of type (2.4) arise in physics. For instance:

(a) If $\beta(r) = |r|^{p-1}r$ — or, more generally, $\beta(r) = \int_0^r \gamma(s)\mathrm{d}s$ where γ is a positive and continuous function on R^1 — the equation arises in the study of a homogeneous gas flowing through a homogeneous porous medium (cf. W. F. Ames [1]).

(b) After some transformations (see J. L. Lions [1]), certain free boundary problems can be written in a form such as seen in problem (2.50).

We observe that in virtue of Proposition 2.10, proved in Chapter II, problem (2.50) can be written as

$$\begin{aligned} &\frac{\mathrm{d}u}{\mathrm{d}t} + \partial\varphi(u) \ni f, \text{ a.e. on }]0, T[\\ &u(0) = u_0 \end{aligned} \tag{2.51}$$

where u: $[0, T] \to H^{-1}(\Omega)$ and φ: $H^{-1}(\Omega) \to]-\infty, +\infty]$ is given by

$$\varphi(u) = \begin{cases} \int_\Omega j(u(x))\mathrm{d}x & \text{if } u \in L^1(\Omega),\ j(u) \in L^1(\Omega) \\ +\infty & \text{otherwise} \end{cases}$$

Since $L^\infty(\Omega) \subset D(\varphi)$ we infer that $D(\varphi)$ is a dense subset of $H^{-1}(\Omega)$. Thus applying Theorems 2.1 and 2.2 with $H = H^{-1}(\Omega)$ and function φ as above, we obtain the following existence result.

PROPOSITION 2.5. *Let $u_0 \in H^{-1}(\Omega)$ and let $f \in L^2(0, T; H^{-1}(\Omega))$. Then problem* (2.50) *has a unique solution $u(t, x)$ satisfying*

$$u \in C(0, T; H^{-1}(\Omega)), \quad \sqrt{t}\,\beta(u) \in L^2(0, T; H_0^1(\Omega)). \tag{2.52}$$

$$\sqrt{t}\,\frac{\partial u}{\partial t} \in L^2(0, T; H^{-1}(\Omega)), \quad j(u) \in L^1(]0, T[\times \Omega). \tag{2.53}$$

Moreover, if u_0, $j(u_0) \in L^1(\Omega)$ then $j(u) \in L^\infty(0, T; L^1(\Omega))$ and

$$\frac{\partial u}{\partial t} \in L^2(0, T; H^{-1}(\Omega)), \quad \beta(u) \in L^2(0, T; H_0^1(\Omega)). \tag{2.54}$$

Suppose further that $u_0 \in H^{-1}(\Omega)$ and f is absolutely continuous from $[0, T]$ into $H^{-1}(\Omega)$. Then

$$t\,\beta(u) \in L^\infty(0, T; H_0^1(\Omega)), \quad t\,\frac{\partial u}{\partial t} \in L^\infty(0, T; H^{-1}(\Omega)). \tag{2.55}$$

As an application of Theorem 2.4, we obtain the following result concerning the asymptotic behaviour as $t \to \infty$ for the solutions of Eq. (2.50).

COROLLARY 2.1. *Assume that $\dfrac{\mathrm{d}f}{\mathrm{d}t} \in L^1(0, \infty; H^{-1}(\Omega))$ and denote $f_\infty = \lim\limits_{t\to\infty} f(t)$ in $H^{-1}(\Omega)$. Suppose further that $f - f_\infty \in L^1(0, \infty; H^{-1}(\Omega))$. Then*

$$\lim_{t\to\infty} \frac{\partial u}{\partial t}(t) = 0 \quad \text{in } H^{-1}(\Omega), \tag{2.56}$$

$$\lim_{t\to\infty} u(t) = u_\infty \quad \text{in } H^{-1}(\Omega), \tag{2.57}$$

where $u_\infty \in H^{-1}(\Omega)$ is the solution of equation

$$-\Delta\,\beta(u_\infty) = f. \tag{2.58}$$

Taking in Proposition 2.5 $\beta(r) = \dfrac{1}{p-1}|r|^{p-2}r$ where $p > 1$, we obtain the following existence result for the nonlinear equation of the diffusion, namely,

COROLLARY 2.2. *Let $u_0 \in H^{-1}(\Omega)$ and let $f \in L^2(0, T; H^{-1}(\Omega))$. Then there exists a unique function $u(t, x)$ satisfying*

$$\frac{\partial u}{\partial t} - \sum_{i=1}^{n} \frac{\partial}{\partial x_i}\left(|u|^{p-2}\frac{\partial u}{\partial x_i}\right) = f \quad \text{on }]0, T[\times\Omega. \tag{2.59}$$

$$u(t, x) = 0 \qquad \text{on }]0, T[\times\Gamma.$$

$$u(0, x) = u_0(x) \qquad \text{on } \Omega.$$

$$u \in C(0, T;\ H^{-1}(\Omega)),\ \sqrt{t}\,|u|^{p-2}u \in L^2(0, T;\ H_0^1(\Omega)). \tag{2.60}$$

$$\sqrt{t}\,\frac{\partial u}{\partial t} \in L^2(0, T; H^{-1}(\Omega)),\ u \in L^p(]0, T[\times\Omega). \tag{2.61}$$

Moreover, if $u_0 \in L^p(]0, T[\times\Omega)$ then

$$|u|^{p-2}u \in L^2(0, T; H_0^1(\Omega)),\ \frac{\partial u}{\partial t} \in L^2(0, T; H^{-1}(\Omega)), \tag{2.62}$$

$$u \in L^\infty(0, T; L^p(\Omega)). \tag{2.63}$$

Finally, suppose that $u_0 \in H^{-1}(\Omega)$ and f is absolutely continuous from $[0, T]$ into $H^{-1}(\Omega)$. Then

$$t|u|^{p-2}u \in L^\infty(0, T; H_0^1(\Omega)),\ t\frac{\partial u}{\partial t} \in L^\infty(0, T;\ H^{-1}(\Omega)). \tag{2.64}$$

§ 3. Variational evolution inequations

The theory of variational evolution inequations is one of the most important domains of application of the ideas and techniques of differential equations associated with maximal monotone operators and with semigroups of nonlinear contractions.

Some of the existence results we shall include in this paragraph rely on the regularity theorem previously established.

3.1. Unilateral conditions on $u(t)$

We are given two Hilbert spaces V and H such that $V \subset H$ and the inclusion mapping of V into H is continuous. Moreover, we asume that V is dense in H. The norms in V and H will be denoted by $\|\cdot\|$ and $|\cdot|$, respectively. The space H is identified with its own dual and we denote the dual space of V by V'. Then the following inclusion relation holds

$$V \subset H \subset V'.$$

Finally, we denote by $\|\cdot\|_*$ the norm of V', and by $(\, , \,)$ the pairing between V and V' (the inner product of H, respectively).

Besides these spaces we are given a linear continuous operator $A\colon V \to V'$, which is assumed to satisfy:

$$(Au, v) = (Av, u) \quad \text{for all } u, v \in V \tag{3.1}$$

$$(Au, u) \geqslant \omega \|u\|^2 \quad \text{for every } u \in V, \tag{3.2}$$

where ω is a positive constant.

Let $\varphi\colon V \to]-\infty, +\infty]$ be a lower-semicontinuous convex function nonidentically $+\infty$. We denote the closure in H of the set $D(\varphi) = \{u \in V; \varphi(u) < +\infty\}$ by $\overline{D(\varphi)}^H$.

THEOREM 3.1. *Let $u_0 \in \overline{D(\varphi)}^H$ and $f \in L^2(0, T; H)$. Then, there exists a unique function $u \in C(0, T; H)$ satisfying*

$$u(t) \in D(\partial\varphi) \text{ a.e. } t \in]0, T[, \; u \in W^{1,2}(\delta, T; H) \textit{ for every } 0 < \delta < T. \tag{3.3}$$

$$\sqrt{t}\, \frac{\mathrm{d}u}{\mathrm{d}t} \in L^2(0, T; H), \;\; \varphi(u) \in L^1(0, T), \;\; u \in L^2(0,T; V). \tag{3.4}$$

$$\left(\frac{\mathrm{d}u}{\mathrm{d}t}(t) + Au(t) - f(t), u(t) - v\right) \leqslant \varphi(v) - \varphi(u(t)), \text{ a.e. } t \in]0, T[\tag{3.5}$$

for every $v \in V$.

$$u(0) = u_0. \tag{3.6}$$

In addition, if $\dfrac{\mathrm{d}f}{\mathrm{d}t} \in L^1(0, T; H)$, then $u(t) \in D(\partial\varphi)$ for every $t \in]0, T[$, $\dfrac{\mathrm{d}^+}{\mathrm{d}t} u(t)$ exists for every $t \in]0, T[$ and

$$t\, \frac{\mathrm{d}u}{\mathrm{d}t} \in L^\infty(0, T; H). \tag{3.7}$$

Proof. Let $L: H \to H$ denote the multivalued operator defined by

$$Lu = [Au + \partial\varphi(u)] \cap H, \quad \forall\, u \in D(L)$$

where

$$D(L) = \{u \in V; [Au + \partial\varphi(u)] \cap H \neq \emptyset\}.$$

Since A is continuous from V to V' and $\partial\varphi$ is maximal monotone in $V \times V'$, we conclude that L is maximal monotone in $H \times H$ (see Chapter II, Corollary 1.1). Obviously, L is cyclically monotone in $H \times H$. Therefore, there exists a lower-semicontinuous convex function $\psi: H \to]-\infty, +\infty]$ such that $L = \partial\psi$. It should be observed that $\psi(u) = \varphi(u) + \frac{1}{2}(Au, u)$.

First we shall prove that

$$\overline{D(L)}^H = \overline{D(\varphi)}^H.$$

To this purpose, we consider x to be arbitrary but fixed in $D(\varphi)$. We set $x_n = (1 + n^{-1}L)^{-1}x$, $n = 1,2, \ldots$. We have

$$x_n + n^{-1}Ax_n + n^{-1}\partial\varphi(x_n) \ni x$$

which implies

$$(x_n + n^{-1}Ax_n - x,\ x_n - v) \leqslant n^{-1}(\varphi(v) - \varphi(x_n)) \text{ for all } v \in V.$$

Taking $v = x$ and using condition (3.2), we get

$$|x_n - x|^2 + n^{-1}\omega\|x_n - x\|^2 \leqslant n^{-1}(\varphi(x) - \varphi(x_n)) + n^{-1}\|Ax\|_*\|x_n - x\|. \tag{3.8}$$

Let $[u, v]$ be arbitrary in $\partial\varphi$. We have

$$\varphi(u) - \varphi(x_n) \leqslant (v, u - x_n) \leqslant \|v\|_*\|u - x_n\|. \tag{3.9}$$

By relations (3.8) and (3.9) we deduce that $\lim_{n\to\infty}|x_n - x| = 0$, i.e., $x \in \overline{D(L)}^H$. Therefore, we have proved that $D(\varphi) \subset \overline{D(L)}^H$ which clearly implies the desired relation. On the other hand, problem (3.5) can be written as

$$\frac{du}{dt} + Lu \ni f, \quad \text{a.e. } t \in]0, T[.$$

Thus we conclude the proof by applying Theorems 2.1 and 2.2.

COROLLARY 3.1. *Let A be a linear continuous operator from V into V' satisfying conditions* (3.1) *and* (3.2). *Let K be a closed convex subset of V. In addition,* *we are given $u_0 \in \overline{K}^H$ and $f \in L^2(0, T; H)$. Then there exists a unique $u \in C(0, T; H)$ satisfying*

$$u(t) \in K, \quad \text{a.e.} \quad t \in]0, T[, \quad u \in L^2(0, T; V). \tag{3.10}$$

$$\sqrt{t}\,\frac{\mathrm{d}u}{\mathrm{d}t} \in L^2(0, T; H), \; u \in W^{1,2}(\delta, T; H), \quad \forall \delta \in]0, T[. \tag{3.11}$$

$$\left(\frac{\mathrm{d}u}{\mathrm{d}t}(t) + Au(t) - f(t), u(t) - v\right) \leqslant 0, \text{ a.e.} \quad t \in]0, T[. \tag{3.12}$$

for all $v \in K$.

$$u(0) = u_0. \tag{3.13}$$

If $\dfrac{\mathrm{d}f}{\mathrm{d}t} \in L^1(0, T; H)$ then $u(t) \in K$ for every $t \in]0, T[$, $\dfrac{\mathrm{d}^+}{\mathrm{d}t} u(t)$

exists at every $t \in]0, T[$ and

$$t\,\frac{\mathrm{d}u}{\mathrm{d}t} \in L^\infty(0, T; H). \tag{3.14}$$

Proof. We apply Theorem 3.1 where $\varphi = I_K$ is the indicator function of the convex K.

THEOREM 3.2. *In addition to the assumptions given in Theorem 3.1 we suppose that the following hypothesis holds:*

(a) *There exists $h \in H$ such that for every $\varepsilon > 0$ and any $u \in D(\varphi)$ the solution u_ε of equation $u_\varepsilon + \varepsilon A u_\varepsilon = u + \varepsilon h$, belongs to $D(\varphi)$ and $\varphi(u_\varepsilon) \leqslant \varphi(u)$.*

Let $f \in L^2(0, T; H)$ and let $u_0 \in \overline{D(\varphi)}^H$ be such that $u_0 \in V$. Then solution $u(t)$ for problems (3.5) *and* (3.6) *satisfies*

$$u \in L^\infty(0, T; V), \; Au \in L^2(0, T; H). \tag{3.15}$$

Proof. Denote by A_H the restriction of A to H, i.e.

$$A_H u = Au \cap H, \quad \forall u \in D(A_H).$$

We observe that assumptions (3.1) and (3.2) imply that A_H is maximal monotone in $H \times H$ and densely defined. We set

$$A_\varepsilon u = \varepsilon^{-1}(u - (1 + \varepsilon A_H)^{-1} u), \quad \varepsilon > 0.$$

Let $h \in H$ be chosen as in hypothesis (a) given above and let $u_\varepsilon(t) = (1 + \varepsilon A_H)^{-1} u(t) + \varepsilon(1 + \varepsilon A_H)^{-1}h$. In (3.5) we take $v = u_\varepsilon(t)$, to obtain

$$\left(\frac{\mathrm{d}u}{\mathrm{d}t} + Au - f(t),\ u(t) - u_\varepsilon(t)\right) \leqslant 0, \text{ a.e. } t \in]0, T[.$$

Hence

$$\left(\frac{\mathrm{d}u}{\mathrm{d}t}, A_\varepsilon u\right) + (Au,\ A_\varepsilon u) \leqslant (f, A_\varepsilon u) + \left(\frac{\mathrm{d}u}{\mathrm{d}t} + Au - f,\ (1 + \varepsilon A_H)^{-1} h\right).$$

Since A_ε is self-adjoint, we get

$$\frac{1}{2}(A_\varepsilon u(t), u(t)) + \int_0^t (Au(s),\ A_\varepsilon u(s))\, \mathrm{d}s \leqslant \int_0^t |f(s)|\, |A_\varepsilon u(s)| \mathrm{d}s +$$

$$+ \frac{1}{2}(A_\varepsilon u_0,\ u_0) + |h|\, (|u(t) - u_0| + \int_0^t (|A_\varepsilon u(s)| + |f(s)|)\, \mathrm{d}s,$$

but

$$(Au - A_\varepsilon u,\ A_\varepsilon u) \geqslant 0 \quad \text{for all } \varepsilon > 0 \text{ and } u \in H$$

or,

$$|A_\varepsilon u|^2 \leqslant (A_\varepsilon u,\ Au).$$

Summarizing to this point, we have obtained the inequality

$$(A_\varepsilon u(t),\ u(t)) + \int_0^t |A_\varepsilon u(s)|^2\, \mathrm{d}s \leqslant C(\|A_\varepsilon u_0\|_* \|u_0\| + \int_0^t |f(s))|^2 \mathrm{d}s + |u(t) - u_0|), \quad 0 \leqslant t \leqslant T. \tag{3.16}$$

Let $\tilde{A}$: $L^2(0, T; H) \to L^2(0, T; H)$ be the realization of A_H in $L^2(0, T; H)$. In other words,

$$(\tilde{A}u)(t) = Au(t), \quad \text{a.e.} \quad t \in]0, T[, \quad u \in D(\tilde{A}),$$

while

$$D(\tilde{A}) = \{u \in L^2(0, T;\ H);\ Au \in L^2(0,\ T;\ H)\}.$$

Obviously, $\tilde{A}$ is maximal monotone and the following equality holds

$$((1 + \varepsilon\tilde{A})^{-1}u)(t) = (1 + \varepsilon A_H)^{-1} u(t), \quad \text{a.e. } t \in]0, T[.$$

In particular, estimate (3.16) implies that $\tilde{A}_\varepsilon u$ is bounded in $L^2(0, T; H)$. Therefore, $u \in D(\tilde{A})$ and

$$\lim_{\varepsilon \to 0} \tilde{A}_\varepsilon u = \tilde{A}u \text{ in } L^2(0, T;\ H).$$

Hence $Au(t) \in H$, a.e. on $]0, T[$ and the following inequality holds

$$(Au(t),\ u(t)) + \int_0^t |Au(s)|^2 \,\mathrm{d}s \leqslant C_1(+ \|u_0{}^2\| + \int_0^t |f(s)|^2 \,\mathrm{d}s)$$

for all $t \in [0, T]$. This completes the proof.
As an immediate consequence of Theorem 3.2 one obtains

COROLLARY 3.2 *In Corollary* 3.1 *we assumed that the following assumption held:*

(b) *There exists* $h \in H$ *such that for every* $\varepsilon > 0$ *and* $u \in K$, $(1 + \varepsilon A)^{-1}(u + \varepsilon h) \in K$.

Let $f \in L^2\ (0,\ T; H)$ *and* $u_0 \in \bar{K}^H$ *be such that* $u_0 \in V$. *Then* $u \in L^\infty(0, T;\ V)$ *and* $Au \in L^2(0,\ T;\ H)$.

3.2. Unilateral conditions on $\dfrac{\mathrm{d}u}{\mathrm{d}t}(t)$

The notations and the assumptions on V, H and A are those used in the first section.

Let K be a closed convex subset of V and let $\partial I_K \subset V \times V'$ be the subdifferential of the indicator function I_K, i.e.

$$\partial I_K(u) = \{y \in V'; (y,\ u - v) \geqslant 0, \quad \forall v \in K\} \quad u \in K. \tag{3.17}$$

Let R_K denote the set of all $y \in V'$ such that there exists $u \in K$ satisfying

$$(y - u,\ u - v) \geqslant 0, \quad \forall v \in K.$$

In other words

$$R_K = R(1 + \partial I_K). \tag{3.18}$$

We consider the following variational problem:

$$\begin{gathered}\left(\frac{\mathrm{d}u}{\mathrm{d}t}(t) + Au(t) - f(t),\ v - \frac{\mathrm{d}u}{\mathrm{d}t}(t)\right) \geqslant 0, \quad \text{a.e. } t \in]0, T[\text{ for all } v \in K \\ u(0) = u_0, \quad \frac{\mathrm{d}u}{\mathrm{d}t}(t) \in K, \quad \text{a.e. } t \in]0, T[.\end{gathered} \tag{3.19}$$

THEOREM 3.3. *We are given*

$$u_0 \in V, f \in W^{1,2}(0, T; V').$$

Then problem (3.19) *has a unique solution* $u \in C(0, T; V)$ *which satisfies*

$$\sqrt{t}\,\frac{du}{dt} \in L^2(0, T; V), \quad \frac{du}{dt} \in L^2(0, T; H). \tag{3.20}$$

If in addition $\frac{d^2f}{dt^2} \in L^2(0, T; V')$ *then*

$$t\,\frac{du}{dt} \in L^\infty(0, T; V). \tag{3.21}$$

Finally, if $f(0) - Au_0 \in R_K$ *and* $f \in W^{2,2}(0, T; V')$ ($f \in W^{1,2}(0, T; H)$, *respectively) then* $\frac{du}{dt} \in L^\infty(0, T; V)$ $\left(\frac{du}{dt} \in L^2(0, T; V), \textit{respectively}\right)$ *and*

$$\frac{d^2u}{dt^2} \ni L^2(0, T; H). \tag{3.22}$$

Proof. The mapping $u \to u + \partial I_K(u)$ from K to V' is monotone. In particular this implies that operator $(1 + \partial I_K)^{-1}\ V' \to V$ is single-valued. Define on V an equivalent norm given by the inner product

$$\langle u, v\rangle = (Au, v); \quad u, v \in V,$$

and consider the operator $\Lambda = (1 + \partial I_K)^{-1} A$ where

$$D(\Lambda) = \{u \in V;\ Au \in R_K\}.$$

We observe that problem (3.19) can be written in the following form

$$\begin{gathered}\frac{du}{dt} + Au + \partial I_K\left(\frac{du}{dt}\right) \ni f, \quad \text{a.e. on }]0, T[\\ u(0) = u_0, \quad \frac{du}{dt} \in K, \quad \text{a.e. on }]0, T[.\end{gathered} \tag{3.23}$$

We set $g = A^{-1} f$ and $v = g - u$. Then problem (3.23) becomes

$$\begin{gathered}\frac{dv}{dt} + \Lambda v = \frac{dg}{dt}, \quad \text{a.e.} \quad t \in]0, T[,\\ v(0) = A^{-1} f(0) - u_0.\end{gathered} \tag{3.24}$$

We mention that assumptions (3.1) and (3.2) imply that A^{-1} is everywhere defined and continuous on V'.

LEMMA 3.1. *Λ is a cyclically maximal monotone subset of $V \times V$. Moreover, $D(\Lambda)$ is dense in V.*

Proof. The first part of this lemma follows immediately by the definition of Λ. Let y be arbitrary in V. Equation

$$v + \Lambda v \ni y$$

is equivalent to

$$Ay \in Au + \partial I_K(u) + u. \tag{3.25}$$

Since ∂I_K is maximal monotone in $V \times V'$ and A is continuous and coercive from V to V', it follows that $R(1 + A + \partial I_K) = V'$. Hence Eq. (3.25) has a unique solution $u \in K$. Thus we have proved that $R(1 + \Lambda) = V$, i.e., Λ is maximal monotone in $V \times V$.

In order to prove that $D(\Lambda)$ is dense in V it suffices to show that R_K is a dense subset of V'. Let y be arbitrary in V. We have already observed that for every $\varepsilon > 0$ equation

$$u_\varepsilon + \varepsilon A u_\varepsilon + \partial I_K(u_\varepsilon) \ni y \tag{3.26}$$

has a unique solution $u_\varepsilon \in K$. We have,

$$(y - u_\varepsilon - \varepsilon A u_\varepsilon,\ u_\varepsilon - v) \geqslant 0 \quad \text{for all } v \in K.$$

The latter implies that $\{\sqrt{\varepsilon}\, u_\varepsilon\}$ is bounded in V as $\varepsilon \to 0$. Thus $\lim\limits_{\varepsilon \to 0} \varepsilon A u_\varepsilon = 0$ in V' and from Eq. (3.26) we deduce that y belongs to the closure of R_K in V'. Since V is dense in V' we finally conclude that R_K is dense in V'.

Proof of Theorem 3.3 (continued): Let $\psi\colon V \to]-\infty, +\infty]$ be a lower-semi-continuous convex function such that $\Lambda = \partial\psi$. According to Lemma 3.1, $\overline{D(\psi)} = V$. Applying Theorem 2.1 to ψ and V we conclude that Eq. (3.19) has a unique solution $u \in C(0, T; V)$ such that $\sqrt{t}\,\dfrac{du}{dt} \in L^2(0, T; V)$. On the other hand, using Eq. (3.19) we get $\left|\dfrac{du}{dt}\right|^2 + \dfrac{1}{2}\dfrac{d}{dt}\|u\|^2 \leqslant \|v\|\,(\|f\|_* + \|Au\|_*) + \dfrac{du}{dt}|v| + \dfrac{d}{dt}(f, u) - \left(\dfrac{df}{dt}, u\right)$, $\forall v \in K$. Since $u \in C(0, T; V)$ and $\dfrac{df}{dt} \in L^2(0, T; V')$, the latter implies that $\dfrac{du}{dt} \in L^2(0, T; H)$. If $f(0) - Au_0 \in R_K$ and $f \in W^{2,2}(0, T; V')$ then in Eq. (3.24), $\dfrac{dg}{dt} \in W^{1,2}(0, T; V)$ and

$v(0) \in D(\Lambda)$. From Theorem 2.2 in Chapter III, this implies that $\frac{dv}{dt} = \frac{dg}{dt} - \frac{du}{dt} \in L^\infty(0, T; V)$. If $f \in W^{1,2}(0, T; H)$ then $\frac{dg}{dt} \in L^2(0, T; V)$ and, by Theorem 2.2, we deduce that $\frac{dv}{dt} \in L^2(0, T; V)$. To conclude the proof it remains to prove (3.22). In problem (3.19) we put $v = \frac{d}{dt} u(t + h)$ to obtain

$$\left(\frac{du}{dt}(t), \frac{d}{dt}(u(t) - u(t + h)\right) + \left(Au(t), \frac{d}{dt}(u(t) - u(t + h))\right) \leqslant \left(f(t), \frac{d}{dt}(u(t) - u(t + h))\right), \quad \text{a.e. on }]0, T[.$$

Next in problem (3.19) we take $t = t + h$ and $v = \frac{du}{dt}(t)$. Subtracting the result from the preceding inequality, one obtains:

$$\left|\frac{d}{dt}(u(t) - u(t + h))\right|^2 + \frac{1}{2}\frac{d}{dt}(A(u(t) - u(t + h)), u(t) - u(t + h)) \leqslant$$
$$\leqslant \left(f(t) - f(t + h), \frac{d}{dt}(u(t) - u(t + h))\right), \quad \text{a.e. on }]0, T[.$$

Assuming that $f \in W^{2,2}(0, T; V')$ this yields

$$\int_0^t \left|\frac{d}{ds}(u(s) - u(s + h))\right|^2 ds \leqslant \frac{1}{2}(Au_0 - Au(h), u_0 - u(h)) +$$
$$+ (f(t) - f(t + h), u(t) - u(t + h)) - (f(0) - f(h), u_0 - u(h)) + \tag{3.27}$$
$$+ \int_0^t \left\|\frac{d}{ds}(f(s) - f(s + h))\right\|_* \|u(s) - u(s - h)\| \, ds.$$

Since $\frac{du}{dt} \in L^2(0, T; V)$ and $f \in W^{2,2}(0, T; V')$ we deduce that $\frac{d^2u}{dt^2} \in L^2(0, T; H)$. The case $f \in W^{1,2}(0, T; H)$ follows by a similar argument.

Next, we shall consider the following problem

$$\left(Au(t) - f(t), \frac{du}{dt}(t) - v\right) \leqslant 0, \ \forall v \in K, \quad \text{a.e. } t \in]0, T[\tag{3.28}$$

$$u(0) = u_0, \frac{du}{dt} \in K, \quad \text{a.e. } t \in]0, T[, \tag{3.29}$$

where K is a closed convex subset of V satisfying

$0 \in K.$

THEOREM 3.4. *Let $u_0 \in V$ and $f \in W^{2,2}(0, T; V')$ be given such that*

$$(f(0) - Au_0, v) \leqslant 0, \quad \forall v \in K. \tag{3.30}$$

Then problems (3.28), (3.29) *have a unique solution $u \in C(0, T; V)$ satisfying*

$$\frac{du}{dt} \in L^\infty(0, T; V).$$

Proof. We shall use the method of parabolic regularization. According to Theorem 3.3, for any $\varepsilon > 0$ there exists a unique function $u_\varepsilon \in C(0, T; V)$ such that

$$\frac{du_\varepsilon}{dt} \in L^\infty(0, T; V); \ \frac{d^2u_\varepsilon}{dt^2} \in L^2(0, T; H) \tag{3.31}$$

and satisfies

$$\varepsilon \frac{du_\varepsilon}{dt} + Au_\varepsilon + \partial I_K\left(\frac{du_\varepsilon}{dt}\right) \ni f(t), \quad \text{a.e.} \quad t \in]0, T[, \tag{3.32}$$

$$u_\varepsilon(0) = u_0, \frac{du_\varepsilon}{dt} \in K, \ \text{a.e.} \quad t \in]0, T[. \tag{3.33}$$

We observe that condition (3.30) implies that

$$f(0) - Au_0 \in \partial I_K(0) \subset R_K. \tag{3.34}$$

We multiply Eq. (3.32) by $\frac{du_\varepsilon}{dt}$. Since ∂I_K is monotone in $V \times V'$, it follows from formula (3.34) that

$$\frac{1}{2} \frac{d}{dt} (A(u_\varepsilon - u_0), u_\varepsilon - u_0) + \left|\frac{du_\varepsilon}{dt}\right|^2 \varepsilon \leqslant \left(f(t) - f(0), \frac{du_\varepsilon}{dt}\right).$$

Hence

$$\frac{1}{2} \omega \|u_\varepsilon(t) - u_0\|^2 + \varepsilon \int_0^t \left|\frac{du_\varepsilon}{ds}\right|^2 ds \leqslant \|f(t) - f(0)\|_* \|u_\varepsilon(t) - u_0\| +$$

$$+ \int_0^t \left(\frac{df}{ds}(s), u_\varepsilon(s) - u_0\right) ds, \quad 0 \leqslant t \leqslant T.$$

Finally,

$$\|u_\varepsilon(t) - u_0\|^2 + \varepsilon \int_0^t \left|\frac{du_\varepsilon}{ds}\right|^2 ds \leqslant C_1\left(\|f(t)-f(0)\|_*^2 + \int_0^t \left\|\frac{df}{ds}\right\|_*^2 ds\right). \quad (3.35)$$

Similarly,

$$\|u_\varepsilon(t+h) - u_\varepsilon(t)\|^2 + \varepsilon \int_0^t \left|\frac{du_\varepsilon}{ds}(s+h) - \frac{du_\varepsilon}{ds}(s)\right|^2 ds \leqslant$$

$$\leqslant C_2\left(\|f(t+h) - f(t)\|_*^2 + \|f(h) - f(0)\|_*^2 + \|u_\varepsilon(h) - u_0\|^2 + \right.$$

$$\left. + \left|\int_0^t \left\|\frac{df}{ds}(s+h) - \frac{df}{ds}(s)\right\|_*^2 ds\right|\right). \quad (3.36)$$

Consequently, we conclude from (3.35) and (3.36) that

$$\left\|\frac{du_\varepsilon}{dt}(t)\right\|^2 + \varepsilon\int_0^T \left|\frac{d^2u_\varepsilon}{dt^2}\right|^2 dt \leqslant C, \qquad 0 \leqslant t \leqslant T, \quad (3.37)$$

where C is a nonnegative constant independent of ε.

Now we may take weakly convergent subsequences. There exists a subsequence denoted again by $\{u_\varepsilon\}$ such that

$$\left.\begin{aligned} u_\varepsilon &\to u \\ \frac{du_\varepsilon}{dt} &\to \frac{du}{dt} \end{aligned}\right\} \text{ in the weak-star topology of } L^\infty(0, T; V) \quad (3.38)$$

as $\varepsilon \to 0$.

We shall prove that u is a solution of problem (3.28). We first observe that $u \in C(0, T; V)$ and

$u_\varepsilon(t) \to u(t)$ weakly in V for all $t \in [0, T]$.

In particular, this implies that $u(0) = u_0$. Moreover, again by (3.38) we easily deduce that $\dfrac{du}{dt}(t) \in K$, a.e. $t \in]0, T[$.

On the other hand, it follows from Eq. (3.32) that

$$\left(\varepsilon\frac{du_\varepsilon}{dt} + Au_\varepsilon - f, \frac{du_\varepsilon}{dt} - v\right) \leqslant 0, \quad \text{a.e. } t \in]0, T[, \ \forall v \in K. \quad (3.39)$$

Taking $v = v(t)$, where $v \in L^2(0, T; V)$ and $v(t) \in K$, a.e. $t \in]0, T[$, and integrating over $]0, T[$, one obtains

$$\int_0^T \left(Au_\varepsilon, \frac{du_\varepsilon}{dt} - v\right) dt + \varepsilon \int_0^T \left(\frac{du_\varepsilon}{dt}, \frac{du_\varepsilon}{dt} - v\right) dt \leqslant \int_0^T \left(f, \frac{du_\varepsilon}{dt} - v\right) dt$$

or

$$\frac{1}{2}((Au_\varepsilon(T), u_\varepsilon(T)) - (Au_0, u_0)) - \int_0^T (Au_\varepsilon, v)\, dt \leqslant$$

$$\leqslant \int_0^T \left(f - \varepsilon \frac{du_\varepsilon}{dt}, \frac{du_\varepsilon}{dt} - v\right) dt. \tag{3.40}$$

Since the function $u \to (Au, u)$ is weakly lower-semicontinuous on V, from formula (3.40) we obtain

$$\frac{1}{2}((Au(T), u(T)) - (Au_0\, u_0)) \leqslant \int_0^T (Au(t), v(t)\, dt +$$

$$+ \int_0^T \left(f(t), \frac{du}{dt}(t) - v(t)\right) dt.$$

Then

$$\int_0^T \left(Au(t) - f(t), \frac{du}{dt}(t) - v(t)\right) dt \leqslant 0, \quad \forall v \in L^2(0, T; K) \tag{3.41}$$

for all $v \in L^2(0, T; V)$ such that $v(t) \in K$, a.e. $t \in]0, T[$. Let v be arbitrary in K and let φ be a C_0^∞ scalar function such that supp $\varphi \subset]0, T[$, $0 \leqslant \varphi(t) \leqslant 1$. Taking in formula (3.41), $v(t) = (1 - \varphi(t)) \dfrac{du}{dt}(t) + \varphi(t)v$, we get

$$\int_0^T \left(Au(t) - f(t), \frac{du}{dt}(t) - v\right) \varphi(t)\, dt \leqslant 0$$

which implies inequation (3.28) by a standard device. The uniqueness follows immediately by using the positivity of the operator A. Thus the proof of Theorem 3.4 is complete.

3.3 A class of nonlinear variational inequations

In the present section, we shall consider the extension of some of the results in Section 2, and in particular in Theorem 3.4, to the case where operator A is nonlinear but still cyclically monotone in $H \times H$.

We now state the basic assumptions which will be in effect throughout this section.

We are given a real Hilbert space H and a real reflexive Banach space V with the following properties:

(i) *$V \subset H$ and the injection mapping of V into H is densely defined and completely continuous.*

Notations for norms and inner product will be the same as in Section 1. Secondly, we are given two functions φ and ψ which are assumed to satisfy:

(ii) *The functions φ and ψ are lower-semicontinuous, convex and proper from H into $]-\infty, +\infty]$. There is $C > 0$ such that*

$$\varphi(u)/|u|^2 \geqslant C \tag{3.42}$$

for sufficiently large $|u|$.

(iii) *There exists $d > 0$ such that*

$$\varphi(\lambda u) = \lambda^d \varphi(u) \text{ for all } u \in H \text{ and } \lambda > 0. \tag{3.43}$$

Moreover, $\partial\psi$ is $\partial\varphi$-monotone, i.e.,

$$\varphi((1 + \lambda\partial\psi)^{-1}x - (1 + \lambda\partial\psi)^{-1}y) \leqslant \varphi(x - y) \tag{3.44}$$

or every $[x, y] \in H \times H$ and all $\lambda > 0$.

(iv) *$D(\psi) \subset V$ and* $\lim\limits_{\|u\| \to \infty} (\psi(u) + |u|^2) = \infty$.

We shall consider first order evolution equations of the type

$$\begin{aligned} &\partial\varphi\left(\frac{\mathrm{d}u}{\mathrm{d}t}(t)\right) + Au(t) \ni f(t), \quad \text{a.e. on }]0, T[\\ &u(0) = u_0 \end{aligned} \tag{3.45}$$

where $A = \partial\psi$.

In particular, for $\varphi = I_K$, Eq. (3.45) is equivalent to Eq. (3.28), but A need not be self-adjoint in H.

We now proceed to our principal result regarding this problem.

THEOREM 3.5. *Let $u_0 \in D(A)$ and $f \in W^{1,2}(0, T; H)$. Under the above hypotheses, there exists at least a solution $u \in W^{1,2}(0, T; H)$ of problem (3.45) which satisfies:*

$$\frac{\mathrm{d}u}{\mathrm{d}t}(t) \in D(\partial\varphi) \text{ and } u(t) \in D(A), \quad \text{a.e. on }]0, T[. \tag{3.46}$$

$$\psi(u) \in L^\infty(0, T), \quad \varphi\left(\frac{\mathrm{d}u}{\mathrm{d}t}\right) \in L^1(0, T). \tag{3.47}$$

$$Au \in L^2(0, T; H). \tag{3.48}$$

Remark. By assumption (iv) it is immediately clear that u also satisfies

$$u \in L^\infty(0, T; V).$$

If A is multivalued, Eq. (3.45) and relation (3.48) have to be interpreted in the sense that there is some $v \in L^2(0, T; H)$ such that $v(t) \in Au(t)$ and $\partial\varphi\left(\frac{du}{dt}\right) + v(t) \ni f(t)$ almost everywhere on $]0, T[$.

Proof of Theorem 3.5. In order to prove the theorem, we begin by considering the following sequence of two-point boundary problems which approximate equation (3.45),

$$\begin{aligned} &-\lambda \frac{d}{dt}\left(\partial\varphi\left(\frac{du}{dt}\right)\right) + \partial\varphi\left(\frac{du}{dt}\right) + A_\lambda u \ni f, \quad \text{a.e. on }]0, T[\\ &\partial\varphi\left(\frac{du}{dt}(0)\right) \ni \lambda^{-1}(u(0) - u_0), \quad \partial\varphi\left(\frac{du}{dt}(T)\right) \ni 0 \end{aligned} \tag{3.49}$$

where $A_\lambda = \lambda^{-1}(1 - (1 + \lambda A)^{-1})$. Secondly, we shall find sufficient *a priori* estimates on the solutions u_λ for problem (3.49) to obtain — via *compactness* arguments — *strong convergence* of u_λ to a solution u of Eq. (3.45).

(a) *Approximate equations.* Let $B_\lambda\colon L^2(0, T; H) \to L^2(0, T; H)$ be the operator defined by

$$(B_\lambda u)(t) = -\lambda \frac{d}{dt}\left(\exp\left(-\frac{t}{\lambda}\right)p_\lambda(t)\right) \quad \text{a.e. on }]0, T[, \quad u \in D(B_\lambda)$$

where $D(B_\lambda)$ consists of all $u \in W^{1,\infty}(0, T; H)$ such that there exist $p_\lambda \in W^{1,2}(0, T; H)$ satisfying

$$p_\lambda(t) \in \partial\varphi\left(\frac{du}{dt}(t)\right), \qquad \text{a.e. on }]0, T[$$

$$p_\lambda(0) = \lambda^{-1}(u(0) - u_0), \quad p_\lambda(T) = 0.$$

Then the boundary value problem (3.49) can be written as

$$B_\lambda u + \exp\left(-\frac{t}{\lambda}\right)A_\lambda u \ni \exp\left(-\frac{t}{\lambda}\right)f. \tag{3.50}$$

We shall prove in the next Chapter (see Lemma 2.1 and the remark which follows) that B_λ is maximal monotone and coercive in $L^2(0, T; H)$. More precisely, we shall see that B_λ is the subdifferential of function

$\Phi_\lambda: L^2(0, T; H) \to]-\infty, +\infty]$ given by

$$\Phi_\lambda(u) = \begin{cases} \lambda \displaystyle\int_0^T \exp\left(-\frac{t}{\lambda}\right) \varphi\left(\frac{du}{dt}\right) dt + \frac{1}{2} |u(0) - u_0|^2 \\ \text{if } u \in W^{1,1}(0,T; H) \quad \text{and } \varphi\left(\dfrac{du}{dt}\right) \in L^1(0, T) \\ +\infty \qquad\qquad \text{otherwise.} \end{cases}$$

Since the operator $u \to \exp\left(-\dfrac{t}{\lambda}\right) A_\lambda u$ is monotone and continuous on $L^2(0, T; H)$, we may conclude as an intermediate step of the proof that Eq. (3.50) has a solution $u_\lambda \in D(B_\lambda)$. In other words, for every $\lambda > 0$ there exists $u_\lambda \in W^{1,\infty}(0, T; H)$ and $p_\lambda \in W^{1,2}(0, T; H)$ satisfying

$$p_\lambda(t) \in \partial\varphi\left(\frac{du_\lambda}{dt}(t)\right), \quad \text{a.e. on }]0, T[. \tag{3.51}$$

$$-\lambda \frac{d}{dt} p_\lambda + p_\lambda + A_\lambda u_\lambda = f, \quad \text{a.e. on }]0, T[\tag{3.52}$$

and

$$\lambda p_\lambda(0) = u_\lambda(0) - u_0, \qquad p_\lambda(T) = 0. \tag{3.53}$$

(b) *A priori estimates.* Firstly, we note that conditions (3.43) and (3.44), Proposition 1.2 and relation (3.51) imply that

$$\left(\frac{d}{dt} A_\lambda u_\lambda, p_\lambda\right) \geqslant 0, \quad \text{a.e. on }]0, T[. \tag{3.54}$$

Secondly, we differentiate relation (3.52), and after we multiply it with p_λ we obtain

$$\frac{1}{2} \frac{d}{dt}\left(|p_\lambda|^2 - \lambda \frac{d}{dt} |p_\lambda|^2\right) + \lambda \left|\frac{dp_\lambda}{dt}\right|^2 \leqslant \left(\frac{df}{dt}, p_\lambda\right), \quad \text{a.e. on }]0, T[.$$

Integrating over $]0, t[$ and using the boundary conditions (3.53), this yields

$$\frac{1}{2} |p_\lambda(t)|^2 + \lambda \int_0^t \left|\frac{dp_\lambda}{ds}\right|^2 ds + (p_\lambda(0) + A_\lambda u_\lambda(0) - f(0), p_\lambda(0)) -$$

$$- \frac{1}{2} |p_\lambda(0)|^2 \leqslant \frac{\lambda}{2} \frac{d}{dt} |p_\lambda(t)|^2 + \int_0^t \left|\frac{df}{ds}\right| |p_\lambda| \, ds, \quad 0 \leqslant t \leqslant T. \tag{3.55}$$

However

$$(A_\lambda u_\lambda(0), p_\lambda(0)) \geqslant (A_\lambda u_0, p_\lambda(0))$$

because A_λ is monotone. Then integrating (3.55) over $]0, T[$, we obtain after some rearrangements

$$\int_0^T |p_\lambda(t)|^2 \,\mathrm{d}t + |p_\lambda(0)|^2 \leqslant C\left(|A^0 u_0|^2 + |f(0)|^2 + \int_0^T \left|\frac{\mathrm{d}f}{\mathrm{d}t}(t)\right|^2 \mathrm{d}t\right). \tag{3.56}$$

Furthermore, taking $t = T$ in (3.55) and using once again the boundary conditions (3.53), we conclude from formula (3.56) that

$$\int_0^T \left(\lambda \left|\frac{\mathrm{d}p_\lambda}{\mathrm{d}t}\right|^2 + |A_\lambda u_\lambda|^2\right) \mathrm{d}t \leqslant C, \quad \forall \lambda > 0. \tag{3.57}$$

We shall denote several constants independent of λ by C.

Comparing inequality (3.56) with (3.53), we conclude that

$$\lim_{\lambda \to 0} u_\lambda(0) = u_0 \quad \text{in } H. \tag{3.58}$$

Let φ^* denote the conjugate function of φ, i.e.,

$$\varphi^*(v) = \sup\{(u, v) - \varphi(u);\ u \in H\}.$$

We recall that $\partial\varphi^* = (\partial\varphi)^{-1}$. Moreover, assumption (ii) implies that $D(\varphi^*) = H$. Since $\dfrac{\mathrm{d}u_\lambda}{\mathrm{d}t} \in \partial\varphi^*(p_\lambda)$, it follows by Lemma 2.1 that

$$\int_0^T \left(\frac{\mathrm{d}p_\lambda}{\mathrm{d}t}, \frac{\mathrm{d}u_\lambda}{\mathrm{d}t}\right) \mathrm{d}t = \varphi^*(0) - \varphi^*(p_\lambda(0)). \tag{3.59}$$

Define

$$\psi_\lambda(u) = \inf\left\{\frac{|u - v|^2}{2\lambda} + \psi(v);\ v \in H\right\}$$

and recall (see Theorem 2.2, Chapter II) that

$$\partial\psi_\lambda = A_\lambda, \tag{3.60}$$

$$\psi_\lambda(u) = \frac{\lambda}{2}|A_\lambda u|^2 + \psi((1 + \lambda A)^{-1}u), \quad \forall u \in H;\ \lambda > 0. \tag{3.61}$$

Let v_0 be arbitrary in $D(\varphi)$. We multiply relation (3.52) with $\frac{du_\lambda}{dt} - v_0$ and integrate over $]0, t[$. After some calculation involving Lemma 2.1 and relations (3.56), (3.57) and (3.59) one obtains

$$\int_0^T \varphi\left(\frac{du_\lambda}{dt}\right) dt + \psi_\lambda(u_\lambda(T)) \leqslant \psi_\lambda(u_\lambda(0)) + T\varphi(v_0) +$$

$$+ \lambda(\varphi^*(0)) - \varphi^*(p_\lambda(0)) + \int_0^T \left(f, \frac{du_\lambda}{dt}\right) dt + C$$

and therefore

$$\int_0^T \varphi\left(\frac{du_\lambda}{dt}\right) dt + \psi_\lambda(u_\lambda(T)) - \psi_\lambda(u_\lambda(0)) \leqslant C + \int_0^T \left(f, \frac{du_\lambda}{dt}\right) dt. \tag{3.62}$$

Furthermore, we have

$$\psi_\lambda(u_\lambda(t)) = \psi_\lambda(u_\lambda(0)) + \int_0^T \left(\frac{du_\lambda}{ds}, A_\lambda u_\lambda\right) ds, \quad 0 \leqslant t \leqslant T. \tag{3.63}$$

Comparing this with relations (3.57) and (3.62), and using hypothesis (ii), it follows that

$$\left\{\frac{du_\lambda}{dt}\right\} \text{ remains bounded in } L^2(0, T; H). \tag{3.64}$$

However

$$\psi_\lambda(u_\lambda(0)) \leqslant \psi_\lambda(v_0) + (A_\lambda u_\lambda(0), u_\lambda(0) - v_0), \quad \forall v_0 \in H.$$

Since $|A_\lambda u_\lambda(0) - A_\lambda u_0| \leqslant \lambda^{-1}|u_\lambda(0) - u_0| \leqslant |p_\lambda(0)|$, and $|A_\lambda u_0| \leqslant |A^0 u_0|$), it follows that $\psi_\lambda(u_\lambda(0))$ are bounded. Consequently, we conclude from Eqs. (3.63) and (3.64) that

$$\psi_\lambda(u_\lambda(t)) \leqslant C \text{ for all } \lambda > 0 \text{ and } t \in [0, T]. \tag{3.65}$$

Letting v_λ be defined by

$$v_\lambda(t) = (1 + \lambda A)^{-1} u_\lambda(t)$$

relation (3.61) implies that

$$\psi(v_\lambda(t)) \leqslant C \quad \text{uniformly on } [0, T]. \tag{3.66}$$

Thus, by hypothesis (iv) we infer that

$$\{v_\lambda\} \text{ remains bounded in } L^\infty(0, T; V). \tag{3.67}$$

Finally, relation (3.64) implies that

$$\left\{\frac{dv_\lambda}{dt}\right\} \text{ remains bounded in } L^2(0, T; H). \tag{3.68}$$

Here we have used, in particular, the fact that $(1 + \lambda A)^{-1}$ is nonexpansive on H.

(c) *Convergence of the approximate solutions.* Since the injection of V into H is compact, it follows from (3.67) and (3.68) and from a well-known compactness theorem (see e.g. J. L. Lions [1], Chapter 1, Th. 5.1), that there exists a subsequence of $\{v_\lambda\}$ which converges *strongly* in $L^2(0, T; H)$ as $\lambda \to 0$. On the other hand, it follows from relations (3.61) and (3.65) that

$$\lim_{\lambda\to 0} |u_\lambda(t) - v_\lambda(t)| = \lim_{\lambda\to 0} \lambda |A_\lambda u_\lambda(t)| = 0 \quad \text{on } [0, T]. \tag{3.69}$$

Thus extracting further subsequences if necessary we may assume that

$$\begin{aligned} &u_\lambda \to u && \text{strongly in } L^2(0, T; H).\\ &\frac{du_\lambda}{dt} \to \frac{du}{dt} && \text{weakly in } L^2(0, T; H)\cdot\\ &A_\lambda u_\lambda \to y && \text{weakly in } L^2(0, T; H).\\ &p_\lambda \to p && \text{weakly in } L^2(0, T; H) \end{aligned} \tag{3.70}$$

Next, from relation (3.57), we conclude that

$$\lambda \frac{dp_\lambda}{dt} \to 0 \quad \text{strongly in } L^2(0, T; H). \tag{3.71}$$

We observe from (3.51), (3.52), (3.70), (3.71) and the definition of "subgradient" that

$$\liminf_{\lambda\to 0} \left[\int_0^T \varphi\left(\frac{du_\lambda}{dt}\right) dt + \psi_\lambda(u_\lambda(T)) - \psi_\lambda(u_\lambda(0))\right] \leqslant \int_0^T \varphi(v)\, dt +$$

$$+ \int_0^T \left(f, \frac{du_\lambda}{dt} - v\right) dt + \int_0^T (y, v)\, dt, \quad \forall v \in L^2(0, T; H). \tag{3.72}$$

On the other hand, the inequality

$$\psi_\lambda(u_\lambda(0)) \leqslant \psi_\lambda(u_0) + (A_\lambda u_\lambda(0),\ u_\lambda(0) - u_0)$$

implies that

$$\limsup_{\lambda\to 0}\ \psi_\lambda(u_\lambda(0)) \leqslant \psi(u_0) \tag{3.73}$$

because $\lambda^{-1}|u_\lambda(0) - u_0| = |p_\lambda(0)|$ are bounded as $\lambda \to 0$. We note that in particular, hypothesis (iii) implies that

$$\varphi\left(\frac{\mathrm{d}v_\lambda}{\mathrm{d}t}\right) \leqslant \varphi\left(\frac{\mathrm{d}u_\lambda}{\mathrm{d}t}\right), \quad \text{a.e. on }]0, T[$$

while as a consequence of (3.61) we have

$$\psi(v_\lambda(T)) \leqslant \psi_\lambda(u_\lambda(T)), \quad \forall\lambda > 0.$$

Putting all these estimates into formula (3.72), we obtain

$$\liminf_{\lambda\to 0}\left[\int_0^T \varphi\left(\frac{\mathrm{d}v_\lambda}{\mathrm{d}t}\right)\mathrm{d}t + \psi(v_\lambda(T))\right] \leqslant \psi(v_0) + \int_0^T (y, v)\,\mathrm{d}t$$

$$+ \int_0^T \left(f, \frac{\mathrm{d}u_\lambda}{\mathrm{d}t} - v\right)\mathrm{d}t + \int_0^T \varphi(v)\,\mathrm{d}t. \tag{3.74}$$

On the other hand, it follows by (3.68), (3.69) and (3.70) that

$$\frac{\mathrm{d}v_\lambda}{\mathrm{d}t} \to \frac{\mathrm{d}u}{\mathrm{d}t} \quad \text{weakly in } L^2(0, T; H)$$

$$v_\lambda(T) \to u(T) \quad \text{weakly in } H.$$

The functions $v \to \int_0^T \varphi(v)\,\mathrm{d}t$ and ψ being lower-semicontinuous and convex on $L^2(0, T; H)$ and H, respectively, they are also weakly lower-semicontinuous. We may therefore conclude that

$$\int_0^T \varphi\left(\frac{\mathrm{d}u}{\mathrm{d}t}\right)\mathrm{d}t + \psi(u(T)) \leqslant \psi(u_0) + \int_0^T (y, v)\,\mathrm{d}t + \int_0^T \left(f, \frac{\mathrm{d}u}{\mathrm{d}t} - v\right)\mathrm{d}t +$$

$$+ \int_0^T \varphi(v)\,\mathrm{d}t, \quad \text{for every } v \in L^2(0, T; H).$$

An equivalent form of this inequality is

$$\int_0^T \varphi\left(\frac{du}{dt}\right) dt \leqslant \int_0^T \varphi(v)\,dt + \int_0^T \left(f - y, \frac{du}{dt} - v\right) dt$$

because $y(t) \in Au(t)$, a.e. on $]0, T[$ and

$$\frac{d}{dt}\psi(u) = \left(y, \frac{du}{dt}\right), \quad \text{a.e. on }]0, T[.$$

This implies, by using a standard device, that

$$\varphi\left(\frac{du}{dt}(t)\right) \leqslant \varphi(v) + \left(f(t) - y(t), \frac{du}{dt}(t) - v\right), \quad \text{a.e. on }]0, T[$$

for all $v \in H$, i.e.,

$$f(t) - y(t) \in \partial\varphi\left(\frac{du}{dt}(t)\right), \quad \text{a.e. on }]0, T[$$

as claimed. We observe that (3.47) is implied by estimates (3.62) and (3.66). Thus the proof of Theorem 3.5 is complete.

COROLLARY 3.3. *Let C be a closed convex cone of H and let $A = \partial\psi$ satisfy hypothesis* (iv). *Suppose further that*

$$(1 + \lambda A)^{-1}x - (1 + \lambda A)^{-1}y \in C \tag{3.75}$$

for every $\lambda > 0$ and all $x, y \in H$ such that $x - y \in C$.

Then for each u_0 in $D(A)$ and any f in $W^{1,2}(0, T; H)$, there exists at least a function $u \in W^{1,2}(0, T; H) \cap L^\infty(0, T; V)$ satisfying

$$\left(\frac{du}{dt}(t) + Au(t) - f(t), \frac{du}{dt}(t) - v\right) \leqslant 0, \text{ a.e. } on\ ]0, T[,\ \forall v \in C. \tag{3.76}$$

$$u(0) = u_0, \quad \frac{du}{dt}(t) \in C, \text{ a.e. } on\ ]0, T[. \tag{3.77}$$

$$Au \in L^2(0, T; H). \tag{3.78}$$

Proof. One applies Theorem 3.5 where $\varphi(u) = \frac{1}{2}|u|^2 + I_C(u)$.

3.4 Applications

In this section we shall present several examples of unilateral problems of physical interest which can be formulated as evolution variational inequations. These examples are chosen mainly to illustrate how the theory is applicable in some important cases. We shall omit all discussions concerning physical motivation for the problems studied here. Comments on those aspects of these problems, along with further solved examples, can be found in the book by G. Duvaut and T. L. Lions [1] and the survey of J. L. Lions [5] (see also J. L. Lions [3]).

As usual Ω will denote a bounded open subset of R^n having a sufficiently smooth boundary Γ. We set $Q =]0, T[\times \Omega$ and $\Sigma =]0, T[\times\Gamma$.

Example 1. Let $\psi \in H^2(\Omega)$ be given such that $\psi(x) \leqslant 0$ a.e. on Γ. We shall now be concerned with the following unilateral problem

$$\begin{aligned} & u(t, x) \geqslant \psi(x) && \text{a.e. } x \in Q, \\ & \frac{du}{dt} = \Delta u + f && \text{on } \{(t, x) \in Q;\ u > \psi\}, \\ & \frac{du}{dt} \geqslant \Delta u + f && \text{on } \{(t, x) \in Q;\ u = \psi\}, \\ & u(t, x) = 0 && \text{on } \Sigma \\ & u(0, x) = u_0(x) && \text{a.e. } x \in \Omega. \end{aligned} \tag{3.79}$$

COROLLARY 3.4. *Let* $u_0 \in L^2(\Omega)$ *and* $f \in L^2(Q)$ *be given such that* $u_0(x) \geqslant \psi(x)$ a.e. *on* Ω. *Then problem* (3.79) *has a unique solution* $u \in C(0, T; L^2(\Omega))$ *satisfying*

$$\sqrt{t}\ \frac{\partial u}{\partial t} \in L^2(Q),\ u \in L^2(0, T;\ H_0^1(\Omega)).$$

Suppose further that $f(t)$ *is absolutely continuous from* $[0, T]$ *into* $L^2(\Omega)$. *Then*

$$t\ \frac{\partial u}{\partial t} \in L^\infty(0, T; L^2(\Omega)).$$

Finally, if $u_0 \in H_0^1(\Omega)$ *and* $f \in L^2(Q)$ *then*

$$u \in L^2(0, T;\ H_0^1(\Omega) \cap H^2(\Omega)).$$

Proof. We apply Corollary 3.1 (respectively Corollary 3.2) where $V = H_0^1(\Omega)$, $H = L^2(\Omega)$, $K = \{u \in K_0^1(\Omega);\ u(x) \geqslant \psi(x)$ a.e. $x \in \Omega\}$ and $A = -\Delta$ is defined from $H_0^1(\Omega)$ into $H^{-1}(\Omega)$ by

$$(Au, v) = a(u, v) = \int_\Omega \operatorname{grad} u \cdot \operatorname{grad} v \mathrm{d}x;\ u, v \in H_0^1(\Omega).$$

Obviously

$$\overline{K}^H = \{u \in L^2(\Omega);\ u \geqslant \psi,\ \text{ a.e. in } \Omega\}.$$

We shall verify condition (b) in Corollary 3.2 where $h = -\Delta\psi$. Let u be arbitrary in K and let $u_\varepsilon \in H_0^1(\Omega) \cap H^2(\Omega)$ be the solution of the Dirichlet problem

$$u_\varepsilon - \varepsilon\Delta u_\varepsilon = u - \varepsilon\Delta\psi \quad \text{on } \Omega$$

$$u_\varepsilon = 0 \quad \text{on } \Gamma.$$

We set $v_\varepsilon = u_\varepsilon - \psi$. Clearly, $v_\varepsilon - \varepsilon\Delta v_\varepsilon = u - \psi \geqslant 0$ a.e. on Ω and $v_\varepsilon \geqslant -\psi \geqslant 0$ on Γ. By using the maximum principle we deduce that $v_\varepsilon \geqslant 0$ a.e. on Ω. Thus, we have proved that $u_\varepsilon \in K$.

Consequently, for every $u_0 \in \overline{K}^H$ and any $f \in L^2(Q)$ there exists a unique function $u \in C(0, T; L^2(\Omega))$ such that $u(t, x) \geqslant \psi(x)$ a.e. on Q, $\sqrt{t}\ \dfrac{\partial u}{\partial t} \in L^2(Q)$ and satisfying the inequality

$$a(u(t), u(t) - v) + \int_\Omega \left(\frac{\partial u}{\partial t}(t, x) - f(t, x)\right)(u(t, x) - v(x))\,\mathrm{d}x \leqslant 0 \qquad (3.80)$$

for all $v \in K$ and almost everywhere on $]0, T[$.

Moreover, if $u_0 \in V = H_0^1(\Omega)$, then in view of Corollary 3.2, $u \in L^2(0, T; H_0^1(\Omega) \cap H^2(\Omega))$.

Now our task is to interpret inequality (3.80). Let $t_0 \in]0, T[$ be a differentiability point for the function $u(t)\colon [0, T] \to L^2(\Omega)$. We set

$$\Omega_0 = \{x \in \Omega;\ u(t_0, x) > \psi(x)\}$$

where the inequality $u > \psi$ is taken in the sense of distributions on Ω. Thus Ω_0 is open and for every $\beta \in \mathscr{D}(\Omega_0)$ there exists $\rho > 0$ such that

$$u(t_0, x) + \rho\beta(x) \geqslant \psi(x), \quad \text{a.e. } x \in \Omega.$$

Taking $v = u \pm \rho\beta$ in inequality (3.80), we get

$$a(u(t_0), \beta) + \int_\Omega \left(\frac{\partial u}{\partial t}(t_0, x) - f(t_0, x)\right)\beta(x)\,\mathrm{d}x = 0.$$

Formally, this means that in the sense of distributions, $u(t_0)$ satisfies

$$\frac{\partial u}{\partial t}(t_0) - \Delta u(t_0) = f(t_0).$$

Next, we take $v = u + \beta$ where $\beta \in \mathscr{D}(\Omega_0)$ and $\beta \geqslant 0$ on Ω_0. One obtains

$$a(u(t_0), \beta) + \int_\Omega \left(\frac{\partial u}{\partial t}(t_0, x) - f(t_0, x) \right) \beta(x) \, \mathrm{d}x \geqslant 0.$$

Hence

$$\frac{\partial u}{\partial t}(t_0) - \Delta u(t_0) \geqslant f(t_0) \quad \text{in } \mathscr{D}'(\Omega).$$

Let us formally define

$$\Omega_t = \{x \in \Omega;\ u(t, x) = \psi(x)\}.$$

More precisely, Ω_t is defined as the complement in Ω of the set Ω_0, since we do not know what $u(x) = \psi(x)$ means for two functions in $H^1(\Omega)$. Then we have shown that there is a nonnegative distribution $\mu(t)$ with supp $\mu(t) \subset \Omega_t$ such that

$$\frac{\partial u}{\partial t} - \Delta u = \mu(t), \quad \text{a.e.} \quad \text{on }]0, T[.$$

In this generalised sense, the solution $u(t, x)$ of the variational inequality (3.80) can be regarded as a solution of problem (3.79).

Now we shall consider a problem related to (3.79), namely,

$$\begin{aligned} &\frac{\partial u}{\partial t}(t, x) \geqslant \psi(x), \quad \text{a.e.} \quad (t, x) \in Q, \\ &\frac{\partial u}{\partial t} = \Delta u + f \quad \text{on } \left\{(t, x);\ \frac{\partial u}{\partial t} > \psi\right\}, \\ &\frac{\partial u}{\partial t} \geqslant \Delta u + f \quad \text{on } \left\{(t, x);\ \frac{\partial u}{\partial t} = \psi\right\}, \\ &u(t, x) = 0 \qquad \text{on }]0, T[\times \Gamma, \\ &u(0, x) = u_0(x), \quad \text{a.e.} \quad x \in \Omega, \end{aligned} \tag{3.81}$$

where $\psi \in H^2(\Omega)$ and $\psi(x) \leqslant 0$, a.e. on Γ.

Applying Theorem 3.3 one obtains in much the same manner as in the preceding corollary the following existence result.

COROLLARY 3.5. *We are given $u_0 \in H_0^1(\Omega)$ and $f \in L^2(0, T; H^{-1}(\Omega))$ such that $\frac{\partial f}{\partial t} \in L^2(0, T; H^{-1}(\Omega))$. Then problem (3.81) has a unique solution $u \in C(0, T; H_0^1(\Omega))$ satisfying*

$$\frac{\partial u}{\partial t} \in L^2(0, T; L^2(\Omega)). \quad \sqrt{t}\ \frac{\partial u}{\partial t} \in L^2(0, T; H_0^1(\Omega)). \tag{3.82}$$

Moreover, if $\frac{\partial^2 f}{\partial t^2} \in L^2(0, T; H^{-1}(\Omega))$, then

$$t\ \frac{\partial u}{\partial t} \in L^\infty(0, T; H_0^1(\Omega)). \tag{3.83}$$

In addition, if $f(0) + \Delta u_0 \in H_0^1(\Omega)$ and

$$f(0, x) + \Delta u_0(x) \geqslant \psi(x), \quad \text{a.e.} \quad x \in \Omega, \tag{3.84}$$

then

$$\frac{\partial u}{\partial t} \in L^\infty(0, T; H_0^1(\Omega)), \quad \frac{\partial^2 u}{\partial t^2} \in L^2(0, T; L^2(\Omega)). \tag{3.85}$$

Now we consider the quasi-stationary problem

$$\begin{aligned}
&\frac{\partial u}{\partial t}(t, x) \geqslant 0, \quad \text{a.e.} \quad (t, x) \in Q \\
&\Delta u + f = 0 \quad \text{in} \left\{(t, x); \ \frac{\partial u}{\partial t} > 0\right\} \\
&\Delta u + f \leqslant 0 \quad \text{in} \left\{(t, x); \ \frac{\partial u}{\partial t} = 0\right\} \\
&u(t, x) = 0, \quad \text{a.e.} \quad (t, x) \in \Sigma \\
&u(0, x) = u_0(x), \quad \text{a.e.} \quad x \in \Omega.
\end{aligned} \tag{3.86}$$

COROLLARY 3.6. *Let $u_0 \in H_0^1(\Omega)$ and $f \in L^2(0, T; H^{-1}(\Omega))$ be such that*

$$\frac{\partial f}{\partial t}, \ \frac{\partial^2 f}{\partial t^2} \in L^2(0, T; H^{-1}(\Omega))$$

and

$$f(0, x) + \Delta u_0 \leqslant 0 \ \textit{in} \ \mathscr{D}'(\Omega). \tag{3.87}$$

Then problem (3.86) *has a unique solution* $u \in C(0, T; H_0^1(\Omega))$ *such that* $\frac{\partial u}{\partial t} \in L^\infty(0, T; H_0^1(\Omega))$.

Proof. We apply Theorem 3.4 where $V = H_0^1(\Omega)$, $H = L^2(\Omega)$, $K = \{u \in H_0^1(\Omega), u(x) \geqslant 0, \text{ a.e. } x \in \Omega\}$ and A defined as above. We conclude the proof by observing that the compatibility assumptions (3.87) imply condition $f(0) + \Delta u_0 \in R(\partial I_K)$ needed in Theorem 3.4.

Example 2. We consider now the following variational problem

$$\frac{\partial u}{\partial t} - \Delta u = f \text{ in } Q =]0, T[\times \Omega$$

$$\left|\frac{\partial u}{\partial n}\right| \leqslant h, \; u\frac{\partial u}{\partial n} + h\,|u| = 0, \quad \text{a.e.} \quad \text{on }]0, T[\times \Gamma \tag{3.88}$$

$$u(0, x) = u_0(x), \quad \text{a.e.} \quad \text{on } \Omega$$

where h is a positive constant.

COROLLARY 3.7. *We are given* $u_0 \in L^2(\Omega)$ *and* $f \in L^2(Q)$. *Then, problem* (3.88) *has a unique solution* $u \in C(0, T; L^2(\Omega))$ *such that*

$$\sqrt{t}\,\frac{\partial u}{\partial t} \in L^2(0, T; L^2(\Omega)), \; u(t) \in L^2(0, T; H^1(\Omega)). \tag{3.89}$$

In addition, if $\frac{\partial f}{\partial t} \in L^1(0, T; L^2(\Omega))$, *then* $t\frac{\partial u}{\partial t} \in L^\infty(0, T; L^2(\Omega))$.

Proof. Let ε be positive and arbitrarily small. We apply Theorem 3.1 where $V = H^1(\Omega)$, $H = L^2(\Omega)$,

$$\varphi(u) = h \int_\Gamma |u| \, d\sigma, \; u \in H^1(\Omega)$$

and A is defined on $H^1(\Omega)$ by

$$(Au, v) = a(u, v) + \varepsilon(u, v), \quad \forall u, \; v \in H^1(\Omega),$$

where $a(u, v) = \int_\Omega \operatorname{grad} u \operatorname{grad} v \, dx$.

Therefore, there exists a unique function $u \in C(0, T; H^1(\Omega))$ such that $u(0) = u_0$, $u(t) \in H^1(\Omega)$, a.e. $t \in]0, T[$ and

$$a(u(t), u(t) - v) + \left(\frac{\partial u}{\partial t}(t) + \varepsilon u(t) - f(t), u(t) - v\right) + \varphi(u(t)) - \\ - \varphi(v) \leqslant 0 \tag{3.90}$$

a.e. $t \in]0, T[$ and for all $v \in H^1(\Omega)$. Taking $v = \pm \rho v$ one obtains

$$\pm \rho \left(\left(\frac{\partial u}{\partial t} + \varepsilon u - f, v \right) + a(u, v) \right) + \rho\varphi(v) \geqslant \varphi(u) + a(u, u) + $$

$$+ \left(\frac{\partial u}{\partial t} + \varepsilon u - f, u \right),$$

and therefore,

$$\left(\frac{\partial u}{\partial t}(t) + \varepsilon u(t) - f(t), u(t) \right) + a(u(t), u(t)) + \varphi(u(t)) = 0, \text{ a.e. } t \in]0, T[$$

$$\left| \left(\frac{\partial u}{\partial t}(t) + \varepsilon u(t) - f(t), v \right) + a(u(t), v) \right| \leqslant \varphi(v), \text{ a.e. } t \in]0, T[\tag{3.91}$$

for any $v \in H^1(\Omega)$.

Consequently, almost everywhere on $]0, T[$, we have

$$\left(\frac{\partial u}{\partial t}(t) + \varepsilon u(t), v \right) + a(u(t), v) = (f(t), v), \quad \forall v \in C_0^\infty(\Omega). \tag{3.92}$$

In other words,

$$\frac{\partial u}{\partial t}(t) + \varepsilon u(t) - \Delta u(t) = f(t) \text{ in } \mathscr{D}'(\Omega). \tag{3.93}$$

Since $\dfrac{\partial u}{\partial t} \in L^2(\Omega)$, a.e. on $]0, T[$ we may conclude that $\Delta u \in L^2(\Omega)$ a.e. on $]0, T[$. In particular, this implies that $\dfrac{\partial u}{\partial n} \in H^{-1/2}(\Gamma)$, a.e. on $]0, T[$. Now we take the inner product of Eq. (3.93) with $v \in H^1(\Omega)$. By Green's formula we obtain

$$\left(\frac{\partial u}{\partial t} + \varepsilon u - f, v \right) + a(u, v) - \int_\Gamma v \frac{\partial u}{\partial n} \, d\sigma = 0, \quad \text{a.e. on }]0, T[$$

where as usual $\int_\Gamma v \dfrac{\partial u}{\partial n} \, d\sigma$ denotes the value of $\dfrac{\partial u}{\partial n} \in H^{-1/2}(\Gamma)$ in $v|_\Gamma \in$ $\in H^{1/2}(\Gamma)$.

But relation (3.91) implies that

$$\left| \int_\Gamma v \frac{\partial u}{\partial n} \, d\sigma \right| \leqslant h \int_\Gamma |v| \, d\sigma, \quad \text{a.e. } t \in]0, T[$$

Hence

$$\int_{\Gamma}\left(u\frac{\partial u}{\partial n}+h\,|\,u\,|\right)\mathrm{d}\sigma=0,\quad \text{a.e.}\quad t\in\,]0,T[.$$

Making use of these relations, we conclude that u satisfies

$$u\frac{\partial u}{\partial n}+h\,|\,u\,|=0,\quad \left|\frac{\partial u}{\partial n}\right|\leqslant h,\quad \text{a.e.}\quad \text{on } \Sigma. \tag{3.94}$$

Let us denote by u_ε the solution of problems (3.93) and (3.94). Equivalently,

$$\frac{\mathrm{d}u_\varepsilon}{\mathrm{d}t}+Au_\varepsilon+\varepsilon u_\varepsilon+\partial\varphi(u_\varepsilon)\ni f,\quad \text{a.e.}\quad t\in\,]0,T[. \tag{3.95}$$

Multiplying formula (3.95) by u_ε and integrating over $]0, t[$, we see that

u_ε remain in a bounded set of $L^2(Q)\cap L^2(0,T;H^1(\Omega))$.

$\sqrt{t}\,\dfrac{\partial u_\varepsilon}{\mathrm{d}t}$ remain in a bounded set of $L^2(Q)$

$\lim\limits_{\varepsilon\to 0}\varepsilon\, u_\varepsilon=0$ in $L^2(Q)$.

Taking weakly convergent subsequences we obtain a solution u for the equation

$$\frac{\mathrm{d}u}{\mathrm{d}t}+Au+\partial\varphi(u)\ni f$$

$$u(0)=u_0$$

which as seen before satisfies problem (3.88) in a certain generalized sense. The details have been omitted.

Example 3. Let γ be a maximal monotone graph in $R^1\times R^1$ such that $\overline{D(\gamma)}=R$. In Corollary 3.3 we take

$$H=L^2(\Omega).\ V=H^1(\Omega),\ C=\{u\in L^2(\Omega);\ u(x)\geqslant 0,\ \text{a.e. on } \Omega\}$$

and $A=\partial\psi$ is defined by

$$Au = -\Delta u \quad \text{for } u \in D(A)$$

where

$$D(A) = \left\{ u \in H^2(\Omega);\ \frac{\partial u}{\partial n} \in -\gamma(u), \quad \text{a.e.} \quad \text{on } \Gamma \right\}.$$

We recall that

$$\psi(u) = \begin{cases} \dfrac{1}{2} \displaystyle\int_\Omega |\operatorname{grad} u|^2 \, dx + \int_\Gamma j(u)\, d\sigma, & \text{if } u \in H^1(\Omega) \text{ and } j(u) \in L^1(\Gamma) \\ +\infty & \text{otherwise.} \end{cases}$$

where $\partial j = \gamma$. Assumptions (i) and (iii) follow easily from the definition of ψ. Let us verify condition (3.75). From Proposition 1.2 it is equivalent to

$$((\partial I_C)_\lambda(u - v),\ \partial\psi(u) - \partial\psi(v)) \geqslant 0 \text{ for all } u, v \in D(\partial\psi) \tag{3.96}$$

while

$$(\partial I_C)_\lambda = \frac{1}{\lambda}(1 - P_C)$$

where P_C is the projection mapping of H into C. Noting that $P_C u = u^+ = \max\{u, 0\}$, we observe that formula (3.96) can be written as

$$\int_\Omega (u - v)^- \Delta(u - v)\, dx \leqslant 0 \quad \text{for all } u, v \in D(A)$$

where $u^- = \inf\{u, 0\}$. But this inequality follows immediately by using Green's formula and the monotonicity of γ.

The general theory can therefore be applied to the present situation.

COROLLARY 3.8. *Let $u_0 \in H^2(\Omega)$ and $f \in L^2(Q)$ be such that*

$$\frac{\partial f}{\partial t} \in L^2(Q), \quad \frac{\partial u_0}{\partial n} \in -\gamma(u_0), \quad \text{a.e.} \quad \text{on } \Gamma. \tag{3.97}$$

Then there exists at least a function u such that

$$u \in C(0, T; L^2(\Omega)) \cap L^2(0, T; H^2(\Omega)). \tag{3.98}$$

$$u \in L^\infty(0, T; H^1(\Omega)), \quad \frac{\partial u}{\partial t} \in L^2(Q). \tag{3.99}$$

$$\int_\Omega \left(\frac{\partial u}{\partial t} - \Delta u - f\right)\left(v - \frac{\partial u}{\partial t}\right) \mathrm{d}x \geqslant 0, \quad \text{a.e. } \textit{on }]0, T[, \quad \forall\, v \in C.$$

$$-\frac{\partial u}{\partial n} \in \gamma(u) \qquad \textit{on } \Sigma. \tag{3.100}$$

$$u(0, x) = u_0(x) \qquad \textit{on } \Omega.$$

$$\frac{\partial u}{\partial t} \geqslant 0 \qquad \text{a.e. } \textit{on } Q.$$

By the same procedure as in *Example* 1, it follows that $u(t)$ satisfies in a generalized sense the unilateral problem

$$\frac{\partial u}{\partial t} - \Delta u = f \quad \text{on} \quad \left\{(t, x);\ \frac{\partial u}{\partial t}(t, x) > 0\right\}$$

$$\frac{\partial u}{\partial t} - \Delta u \geqslant f \quad \text{on} \quad \left\{(t, x);\ \frac{\partial u}{\partial t}(t, x) = 0\right\}$$

$$\frac{\partial u}{\partial t} \geqslant 0 \qquad \text{on } Q$$

$$\frac{\partial u}{\partial n} \in -\gamma(u) \quad \text{on } \Sigma.$$

$$u(0, x) = u_0(x) \quad \text{on } \Omega.$$

§ 4. Nonlinear Volterra equations with positive kernels in Hilbert spaces

This paragraph concerns integral equations of the form

$$u(t) + \int_0^t a(t - s)Au(s)\, \mathrm{d}s \ni f(t), \quad 0 \leqslant t < \infty \tag{4.1}$$

on a Hilbert space H. Here $a(t)$ is a prescribed real function on R^+ while A is a monotone (multivalued) operator in H.

Equations of this type arise in the study of elastic materials with memory and in several problems in continuum mechanics (see B.D. Coleman and M.E. Gurtin [1], and R.C. MacCamy [1]).

The results which we shall give below will be divided into two categories: some of them are asymptotic stability criteria for solutions (the existence of which is assumed) and others ensure the existence of global solutions under corresponding conditions. The method, while technically different from that used in the preceeding paragraphs, follows roughly the same pattern.

4.1. Positive kernels

This section provides some auxiliary results concerning the integral operator

$$(Lu)(t) = \int_0^t a(t-s)u(s)\,\mathrm{d}s, \quad u \in L^2_{\mathrm{loc}}(0, \infty; H) \tag{4.2}$$

where $a \in L^1_{\mathrm{loc}}(0, \infty; H)$. We use the symbol $L^p_{\mathrm{loc}}(0, \infty; H)$, $1 \leqslant p < \infty$ to denote the space consisting of all strongly measurable functions $u :]0, \infty[\to H$, with the property that $u \in L^p(0, T; H)$ for every $0 < T < \infty$.

It is well known that under the above assumption, for every $0 < T < \infty$, the operator L is continuous from $L^2(0, T; H)$ into itself.

DEFINITION 4.1. The function a is called *positive kernel* if the operator L is positive on $L^2(0, T; H)$ for all T. In other words,

$$\int_0^T (Lu(t), u(t))\,\mathrm{d}t \geqslant 0 \text{ for all } u \in L^2(0, T; H) \text{ and every } T > 0. \tag{4.3}$$

Then a is called a *strongly positive kernel* if there exist constants $\varepsilon > 0$, $\alpha > 0$ such that $a(t) - \varepsilon e^{-\alpha t}$ is a positive kernel.

Let $\hat{a}(\lambda)$ be the *Laplace transform* of $a(t)$, i.e.,

$$\hat{a}(\lambda) = \int_0^\infty \mathrm{e}^{-\lambda t} a(t)\,\mathrm{d}t, \quad \lambda \in C. \tag{4.4}$$

It is well known that positivity is closely connected to the Laplace transform of a. In the sequel we shall use the following positivity criterion.

LEMMA 4.1. *Let* $a \in L^\infty(0, \infty)$ *be such that*

$$Re\,\hat{a}(\lambda) > 0 \quad in\ Re\,\lambda > 0. \tag{4.5}$$

Then $a(t)$ defines a positive kernel.

Proof. Let $T > 0$ be arbitrary but fixed. For any $u \in L^2(0, T; H)$ define

$$u_T(t) = \begin{cases} u(t) & \text{if } t \in]0, T[\\ 0 & \text{otherwise.} \end{cases}$$

Let $\varepsilon > 0$. We have

$$\int_0^T e^{-2\varepsilon t}(Lu(t), u(t))\, dt = \int_0^\infty (L_\varepsilon(e^{-\varepsilon t}u_T(t)), e^{-\varepsilon t}u_T(t)\, dt$$

where

$$(L_\varepsilon v)(t) = \int_0^t e^{-\varepsilon(t-s)}\, a(t-s)\, v(s)\, ds.$$

One sees that

$$\widehat{L_\varepsilon(e^{-\varepsilon t}u_T)}\,(i\tau) = \hat{a}(\varepsilon + i\tau)\, \hat{u}_T(\varepsilon + i\tau), \quad \forall \tau \in R^1.$$

Thus the Parseval equality leads to

$$\int_0^T e^{-2\varepsilon t}(Lu, u)\, dt = \frac{1}{2\pi}\int_{-\infty}^{\infty} Re\, \hat{a}(\varepsilon + i\tau)\, |\hat{u}_T(\varepsilon + i\tau)|^2\, d\tau.$$

Hence

$$\int_0^T e^{-2\varepsilon t}(Lu, u)\, dt \geqslant 0 \quad \text{for any } \varepsilon > 0 \tag{4.6}$$

which clearly implies that $\int_0^T (Lu, u)\, dt \geqslant 0$ as claimed.

Remark. If $a \in L^1(0, \infty)$ then the argument just used in the proof of Lemma 3.1 shows that the positivity of kernel $a(t)$ is implied by the following condition

$$Re\, \hat{a}(i\tau) \geqslant 0 \quad \text{for all } \tau \in R^1. \tag{4.7}$$

We are ready now to discuss strong positivity.

PROPOSITION 4.1. *Let $a(t)$ satisfy the following conditions:*

(i) $a \in C([0, \infty[) \cap C^2(]0, \infty[)$.

(ii) $(-1)^k \dfrac{d^k}{dt^k} a(t) \geqslant 0$ *for* $t > 0$, $k = 0, 1, 2$.

(iii) $a(t) \not\equiv$ *constant*.

Then $a(t)$ defines a strongly positive kernel.

Proof. We first note that (ii) implies that $\lim_{t\to\infty} a(t) = a(\infty)$ exists and is greater or equal to zero, and that $b(t) = a(t) - a(\infty)$ again satisfies assumptions (i) ~ (iii). To prove that $a(t)$ defines a strongly positive kernel, it suffices to show that there exist constants $\varepsilon, \alpha > 0$ such that

$$Re\left(\hat{b}(\lambda) - \frac{\varepsilon}{\lambda + \alpha}\right) > 0, \quad \text{on } Re\,\lambda > 0. \tag{4.8}$$

We observe that for $Re\,\lambda > 0$,

$$\hat{b}(\lambda) = \frac{b(0)}{\lambda} + \frac{1}{\lambda}\int_0^\infty e^{-\lambda t}\frac{db}{dt}\,dt.$$

Assumption (ii) implies that $\dfrac{db}{dt} \in L^1(0, \infty)$ and therefore, $\hat{b}(\lambda)$ exists for $Re\,\lambda \geqslant 0$, $\lambda \neq 0$. Moreover, in view of (ii) and (iii)

$$Re\,\hat{b}(i\eta) = \int_0^\infty b(t)\cos\eta t\,dt > 0 \quad \text{for all } \eta \neq 0.$$

Thus if we can show that $Re\,\hat{b}(i\eta) \geqslant C/\eta^2$ for all $|\eta|$ sufficiently large, then it will follow that we can choose α and ε so that relation (4.8) holds for all $Re\,\lambda = 0$.
Then the maximum principle implies that relation (4.8) holds for $Re\,\lambda \geqslant 0$.

Summarizing to this point, we need to show that

$$Re\,\hat{b}(i\eta) \geqslant C/\eta^2, \quad C > 0 \tag{4.9}$$

for all $|\eta| \geqslant N$, where N is some large number. Without loss of generality, we need only to consider $\eta \geqslant N$. We note that

$$Re\,\hat{b}\,(i\eta) = \int_0^\infty b(t)\cos\eta t\,dt = \frac{1}{\eta}\int_0^\infty b\left(\frac{t}{\eta}\right)\cos t\,dt. \tag{4.10}$$

In order to establish relation (4.9) it suffices to show that the last integral in Eq. (4.10) is nondecreasing as η increases. Integrating by parts, we have

$$\int_0^\infty b\left(\frac{t}{\eta}\right)\cos t\,dt = -\int_0^\infty \frac{d}{dt}\,b(t)\sin t\eta\,dt,$$

which is uniformly convergent because $\dfrac{d}{dt}\,b \in L^1(0, \infty)$. Differentiating under integral sign, we obtain

$$\frac{d}{d\eta}\int_0^\infty b\left(\frac{t}{\eta}\right)\cos t\,dt = -\int_0^\infty t\,\frac{d}{dt}\,b(t)\cos\eta t\,dt \geqslant 0 \quad \text{for all } \eta > N. \tag{4.11}$$

The above calculation will be justified if the integral on the right of Eq. (4.11) is uniformly convergent. This is in fact true because

$$-\int_0^\infty t \frac{d}{dt} b(t) \cos \eta t \, dt = \frac{1}{\eta} \left[\int_0^\infty \frac{d}{dt} b(t) \sin \eta t \, dt + \right.$$

$$\left. + \int_0^\infty t \sin \eta t \frac{d^2}{dt^2} b(t) \, dt \right]$$

and $t \dfrac{d^2}{dt^2} b \in L^1(0, \infty)$ in consequence of assumption (ii). This completes the proof.

Remark. Conditions (i) and (ii) are stronger than really needed. It turns out that the conclusions of Proposition 4.1 hold if we merely assume that: (a) $a \in C^1(0, \infty) \cap L^1(0, 1)$ (b) $a(t) \geqslant 0$, $\dfrac{d}{dt} a(t) \leqslant 0$ (c) $\dfrac{d}{dt} a(t)$ is non-decreasing (d) $\dfrac{da}{dt}(t) \not\equiv 0$. The paper by R.C MacCamy and J. S. Wong [1] contains the proof of Proposition 4.1 under these general conditions as well as other positivity results.

4.2. Equation (4.1) with $A = \partial\varphi$

We shall discuss in this section Eq. (4.1) when

$$A = \partial\varphi \tag{4.12}$$

φ being a lower-semicontinuous, proper and convex function from H into $]-\infty, +\infty]$. Concerning $a(t)$, we will make convenient positivity assumptions below.

(i) *$a(t)$ is continuous on* $[0, \infty[$ *and absolutely continuous on every compact interval of* $]0, \infty[$. *Moreover,*

$$(-1)^k \frac{d^k}{dt^k} a(t) \geqslant 0, \qquad \text{a.e. } on \ ]0, \infty[, \quad k = 0, 1. \tag{4.13}$$

(ii) *$a(t)$ is a positive kernel*, i.e., $Re\, \hat{a}(\lambda) > 0$ *on* $Re\, \lambda > 0$.

A function $u \in L^2_{loc}(0, \infty; H)$ is said to be a solution of Eq. (4.1) if there exists a function $w \in L^2_{loc}(0, \infty; H)$ such that

$$w(t) \in D(A), \quad \text{a.e. on } t > 0. \tag{4.14}$$

$$u(t) + \int_0^t a(t-s)\, w(s)\, ds = f(t), \quad \text{a.e. } t > 0. \tag{4.15}$$

For simplicity we shall write Au instead of w.
We can state the following result.

THEOREM 4.1. *Let* $A = \partial\varphi$ *and* $a(t)$ *satisfy the above assumptions. We are given* $f \in W^{1,2}_{\text{loc}}(0, \infty; H)$ *such that*

$$f(0) \in D(\varphi). \tag{4.16}$$

Then there exists at least one solution $u(t)$ *of Eq.* (4.1) *such that*

$$u \in W^{1,2}_{\text{loc}}(0, \infty; H) \tag{4.17}$$

$$t \to \varphi(u(t)) \text{ is absolutely continuous on every } [0, T]. \tag{4.18}$$

If A *is strictly monotone, then the solution is unique.*

Remark. Before proceeding to the proof of Theorem 3.1 we shall point out that $W^{1,2}_{\text{loc}}(0, \infty; H)$ denotes the local space

$$W^{1,2}_{\text{loc}}(0, \infty; H) = \left\{ u \in L^2_{\text{loc}}(0, \infty; H);\ \frac{\mathrm{d}u}{\mathrm{d}t} \in L^2_{\text{loc}}(0, \infty; H) \right\}$$

where $\frac{\mathrm{d}}{\mathrm{d}t}$ is taken in the sense of H-valued vectorial distributions on $]0, \infty[$.
In other words, $W^{1,2}_{\text{loc}}(0, \infty; H)$ is the space of all functions $u\colon [0, \infty] \to H$ with the property that $u(t)$ is absolutely continuous on every finite interval $[0, T]$, and $\frac{\mathrm{d}u}{\mathrm{d}t}$ belongs to $L^2_{\text{loc}}(0, \infty; H)$.

Proof. Let $\varphi_\lambda\colon H \to]-\infty, +\infty]$ be the regularized convex function associated with φ, i.e.

$$\varphi_\lambda(u) = \inf \left\{ \frac{|u - v|^2}{2\lambda} + \varphi(v);\ v \in H \right\}, \quad \lambda > 0.$$

We recall (see Theorem 2.2, in Chapter II) that φ_λ is Fréchet differentiable on H and $\partial\varphi_\lambda = A_\lambda = \lambda^{-1}(1 - (1 + \lambda A)^{-1})$. Inasmuch as A_λ is Lipschitzian on H, the approximating equation

$$u_\lambda(t) + \int_0^t a(t - s)\, A_\lambda u_\lambda(s)\, \mathrm{d}s = f(t), \quad 0 \leqslant t < \infty \tag{4.19}$$

has a unique solution $u_\lambda \in L^2_{\text{loc}}(0, \infty; H)$. Since $f \in W^{1,2}_{\text{loc}}(0, \infty; H)$ and $\frac{\mathrm{d}a}{\mathrm{d}t} \in L^1_{\text{loc}}(0, \infty; H)$, there results that $u \in W^{1,1}_{\text{loc}}(0, \infty; H)$. By differentiating both sides of Eq. (4.19) we get

$$u_\lambda'(t) + a(0)A_\lambda u_\lambda(t) + \int_0^t a'(t-s)\, A_\lambda u_\lambda(s) = f'(t), \text{ a.e. on }]0, \infty[. \quad (4.20)$$

where $'$ stands for $\frac{\mathrm{d}}{\mathrm{d}t}$. Multiply Eq. (4.20) by $A_\lambda u_\lambda$ and use the relation

$$\frac{\mathrm{d}}{\mathrm{d}t}\varphi_\lambda(u_\lambda(t)) = (A_\lambda u_\lambda(t), u_\lambda'(t)), \quad \text{a.e. } t > 0.$$

We get

$$\frac{\mathrm{d}}{\mathrm{d}t}\varphi_\lambda(u_\lambda(t)) + a(0)\,|A_\lambda u_\lambda(t)|^2 \leqslant |f'(t)|\,|A_\lambda u_\lambda(t)| +$$

$$+ |A_\lambda u_\lambda(t)| \int_0^t |a'(t-s)|\,|A_\lambda u_\lambda(s)|\,\mathrm{d}s,$$

and one deduces that

$$\varphi_\lambda(u_\lambda(t)) + a(0)\int_0^t \left|A_\lambda u_\lambda(s)\right|^2 \mathrm{d}s \leqslant \varphi_\lambda(f(0)) + \int_0^t |f'(s)|\,|A_\lambda u_\lambda(s)|\,\mathrm{d}s +$$
$$(4.21)$$
$$+ \int_0^t |A_\lambda u_\lambda(s)|\,\mathrm{d}s \int_0^s |a'(s-\tau)|\,|A_\lambda u_\lambda(\tau)|\,\mathrm{d}\tau, \quad t \geqslant 0.$$

Since the L^2-norm on $]0, t[$ of $\int_0^s |a'(s-\tau)|\,|A_\lambda u_\lambda(\tau)|\,\mathrm{d}\tau$ is majorized by $\int_0^t |a'(\tau)|\,\mathrm{d}\tau \left(\int_0^t |A_\lambda u_\lambda(\tau)|^2\,\mathrm{d}\tau\right)^{1/2}$, it follows by Schwarz's inequality that

$$\int_0^t |A_\lambda u_\lambda(s)|\,\mathrm{d}s \int_0^s |a'(s-\tau)|\,|A_\lambda u_\lambda(\tau)|\mathrm{d}\tau \leqslant \int_0^t |A_\lambda u_\lambda(s)|^2\,\mathrm{d}s \int_0^t |a'(s)|\,\mathrm{d}s.$$

Putting this estimate into formula (4.21), we conclude that

$$\varphi_\lambda(u_\lambda(t)) + a(t)\int_0^t |A_\lambda u_\lambda(s))|^2\,\mathrm{d}s \leqslant \varphi_\lambda(f(0)) + \int_0^t |f'(s)|\,|A_\lambda u_\lambda(s)|\,\mathrm{d}s.$$

Let $0 < T < \infty$ be such that $a(T) > 0$. Then the last estimate yields

$$\varphi_\lambda(u_\lambda(t)) + \frac{1}{2}a(t)\int_0^t |A_\lambda u_\lambda(s)|^2 \,\mathrm{d}s \leqslant \varphi(f(0)) + \frac{1}{2a(t)}\int_0^t |f'(s)|^2 \,\mathrm{d}s \tag{4.22}$$

for all $0 \leqslant t \leqslant T$ and $\lambda > 0$.
Let u_0 be any arbitrary element of $D(A)$. Then the inequality

$$\varphi_\lambda(u_\lambda) \geqslant \varphi_\lambda(u_0) - (A_\lambda u_0, u_0 - u_\lambda)$$

gives with some constants α and β,

$$\varphi_\lambda(u_\lambda(t)) \geqslant \alpha|u_\lambda(t)| + \beta \quad \forall \lambda > 0 \tag{4.23}$$

because $|A_\lambda u_0| \leqslant |A^0 u_0|$ and $\varphi_\lambda(u_0)$ are bounded.
On the other hand, applying Schwarz's inequality in Eq. (4.19), we obtain

$$|u_\lambda(t)| \leqslant |f(t)| + \varepsilon\int_0^t |A_\lambda u_\lambda(s)|^2 \,\mathrm{d}s + C_\varepsilon, \quad \text{for } t \geqslant 0$$

where ε is positive and arbitrarily small. Combining this estimate with (4.22) and (4.23), we find that

$$\varphi_\lambda(u_\lambda(t)) \text{ are uniformly bounded on } [0, T] \tag{4.24}$$

and

$$\{A_\lambda u_\lambda\} \text{ is bounded in } L^2(0, T; H).$$

$$\{u_\lambda(t)\} \text{ is bounded in } L^\infty(0, T; H).$$

Therefore, we may assume that as $\lambda \to 0$,

$$u_\lambda \to u \text{ weakly-star in } L^\infty(0, T; H) \tag{4.25}$$

$$A_\lambda u_\lambda \to g \text{ weakly in } L^2(0, T; H).$$

Let $L: L^2(0, T; H) \to L^2(0, T; H)$ be the integral operator defined by Eq. (4.2). We note that condition (ii) implies that L is positive on the space $L^2(0, T; H)$. Since Eq. (4.19) can also be written in the form

$$u_\lambda + LA_\lambda u_\lambda = f$$

from the positivity of L we find

$$\int_0^T (A_\lambda u_\lambda(t) - A_\mu u_\mu(t), u_\lambda(t) - u_\mu(t))\,\mathrm{d}t \leqslant 0, \text{ for all } \lambda, \mu > 0. \tag{4.26}$$

Since A (more precisely, its realization) is maximal monotone in $L^2(0, T; H) \times L^2(0, T; H)$, from Proposition 1.1 (Chapter II) applied to relation (4.26) we deduce that $u(t) \in D(A)$, a.e. on $]0, T[$ and

$$g(t) \in Au(t), \text{ a.e. } \quad t \in]0, T[.$$

This clearly shows that $u(t)$ is a solution of Eq. (4.1) on the interval $[0, T]$.

Inasmuch as u' and Au belong to $L^2(0, T; H)$, it follows from Lemma 2.1 that function $t \to \varphi(u(t))$ is absolutely continuous on $[0, T]$ and

$$\frac{\mathrm{d}}{\mathrm{d}t} \varphi(u(t)) = (u'(t), Au(t)), \quad \text{a.e.} \quad \text{on }]0, T[. \tag{4.27}$$

To conclude with the existence, it remains to show that $u(t)$ can be continued on $]T, \infty[$. To this purpose, consider the equation

$$v(t) + \int_0^t a(t-s)\, Av(s)\, \mathrm{d}s \ni f(t+T) - \int_0^T a(T+t-s)\, Au(s)\, \mathrm{d}s \tag{4.28}$$

on the interval $[0, T]$. Let $f_1(t)$ denote the right-hand side of Eq. (4.28). Clearly, $f_1 \in W^{1,2}(0, T; H)$ and $f_1(0) = u(T) \in D(\varphi)$. Thus, according to the first part of this proof, there exists at least a solution $v \in W^{1,2}(0, T; H)$ of Eq. (4.28). It now follows readily that function $\tilde{u}(t) : [0, 2T] \to H$ defined by

$$\tilde{u}(t) = \begin{cases} u(t) & \text{if } 0 \leqslant t \leqslant T \\ v(t-T) & \text{if } T < t \leqslant 2T \end{cases}$$

satisfies Eq. (4.1) on $[0, 2T]$. Thus, by repeatedly using this argument one obtains a solution u of Eq. (4.1) defined on the whole half-axis and satisfying the claimed conditions.

If A is strictly monotone, i.e.,

$$(Au - Av, u - v) = 0 \quad \text{iff } u = v$$

the uniqueness of the solution is an immediate consequence of the positivity of L. The theorem is thereby proved.

We shall now investigate the behaviour of the solutions for Eq. (4.1). The result we shall prove can be stated as follows.

THEOREM 4.2. *Let the assumptions of Theorem 4.1 be satisfied. Suppose furthermore that* $\dfrac{\mathrm{d}f}{\mathrm{d}t} \in L^2(0, \infty; H)$ *and*

$$a(\infty) > 0, \lim_{|u| \to \infty} \varphi(u) = \infty. \tag{4.29}$$

Then any solution $u(t)$ of Eq. (4.1) satisfies

$$|u(t)| \text{ bounded on } [0, \infty[. \tag{4.30}$$

$$\frac{du}{dt} \text{ and } Au \text{ belong to } L^2(0, \infty; H). \tag{4.31}$$

$$\lim_{t \to \infty} \varphi(u(t)) = \varphi_\infty = \min \{\varphi(v);\ v \in H\}. \tag{4.32}$$

Assume furthermore that

$$\sqrt{t}\,\frac{da}{dt} \in L^1(0, \infty),\ \varphi(f) \in L^1(0, \infty),\ \sqrt{t}\,\frac{df}{dt} \in L^2(0, \infty; H). \tag{4.33}$$

Then

$$\sqrt{t}\,\frac{du}{dt},\ \sqrt{t}\,Au \in L^2(0, \infty; H) \tag{4.34}$$

and

$$t\varphi(u(t)) \in L^\infty(0, \infty),\ \varphi(u) \in L^1(0, \infty). \tag{4.35}$$

Proof. Let u be any solution of Eq. (4.1). Obviously, $u \in W^{1,2}_{\mathrm{loc}}(0, \infty; H)$ and the following equality

$$u'(t) + a(0)\, Au(t) + \int_0^t a'(t-s)\, Au(s)\, ds = f'(t) \tag{4.36}$$

holds almost everywhere on $]0, \infty[$. Let us now multiply both sides of Eq. (4.36) by $Au(t)$ and integrate over $]0, t[$. Then using the same argument as in the proof of Theorem 4.1, one obtains the estimate

$$\varphi(u(t)) + \frac{1}{2}\,a(t) \int_0^t |Au(s)|^2\, ds \leqslant \varphi(f(0)) + \frac{1}{2a(t)} \int_0^t |f'(s)|^2\, ds, \text{ for } t > 0. \tag{4.37}$$

Thus, it follows from condition (4.29) that

$$|u(t)| \text{ is bounded on } [0, \infty[. \tag{4.38}$$

The function φ being convex and lower-semicontinuous, it is bounded from below by an affine function. Thus, using once again estimate (4.37) and assumption (ii) we conclude that

$$Au \in L^2(0, \infty; H). \tag{4.39}$$

Next from Eq. (4.36) and (4.27) we derive that

$$\frac{du}{dt} \in L^2(0, \infty; H), \quad \frac{d}{dt}\varphi(u) \in L^1(0, \infty).$$

This is in fact true because $a' \in L^1(0, \infty)$ (in consequence of assumption (ii)) and $Au \in L^2(0, \infty; H)$. We may conclude therefore, that $\lim_{t\to\infty} \varphi(u(t)) = \varphi_\infty$. But note that by definition of $\partial\varphi$,

$$\varphi(u(t)) \leqslant \varphi(v) + (Au(t), u(t) - v) \quad \text{for every } v \in H.$$

Then relations (4.38) and (4.39) imply that

$$\varphi_\infty \leqslant \varphi(v), \quad \forall v \in H$$

as claimed.

Now, suppose that condition (4.33) holds.

Multiplying Eq. (4.36) by $t\,Au(t)$ and integrating over $]0, t[$, yields

$$t\varphi(u(t)) + a(0)\int_0^t s|Au(s)|^2\,ds \leqslant \int_0^t \varphi(u(s))\,ds -$$

$$-\int_0^t\int_0^s sa'(s-\tau)\,(Au(s), Au(\tau))\,d\tau\,ds + \tag{4.40}$$

$$+\left(\int_0^t s|f'(s)|^2\,ds\right)^{1/2}\left(\int_0^t s|Au(s)|^2\,ds\right)^{1/2}, \quad \text{for } t \geqslant 0.$$

Next, we multiply Eq. (4.1) by $Au(t)$ and integrate over $]0, t[$. Since $a(t)$ defines a positive kernel, this yields

$$\int_0^t (Au(s), u(s) - f(s))\,ds \leqslant 0 \quad \text{for } t \geqslant 0$$

and therefore

$$\int_0^t \varphi(u(s))\,ds \leqslant \int_0^t \varphi(f(s))\,ds \quad \text{for every } t \geqslant 0. \tag{4.41}$$

We set

$$I(t) = -\int_0^t \int_0^s sa'(s-\tau)\,(Au(s),\ Au(\tau))\,\mathrm{d}\tau\,\mathrm{d}s$$

and

$$I_0(t) = \int_0^t s|Au(s)|^2\,\mathrm{d}s.$$

Interchanging the order of integration, which is easily justified by the above hypotheses and Fubini's theorem, and applying Schwarz's inequality yields

$$|I(t)| \leqslant -\int_0^t a'(\tau)\,\mathrm{d}\tau \left(\int_\tau^t s\,|Au(s-\tau)|^2\,\mathrm{d}s\right)^{1/2} I_0(t),$$

and an explicit calculation shows that

$$|I(t)| \leqslant (a(0)-a(t))\,I_0^2(t) + I_0(t)\left(\int_0^t |Au(s)|^2\,\mathrm{d}s\right)^{1/2} \int_0^t \sqrt{s}\,|a'(s)|\,\mathrm{d}s. \quad (4.42)$$

Putting this into (4.40) and using inequality (4.41), we obtain after some elementary calculations that

$$t\varphi(u(t)) + \int_0^t s|Au(s)|^2\,\mathrm{d}s \leqslant C \quad \text{for } t \geqslant 0$$

as claimed. This completes the proof.

Remark. The result obtained in Theorem 4.1 remains true if conditions (4.13) are replaced by

$$a(0) > 0,\ a'(t) \geqslant 0, \quad \text{a.e. on }]0, \infty[. \quad (4.43)$$

Indeed, instead of inequality (4.22), we obtain in this case

$$\varphi_\lambda(u_\lambda(t)) + \left(a(0) - \frac{1}{2}a(t)\right)\int_0^t |A_\lambda u_\lambda|^2\,\mathrm{d}s \leqslant \varphi(f(0)) +$$

$$+ \frac{1}{a(0) - \dfrac{1}{2}a(t)} \int_0^t |f'(s)|^2\,\mathrm{d}s$$

Afterwards, the proof follows the same lines as above. If we make further assumptions on a, for instance that $a(\infty) < 2a(0)$, then the conclusions of Theorem 4.2 are also valid assuming condition (4.43) instead of condition (4.13).

Before concluding this section, we shall give an example to illustrate the above results.

Let β be a maximal monotone graph in $R^1 \times R^1$ such that $0 \in D(\beta)$ and let $j: R^1 \to]-\infty, +\infty]$ be such that $\partial j = \beta$. We take $H = L^2(\Omega)$ and $\varphi : H \to]-\infty, +\infty]$ defined by

$$\varphi(u) = \begin{cases} \dfrac{1}{2}\displaystyle\int_\Omega |\text{grad}\, u|^2\, dx + \int_\Omega j(u)\, dx & \text{if } u \in H_0^1(\Omega) \text{ and } j(u) \in L^1(\Omega) \\ +\infty & \text{otherwise.} \end{cases}$$

We recall that

$$\partial\varphi(u)(x) = -\Delta u(x) + \beta(u(x)), \quad \text{a.e. on } \Omega$$

and

$$\|\Delta u\|_H \leqslant \|(\partial\varphi)^0\,(u)\|_H + C \quad \text{for every } u \in D(\partial\varphi) \tag{4.44}$$

where

$$D(\partial\varphi) = H_0^1(\Omega) \cap H^2(\Omega) \cap D(\tilde{\beta}),$$

$\tilde{\beta}$ being the realization of β in $L^2(\Omega)$. Eq. (4.1) with $A = \partial\varphi$ can be written as

$$\begin{aligned} & u(t, x) - \int_0^t a(t-s)\, \Delta u(s, x)\, ds + \int_0^t a(t-s)\, \beta(u(s, x))\, ds \ni f(t, x) \\ & \qquad\qquad\qquad\qquad\qquad\qquad \text{on }]0, \infty[\times\Omega \\ & u(t, x) = 0. \qquad\qquad\qquad\qquad\qquad \text{on } [0, \infty[\times\Gamma. \end{aligned} \tag{4.45}$$

Suppose that $a(t)$ satisfies assumptions (i) and (ii). Theorems 4.1 and 4.2 therefore yield the following result. *Given*

$$f \in L^2_{\text{loc}}(0, \infty; L^2(\Omega)), \quad \frac{\partial f}{\partial t} \in L^2(0, \infty; L^2(\Omega))$$

$$f(0,.) \in H_0^1(\Omega), \quad j(f(0,.)) \in L^1(\Omega).$$

There exists a unique solution $u(t, x)$ *for Eq.* (4.45) *satisfying:*

$$u \in L^2(0, \infty; H_0^1(\Omega) \cap H^2(\Omega)), \quad \frac{\partial u}{\partial t} \in L^2(0, \infty; L^2(\Omega)). \tag{4.46}$$

$$\|u(t,.)\|_{H_0^1(\Omega)} \text{ bounded on }]0, \infty[. \tag{4.47}$$

In particular, (4.47) implies that $\{u(t,.);\ t \geqslant 0\}$ is a compact subset of $L^2(\Omega)$. Let $u_\infty \in L^2(\Omega)$ be any limit point of $u(t,.)$ as $t \to \infty$. We note that Theorem 4.2, Eq. (4.32), implies that $u_\infty \in (\partial\varphi)^{-1}(0)$. This shows that $u_\infty \in H_0^1(\Omega) \cap H^2(\Omega)$ is the (unique) solution of the nonlinear boundary problem

$$\begin{aligned} \Delta u_\infty - \beta(u_\infty) &\ni 0 \quad \text{on } \Omega \\ u_\infty &= 0 \quad \text{on } \Gamma \end{aligned}$$

Thus we have shown that

$$\lim_{t\to\infty} u(t, x) = u_\infty(x) \quad \text{strongly in } L^2(\Omega). \tag{4.48}$$

4.3. Equation (4.1) with A demicontinuous

Let V be a real Hilbert space, densely and continuously imbedded in H. We identify H with its own dual and denote by V' the dual of V. Then the following relation holds

$$V \subset H \subset V'$$

Notations for norms and inner products will be the same as those given in Section 3.1

We shall discuss in this section Eq. (4.1) under the main assumption that the operator A is demicontinuous and monotone from V to V'.

THEOREM 4.3. *Consider* Eq. (4.1) *under the following assumptions:*

(j) $a \in L^1(0, \infty)$ *and* $\operatorname{Re} \hat{a}(i\eta) \geqslant 0$ *for every* $\eta \in R^1$.

(jj) *The operator A is monotone, single-valued and demicontinuous from V to V'. There exists $C > 0$ such that*

$$\|Au\|_* \leqslant C\|u\| \quad \text{for all } u \in V. \tag{4.49}$$

(jjj) *There exists $\omega > 0$ such that*

$$(Au, u) \geqslant \omega\|u\|^2 \quad \text{for all } u \in V. \tag{4.50}$$

Then for any $f \in L^2(0, \infty; V)$, Eq. (4.1) *has at least one solution $u \in L^2(0, \infty; H)$.*

Proof. We introduce the notations

$$\mathscr{V} = L^2(0, \infty; V),\ \mathscr{H} = L^2(0, \infty; H),\ \mathscr{V}' = L^2(0, \infty; V)$$

and denote $\langle , \rangle$ the natural pairing between $\mathscr{V}$ and $\mathscr{V}'$, i.e.,

$$\langle u, v\rangle = \int_0^\infty (u(t), v(t))\, dt, \quad \text{for } u \in \mathscr{V},\ v \in \mathscr{V}'.$$

Let us denote $\tilde{A} : \mathscr{V} \to \mathscr{V}'$ the operator defined by

$$(\tilde{A}u)(t) = Au(t), \quad \forall u \in \mathscr{V}, \quad \text{a.e. } t > 0.$$

We note that assumption (jj) implies that $\tilde{A}$ is monotone and demicontinuous from $\mathscr{V}$ to $\mathscr{V}'$. Thus, according to a well-known result proved in Chapter II, Section 1.1, $\tilde{A}$ is maximal monotone in $\mathscr{V} \times \mathscr{V}'$. Moreover, by assumption (jjj) we have

$$\langle \tilde{A}u, u\rangle \geqslant \omega \|u\|_{\mathscr{V}}^2, \quad \forall u \in \mathscr{V}. \tag{4.51}$$

Since $a \in L^1(0, \infty)$, the integral operator $(Lv)(t) = \int_0^t a(t-s)v(s)\, ds$ is well-defined and continuous on $\mathscr{V}, \mathscr{H}$ and $\mathscr{V}'$. On the other hand, as we have already noted, assumption (j) also implies that L is positive on all spaces listed above. In particular this implies that for every $\lambda > 0$, $(1 + \lambda L)^{-1}$ is well-defined and nonexpansive on $\mathscr{V}, \mathscr{H}$ and $\mathscr{V}'$. Moreover, $(1 - \lambda L)^{-1}$ is monotone in $\mathscr{V} \times \mathscr{V}'$. Thus in virtue of Corollary 1.2 (Chapter II), for every $\lambda > 0$ and $f \in \mathscr{V}$, the equation

$$\tilde{A}u_\lambda + (\lambda + L)^{-1} u_\lambda = (\lambda + L)^{-1} f \tag{4.52}$$

has a unique solution $u_\lambda \in \mathscr{V}$.

Observe that Eq. (4.52) can be rewritten as

$$u_\lambda + \lambda \tilde{A}u_\lambda + L\tilde{A}u_\lambda = f. \tag{4.53}$$

Multiplying Eq. (4.53) by $\tilde{A}u_\lambda$ and using relation (4.51), yields

$$\omega \|u_\lambda\|_{\mathscr{V}}^2 + \lambda \|\tilde{A}u_\lambda\|_{\mathscr{H}}^2 \leqslant \langle f, \tilde{A}u_\lambda\rangle.$$

Next, from assumption (4.49)

$$\|u_\lambda\|_{\mathscr{V}}^2 + \lambda \|\tilde{A}u_\lambda\|_{\mathscr{H}}^2 \leqslant C_1 \|f\|_{\mathscr{V}}^2 \tag{4.54}$$

for some positive constant C_1.

We know that

$$\|(\lambda + L)^{-1} u\|_{\mathscr{V}} \leqslant \|u\|_{\mathscr{V}}/\lambda \quad \text{for all } \lambda > 0 \text{ and } u \in \mathscr{V}.$$

Therefore Eq. (4.52) yields

$$\lambda \|\tilde{A}u_\lambda\|_{\mathscr{V}} \leqslant \|u_\lambda\|_{\mathscr{V}} + \|f\|_{\mathscr{V}}, \quad \forall \lambda > 0. \tag{4.55}$$

Now, we consider an auxiliary function v_λ defined by

$$v_\lambda = u_\lambda + \lambda \tilde{A} u_\lambda .$$

Taking into account estimates (4.54) and (4.55) we deduce that

$$\{v_\lambda\} \text{ is bounded in } \mathscr{V}$$

$$\lim_{\lambda \to 0} (u_\lambda - v_\lambda) = 0 \quad \text{in } \mathscr{H}.$$

Therefore, a subsequence (again denoted u_λ) can be extracted from $\{u_\lambda\}$ such that as $\lambda \to 0$,

$$\begin{aligned} u_\lambda &\to u \quad \text{weakly in } \mathscr{V} \\ v_\lambda &\to u \quad \text{weakly in } \mathscr{V} \\ \tilde{A} u_\lambda &\to w \quad \text{weakly in } \mathscr{V}'. \end{aligned} \tag{4.56}$$

To conclude the proof we must show that $w = \tilde{A}u$. Let us define

$$\tilde{A}_0 u = \tilde{A} u \quad \text{for} \quad u \in D(\tilde{A}_0)$$

where

$$D(\tilde{A}_0) = \{u \in \mathscr{V}; \quad \tilde{A}u \in \mathscr{H}\}.$$

We have already seen (e.g. Section 3.3 in Chapter III) that under our conditions $\tilde{A}_0$ is maximal monotone in $\mathscr{H} \times \mathscr{H}$ and $D(\tilde{A}_0)$ is a dense subset of $\mathscr{H}$. Then u_λ can be written as

$$u_\lambda = (1 + \lambda \tilde{A}_0)^{-1} v_\lambda$$

and

$$\tilde{A} u_\lambda = \tilde{A}_\lambda v_\lambda, \text{ where}$$

$$\tilde{A}_\lambda = \tilde{A}_0 (1 + \lambda \tilde{A}_0)^{-1}.$$

Thus, using once again the positivity of L, it follows from Eq. (4.53) that

$$\langle \tilde{A}_\lambda v_\lambda - \tilde{A}_\mu v_\mu, \quad v_\lambda - v_\mu \rangle \leqslant 0 \quad \text{for all} \quad \lambda, \mu > 0. \tag{4.57}$$

Extracting further subsequences if necessary, we may assume that

$$\lim_{\lambda \to 0} \langle \tilde{A}_\lambda v_\lambda, \quad v_\lambda \rangle = l. \tag{4.58}$$

Therefore relations (4.56) yield

$$l \leqslant \langle w, u \rangle. \tag{4.59}$$

On the other hand, $\tilde{A}_\lambda$ being monotone in $\mathscr{H} \times \mathscr{H}$, we have

$$\langle \tilde{A}_\lambda v_\lambda - \tilde{A}_\lambda v, v_\lambda - v \rangle \geqslant 0, \text{ for every } v \in \mathscr{H}.$$

Let v be any fixed element of $D(\tilde{A}_0)$. Then $\tilde{A}_\lambda v \to \tilde{A}_0 v$ strongly in $\mathscr{H}$ (see Proposition 3.6, Chapter II) so that the above inequality implies that (one uses also (4.56) and (4.59))

$$\langle w - \tilde{A}v, u - v \rangle \geqslant 0, \quad \forall v \in D(\tilde{A}_0). \tag{4.60}$$

Now let v be an arbitrary element in $\mathscr{V}$. As mentioned earlier, for every $\varepsilon > 0$ the equation

$$v_\varepsilon + \varepsilon \tilde{A} v_\varepsilon = v \tag{4.61}$$

has a unique solution $v_\varepsilon \in D(\tilde{A}_0)$. Moreover, $\lim\limits_{\varepsilon \to 0} v_\varepsilon = v$ strongly in H because $D(\tilde{A}_0)$ is dense in H. Multiplying this equation by $\tilde{A}v_\varepsilon$ and using hypotheses (jj) and (jjj), we see that

$$\omega \, \|v_\varepsilon\|_{\mathscr{V}}^2 + \varepsilon \, \|\tilde{A}v_\varepsilon\|_{\mathscr{H}}^2 \leqslant (v, \tilde{A}v_\varepsilon).$$

Thus it follows that

$$v_\varepsilon \to v \quad \text{weakly in } \mathscr{V}.$$

We wish to show that $\tilde{A}v_\varepsilon \to \tilde{A}v$ weakly in $\mathscr{V}'$ and

$$\limsup_{\varepsilon \to 0} \langle \tilde{A}v_\varepsilon, v_\varepsilon \rangle \leqslant \langle \tilde{A}v, v \rangle.$$

Extracting a subsequence if necessary, we may assume that

$$\tilde{A}v_\varepsilon \to w_0 \text{ weakly in } \mathscr{V}'.$$

Thus from Eq. (4.61) it follows that

$$\limsup_{\varepsilon \to 0} \langle \tilde{A}v_\varepsilon, \overline{v}_\varepsilon \rangle \leqslant \langle v, w_0 \rangle.$$

Since $\tilde{A}$ is monotone, this yields

$$\langle v, w_0 \rangle \geqslant \langle \tilde{A}\overline{v}, \overline{v} - \overline{v} \rangle + \langle w_0, \overline{v} \rangle \quad \text{for every} \quad \overline{v} \in \mathscr{V}.$$

Equivalently

$$\langle w_0 - \tilde{A}\bar{v}, \bar{v} - v \rangle \geqslant 0 \quad \text{for every} \quad \bar{v} \in \mathscr{V}.$$

The operator $\tilde{A}$ being maximal monotone in $\mathscr{V} \times \mathscr{V}'$, we see that $w_0 = \tilde{A}v$ and $\limsup_{\varepsilon \to 0} \langle \tilde{A}v_\varepsilon, v_\varepsilon \rangle \leqslant \langle \tilde{A}v, v \rangle$ as claimed. Now in inequality (4.60) we choose $v = v_\varepsilon$ and let $\varepsilon \to 0$. The results just proved then imply that this inequality extends to all $\bar{v} \in \mathscr{V}$. Again, since $\tilde{A}$ is maximal monotone in $\mathscr{V} \times \mathscr{V}'$, we may conclude that $w = \tilde{A}u$. Now using estimate (4.54) we see that

$$\lambda \tilde{A} u_\lambda \to 0 \quad \text{strongly in } \mathscr{H} \text{ as } \lambda \to 0.$$

Since L is weakly continuous in $\mathscr{V}'$, we may pass to limit in Eq. (4.53) to obtain that

$$u + L\tilde{A}u = f,$$

where we have employed relation (4.56). Thus our theorem is proven.

Remark. Theorem 3.3 is formulated in such a way that it can be easily applied to partial differential equations in a fashion analogous to that of Section 2.3 in Chapter III. A typical example is the nonlinear boundary value problem

$$u(t, x) + \sum_{|\alpha| \leqslant m} (-1)^\alpha \int_0^t a(t-s) D^\alpha A_\alpha(x, D^\gamma u(s, x))\, ds = f(t, x)$$

$$\text{on }]0, \infty[\times \Omega, \tag{4.62}$$

$$D^\alpha u(t, x) = 0 \qquad \text{on }]0, \infty[\times \Gamma, \quad |\alpha| \leqslant m - 1$$

under appropiate growth and monotonicity conditions on A_α.

In this case $V = H_0^m(\Omega)$, $H = L^2(\Omega)$ and A is defined by

$$(Au, v) = \sum_{|\alpha| \leqslant m} \int_\Omega A_\alpha(x, D^\gamma) D^\alpha v \, dx, \quad \text{for} \quad u, v \in H_0^m(\Omega).$$

For a direct treatment of this problem we refer the reader to R. C. MacCamy [2].

4.4. A class of integro-differential equations

We shall consider now the integro-differential equation

$$\frac{du}{dt}(t) + \partial\varphi(u(t)) + \int_0^t a(t-s) Au(s)\, ds \ni f(t), \quad \text{a.e. } t > 0 \tag{4.63}$$

subject to the initial condition

$$u(0) = u_0. \tag{4.64}$$

We are given reflexive Banach spaces V and W which satisfy

$$V \subset W \subset H,$$

where each space to the left of an inclusion sign is dense in the space to its right and the corresponding inclusion map is continuous.

As usual, we identify H with its own dual and denote by V' and W' the dual space of V and W, respectively. Then the following inclusion relation holds

$$V \subset W \subset H \subset W' \subset V'.$$

In Eq. (4.63), φ is a lower-semicontinuous convex proper function from H to $]-\infty, +\infty]$ and A is a nonlinear monotone operator from W to W'. More precisely, the basic assumptions under which the results of this section will be established are as follows:

(a) The injection map of V into W is compact.

(b) $D(A) = W$ and A is single-valued, demicontinuous and coercive from W to W'. Moreover, A is cyclically monotone in $W \times W'$ and bounded on every bounded subset of W.

(c) $D(\varphi) \subset V$ and

$$\lim_{\|u\|_V \to \infty} (\varphi(u) + |u|^2) = \infty. \tag{4.65}$$

(d) For every $u \in D(\varphi)$ and for any $\varepsilon > 0$, solution u_ε of the equation $u_\varepsilon + \varepsilon A u_\varepsilon = u$ lies in $D(\varphi)$ and satisfies

$$\varphi(u_\varepsilon) \leqslant \varphi(u). \tag{4.66}$$

Concerning the kernel $a(t)$, a hypothesis we shall keep throughout this section is the following one:

(e) $a \in C([0, \infty[) \cap C^2(]0, \infty[)$ and

$$(-1)^k \frac{d^k a}{dt^k} \geqslant 0, \quad k = 0, 1, 2 \quad \text{on }]0, \infty[. \tag{4.67}$$

The main result is

THEOREM 4.4. *Suppose that the above hypotheses hold. Let u_0 and f be given such that*

$$u_0 \in D(\varphi), \quad f \in W^{1,2}_{loc}(0, \infty; H). \tag{4.68}$$

Then Eq. (4.63) *has at least one solution* $u \in W^{1,2}_{\mathrm{loc}}(0, \infty; H)$ *satisfying the initial condition* (4.64). *Moreover,*

$$u \in C(0, \infty; W) \cap L^{\infty}_{\mathrm{loc}}(0, \infty; V). \tag{4.69}$$

Proof. As we have already mentioned, our hypotheses concerning A imply that the operator $A_0 : H \to H$ defined by

$$A_0 u = Au \quad \text{for} \quad u \in D(A_0) = \{u \in W; \quad Au \in H\}$$

is maximal monotone in $H \times H$ (see e.g. Section 2.3 in Chapter III). On the other hand, A_0 being cyclically monotone in $H \times H$, there exists a lower-semi-continuous convex function $\psi : H \to]-\infty, +\infty]$ such that (see Theorem 2.3, Chapter II) $A_0 = \partial\psi$.

Without any loss of generality we may assume that ψ is positive on H. This can be achieved by shifting the range of A.

Let $0 < T < \infty$ be such that $a(T) > 0$. On the interval $[0, T]$ we consider the approximating equation

$$\frac{\mathrm{d}u_\lambda}{\mathrm{d}t}(t) + \partial\varphi(u_\lambda(t)) + \int_0^t a(t-s)A_\lambda u_\lambda(s)\,\mathrm{d}s \ni f(t), \tag{4.70}$$

subject to the initial value condition

$$u_\lambda(0) = u_0 \tag{4.71}$$

where $A_\lambda = \lambda^{-1}(1 - (1 + \lambda A_0)^{-1})$ and $\lambda > 0$.

Since A_λ is Lipschitzian on H, an argument similar to that used in the proof of Theorem 2.1 shows that the problem (4.70) – (4.71) has a unique solution $u_\lambda \in C(0, T; H)$ which satisfies

$$\frac{\mathrm{d}u_\lambda}{\mathrm{d}t} \in L^2(0, T; H), \quad \varphi(u_\lambda) \in L^\infty(0, T)$$

and

$$u_\lambda(t) \in D(\partial\varphi), \quad \text{a.e.} \quad \text{on }]0, T[.$$

In order to obtain the convergence of u_λ as $\lambda \to 0$ some *a priori* estimates are necessary.

LEMMA 4.2. *There exists some positive constant C such that for all* $\lambda > 0$ *and* $t \in [0, T]$,

$$\int_0^t \left|\frac{\mathrm{d}u_\lambda}{\mathrm{d}s}(s)\right|^2 \mathrm{d}s + \left|\int_0^t A_\lambda u_\lambda(s)\,\mathrm{d}s\right|^2 + \varphi(u_\lambda(t)) \leqslant C. \tag{4.72}$$

Proof. For simplicity we shall write u' instead of $\mathrm{d}u/\mathrm{d}t$. We set

$$\psi_\lambda(u) = \inf_{v \in H}\left\{\frac{|u - v|^2}{2\lambda} + \psi(v)\right\}, \quad u \in H, \quad \lambda > 0.$$

Let us consider for $\lambda > 0$, the "energy function"

$$E_\lambda(t) = \psi_\lambda(u_\lambda(t)) + \frac{1}{2} a(t) \left| \int_0^t A_\lambda u_\lambda(s)\, ds \right|^2 - $$

$$- \frac{1}{2} \int_0^t a'(t-s) \left| \int_0^t A_\lambda u_\lambda(\tau)\, d\tau \right|^2 ds, \quad 0 \leqslant t \leqslant T.$$

Since $A_\lambda = \partial\psi_\lambda$, it follows from Lemma 2.1 that

$$(u_\lambda'(t), A_\lambda u_\lambda(t)) = (\psi_\lambda(u_\lambda(t)))', \quad \text{a.e. } 0 < t < T. \tag{4.73}$$

We now differentiate $E_\lambda(t)$, taking into account Eqs. (4.70) and (4.73). There results

$$E_\lambda'(t) = \frac{1}{2} a'(t) \left| \int_0^t A_\lambda u_\lambda(s)\, ds \right|^2 - \frac{1}{2} \int_0^t a''(t-s) \left| \int_s^t A_\lambda u_\lambda(\tau)\, d\tau \right|^2 ds +$$

$$+ (f(t) - y_\lambda(t), A_\lambda u_\lambda(t)), \text{ a. e. } 0 < t < T, \tag{4.74}$$

where

$$y_\lambda(t) = f(t) - u_\lambda'(t) - \int_0^t a(t-s)\, A_\lambda u_\lambda(s)\, ds.$$

The existence of the integrals occurring in Eq. (4.74) is easily justified by the hypothesis. In fact it follows readily from (e) that $\lim_{t \to 0} ta'(t) = 0$ and $ta''(t) \in L^1(0, \infty)$.

From hypotheses (4.67) and (4.74) one observes that

$$E_\lambda'(t) \leqslant (f(t) - y_\lambda(t), A_\lambda u_\lambda(t)), \text{ a.e. } 0 < t < T.$$

Inasmuch as $y_\lambda \in \partial\varphi(u_\lambda)$, condition (d) and Proposition 1.1, part (jj) give on $]0, T[$

$$(y_\lambda(t), A_\lambda u_\lambda(t)) \geqslant 0.$$

Hence

$$E_\lambda(t) \leqslant \psi_\lambda(u_0) + \int_0^t (f(s), A_\lambda u_\lambda(s))\, ds, \quad 0 \leqslant t \leqslant T. \tag{4.75}$$

We set

$$F_\lambda(t) = \int_0^t A_\lambda u_\lambda(s)\, ds, \quad 0 \leqslant t \leqslant T.$$

Integrating by parts in formula (4.75) we obtain after some rearrangements that

$$\psi_\lambda(u_\lambda(t)) + \frac{1}{2}a(t)\,|F_\lambda(t)|^2 \leqslant \psi_\lambda(u_0) + (f(t), F_\lambda(t)) - \int_0^t (f'(s), F_\lambda(s))\,ds$$

and

$$\psi_\lambda(u_\lambda(t)) + \frac{1}{4}a(t)\,|F_\lambda(t)|^2 \leqslant \psi_\lambda(u_0) + \frac{2|f(t)|^2}{a(t)} + \int_0^t |f'(s)|\,|F_\lambda(s)|\,ds.$$

Since $\psi_\lambda(u_0) \leqslant \psi(u_0)$ and $a(t) > a(T) > 0$, the above integral inequality yields

$$\psi_\lambda(u_\lambda(t)) + |F_\lambda(t)|^2 \leqslant C \text{ for } 0 \leqslant t \leqslant T, \tag{4.76}$$

where $C > 0$ is independent of λ.
Write Eq. (4.70) as

$$u'_\lambda(t) + y_\lambda(t) + a(0)\,F_\lambda(t) + \int_0^t a'(t-s)\,F_\lambda(s)\,ds = f(t). \tag{4.77}$$

and multiply it by y_λ. Using once again (4.73) we obtain

$$\varphi(u_\lambda(t)) + \int_0^t |y_\lambda(s)|^2\,ds \leqslant \varphi(u_0) + \int_0^t |y_\lambda(s)|\,(a(0)\,|F_\lambda(s)| +$$

$$+ |f(s)|)ds + \int_0^t\int_0^s |a'(s-\tau)|\,|y_\lambda(s)|\,|F_\lambda(\tau)|\,d\tau\,ds.$$

From inequality (4.76), it follows that

$$\varphi(u_\lambda(t)) + \int_0^t |y_\lambda(s)|^2\,\mathrm{d}s \leqslant \varphi(u_0) + C\int_0^t |y_\lambda(s)|\,\mathrm{d}s$$

because

$$\int_0^t\int_0^s |a'(s-\tau)|\,|y_\lambda(s)|\,\mathrm{d}\tau\,\mathrm{d}s \leqslant |a'|_{L^1(0,\,t)}\int_0^t |y_\lambda(s)|\,\mathrm{d}s.$$

This together with relations (4.76) and (4.77) yields estimate (4.72).

We continue the proof of Theorem 4.4, by noting that (d) and Lemma 4.2 imply that

$$\varphi(v_\lambda(t)) \leqslant \varphi(u_\lambda(t)) \leqslant C,\ 0 \leqslant t \leqslant T;\ \lambda > 0. \tag{4.78}$$

where $v_\lambda(t) = (1 + \lambda A_0)^{-1}u_\lambda(t)$. Taking into account assumption (c), the result is that

$$\|v_\lambda(t)\|_V + \|u_\lambda(t)\|_V \leqslant C_0, \quad 0 \leqslant t \leqslant T. \tag{4.79}$$

where C_0 is some positive constant independent of λ. Moreover, since $(1 + \lambda A_0)^{-1}$ is nonexpansive on H, we have

$$|v_\lambda'(t)| \leqslant |u_\lambda'(t)|, \quad \text{a.e.} \quad \text{on }]0, T[$$

so that estimate (4.72) gives

$$\int_0^T |v_\lambda'(t)|^2 \, dt \leqslant \int_0^T |u_\lambda'(t)|^2 \, dt \leqslant C, \ \lambda > 0. \tag{4.80}$$

Since the injection map of V into W is compact by hypothesis, it follows from relations (4.79) and (4.80) that $\{u_\lambda\}$ and $\{v_\lambda\}$ are compact subsets of $L^2(0, T; W)$ (see J. L. Lions [1], Chapter I, Theorem 5.1).

Hence, there exist subsequences $\{u_{\lambda_n}\}$ and $\{v_{\lambda_n}\}$ such that $u_{\lambda_n} \to u$, $v_{\lambda_n} \to v$ strongly in $L^2(0, T; W)$ as $\lambda_n \to 0$.

On the other hand, the equality

$$v_\lambda(t) + \lambda A v_\lambda(t) = u_\lambda(t)$$

implies

$$|v_\lambda(t) - u_\lambda(t)|^2 \leqslant -\lambda(Au_\lambda(t), v_\lambda(t) - u_\lambda(t))$$

because A is monotone. Hence

$$|v_\lambda(t) - u_\lambda(t)|^2 \leqslant \lambda \, \|Au_\lambda(t)\|_{W'} \|u_\lambda(t) - v_\lambda(t)\|_W.$$

Since A is bounded on the bounded subsets of W, this implies that $u(t) = v(t)$ a.e. on $]0, T[$.

Now, extracting further subsequences if necessary we may assume that

$$u_{\lambda_n} \to u \text{ strongly in } L^2(0, T; W),$$

$$u_{\lambda_n}' \to u' \text{ weakly in } L^2(0, T; H),$$

$$Av_{\lambda_n} \to w \text{ weakly-star in } L^\infty(0, T; W').$$

But A is demicontinuous from W to W', so that we conclude that $w(t) = Au(t)$ a.e. on $]0, T[$.

On the other hand, $\int_0^t a(t - s)A_{\lambda_n} u_{\lambda_n}(s) \, ds$ is bounded in $L^\infty(0, T; H)$ by Lemma 4.2. We may therefore infer that

$$\int_0^t a(t - s) \, A_{\lambda_n} u_{\lambda_n}(s) \, ds \to \int_0^t a(t - s) \, Au(s) \, ds$$

17 – c. 2144

in the weak topology of $L^2(0, T; H)$. In Eq. (4.70) we put $\lambda = \lambda_n$ and let $n \to \infty$. We obtain that $u(t) \in D(\partial\varphi)$ a.e. on $]0, T[$ and

$$u'(t) + \partial\varphi(u(t)) + \int_0^t a(t-s)\, Au(s)\, \mathrm{d}s \ni f(t), \quad \text{a. e. } t \in]0, T[.$$

Here we have used, in particular, the demiclosedness of $\partial\varphi$ in $L^2(0, T; H) \times L^2(0, T; H)$. Moreover, we have

$$u' \in L^2(0, T; H), \quad Au \in L^\infty(0, T; W') \tag{4.81}$$

and

$$\int_0^t a(t-s)\, Au(s)\, \mathrm{d}s \in L^2(0, T; H). \tag{4.82}$$

It also follows that

$$u(0) = u_0.$$

The former relation is true since $u'_{\lambda_n} \to u'$ weakly in $L^2(0, T; H)$ while $u_{\lambda_n}(0) = u_0$.

Finally, in (4.78) we put $\lambda = \lambda_n$ and let $n \to \infty$, thereby obtaining

$$\varphi(u(t)) \leqslant C \quad \text{for} \quad 0 \leqslant t \leqslant T.$$

Here we have used the lower-semicontinuity of φ. By hypothesis (c) the above inequality implies that $u \in L^\infty(0, T; V)$. Since $u(t)$ is H-valued continuous on $[0, T]$ and the injection of V into W is compact we conclude that $u \in C(0, T; W)$. With the same procedure used in the proof of Theorem 4.1, we deduce that the function $u(t)$ can be extended to the whole R^1_+ such that (4.81), (4.82) and continuity are preserved. Theorem 4.4 is thus proved.

Remark. Theorem 4.4 remains valid if assumption (c) is replaced *by* $D(\partial\varphi) \subset V$ *and there are positive constants* α, β *such that*

$$\text{(c)}' \qquad |(\partial\varphi)^0 u|^2 + \beta\|u\|_W^2 \geqslant \alpha\|u\|_V^2, \ \forall u \in D(\partial\varphi).$$

Indeed the coercivity of A implies that $\lim_{\|u\|_W \to \infty} \psi(u) = \infty$ (see Proposition 2.6, Chapter II) so that estimate (4.76) gives

$$\{v_\lambda\}, \text{ and } \{u_\lambda\} \text{ are bounded in } L^\infty(0, T; W).$$

Thus, assumption (c)′ and Lemma 4.2 imply that $\{u_\lambda\}$ is bounded in $L^2(0, T; V)$. Afterwards, the proof follows the same lines as before. However, under the above hypotheses solution $u(t)$ will satisfy

$$u \in C(0, \infty; H) \cap L^\infty_{\mathrm{loc}}(0, \infty; W) \cap L^2_{\mathrm{loc}}(0, \infty; V)$$

instead of (4.69). Let us illustrate with a simple example.

Consider the mixed boundary initial value problem

$$u_t(t, x) = u_{xx}(t, x) + \int_0^t a(t - s)\,(\sigma(u_x))_x\, ds + f(t, x), \quad (t, x) \in\,]0,\infty[\, \times\,]0,1[$$

$$u(t, 0) = u(t, 1) = 0, \quad 0 < t < \infty, \tag{4.83}$$

$$u(0, x) = u_0(x), \qquad 0 < x < 1.$$

We have used here the standard notations: $u_t = \partial u/\partial t$ and $u_x = \partial u/\partial x$.

We suppose that $\sigma(r)$ is a continuously differentiable function on R^1, satisfying

$$\sigma'(r) \geqslant 0, \; r \in R^1,$$

and

$$|\sigma(r)| \leqslant m_1\,|r| + m_2, \; r \in R^1,$$

where m_1, m_2 are positive constants.

Define operator A from $H_0^1(0, 1)$ into its dual space by

$$(Au, v) = \int_0^1 \sigma(u_x) v_x \mathrm{d}x; \quad u, v \in H_0^1(0, 1).$$

It is easy to see that under our assumptions on σ operator A is cyclically monotone and demicontinuous from $H_0^1(0, 1)$ into $(H_0^1(0, 1))'$. If we further impose condition

$$r\sigma(r) \geqslant m(|r|^2 + 1), \; m > 0,$$

then A is coercive on $H_0^1(0,1)$. Let $\varphi\colon L^2(0,1) \to\,]-\infty, +\infty]$ be the convex function defined by

$$\varphi(u) = \begin{cases} \dfrac{1}{2}\displaystyle\int_0^1 |u_x|^2 \,\mathrm{d}x & \text{if } u \in H_0^1(0, 1), \\ +\infty & \text{otherwise.} \end{cases}$$

We have

$$\partial\varphi(u) = -u_{xx}$$

for all $u \in H_0^1(0, 1) \cap H^2(0, 1) = D(\partial\varphi)$. It follows easily that conditions (a), (b), (c)′, (d), are satisfied by $V = H_0^1(0, 1) \cap H^2(0, 1)$, $W = H_0^1(0, 1)$, $H = L^2(0, 1)$ and A as defined above. Furthermore, we assume that $a(t)$ satisfies condition (e).

Theorem 4.4 and the previous remark therefore yield the following result. *Given:*

$$u_0 \in H_0^1(0, 1), \quad f, \frac{\partial f}{\partial t} \in L^2_{\text{loc}}(0, \infty; L^2(0, 1)).$$

There exists at least a function $u(t, x)$ satisfying problem (4.83) *and*

$$u \in L^\infty(0, \infty; H_0^1(0, 1)) \cap L^2_{\text{loc}}(0, \infty; H_0^1(0, 1) \cap H^2(0, 1)) \cap C(0, \infty; L^2(\Omega)), \quad (4.84)$$

$$u_t \in L^2_{\text{loc}}(0, \infty; L^2(0, 1)). \quad (4.85)$$

Remark. Equations of the form (4.83) are of physical interest and have been considered by many authors. For a different approach to the study of related problems, we refer the reader to J. M. Greenberg. R. C. MacCamy and V. J. Mizel [1].

We now return to the discussion of a slightly more general equation than (4.1), namely

$$\frac{\mathrm{d}u}{\mathrm{d}t} + m\,Au + \int_0^t a(t-s)\,Au(s)\,\mathrm{d}s \ni f(t), \quad \text{a.e.} \quad t > 0 \quad (4.86)$$

$$u(0) = u_0$$

where $m > 0$ and $a(t)$ satisfies assumption (e) of Theorem 4.4.

Operator A is assumed to be of the form $A = \partial\varphi$ where φ is a lower-semicontinuous convex function from H to $]-\infty, +\infty]$ satisfying condition (c). In this case V is a real reflexive Banach space such that $V \subset H$ and the inclusion mapping of V into H is densely defined, continuous and compact.

THEOREM 4.5. *Let $V, A, a(t)$ be given, satisfying the above hypotheses. Let u_0, f be given, satisfying*

$$u_0 \in D(\varphi), \quad f \in W^{1,2}_{\text{loc}}(0, \infty; H).$$

Then Eq. (4.86) *has at least one solution u which satisfies*

$$u \in W^{1,2}_{\text{loc}}(0, \infty; H) \cap L^\infty_{\text{loc}}(0, \infty; V)$$

$$Au \in L^\infty_{\text{loc}}(0, T; H).$$

We omit the details of the proof of Theorem 4.5 which follows the pattern of the proof of Theorem 4.4.

4.5. Further investigation of the preceding case

We shall again consider Eq. (4.63), essentially under the same assumptions as in the preceding section. More precisely, we retain conditions (a) $\sim$ (e) from Theorem 4.4 and further suppose that

(f) $a(t) \not\equiv a(0)$ and one of the following two conditions holds

$a(\infty) > 0$; $a \in L^1(0, \infty)$.

(g) $A^{-1}(0)$ consists of a single point.

It turns out that the above conditions ensure the asymptotic stability of solution $u(t)$ of Eq. (4.63). For simplicity we shall restrict ourselves to the discussion of Eq. (4.63) where $f(t) = 0$. We shall prove the following conditional asymptotic result.

THEOREM 4.6. *Assume that conditions (a) $\sim$ (g) hold. If $u \in W^{1,2}_{\text{loc}}(0, \infty; H)$ is any solution of* Eq. (4.63) *satisfying*

$$Au \in L^2_{\text{loc}}(0, \infty; H), \quad u \in C([0, \infty]; W) \tag{4.87}$$

then

$$\lim_{t\to\infty} u(t) = A^{-1}(0) \text{ strongly in } W. \tag{4.88}$$

Remark. $C([0, \infty]; W)$ denotes the space of all uniformly continuous functions u: $[0, \infty[\to W$.

Proof. Let $\psi: H \to]-\infty, +\infty]$ be such that $A_0 = \partial\psi$, where $A_0 u = Au \cap H$ for every $u \in W$. Since $0 \in R(A_0)$, we may choose ψ such that $\min_{u\in H} \psi(u) = 0$.

Let u be any function satisfying the conditions of the theorem. We multiply equation

$$u'(t) + \partial\varphi(u(t)) + \int_0^t a(t-s)\, Au(s)\, ds \ni 0, \quad \text{a. e. } t > 0 \tag{4.89}$$

by $Au(t)$ and integrate over $]0, t[$. The use of Proposition 1.1 part (jj) and of condition (d) leads to

$$\psi(u(t)) + \int_0^t ds \int_0^s a(s-\tau)\; (Au(s), Au(\tau))\, d\tau \leqslant \psi(u(0)), \quad t > 0.$$

Since in view of Proposition 4.1 $a(t)$ defines a strongly positive kernel, there exist positive constants α and ε such that

$$\psi(u(t)) + \varepsilon \int_0^t ds \int_0^s e^{-\alpha(s-\tau)} (Au(s), Au(\tau))\, d\tau \leqslant \psi(u(0)). \tag{4.90}$$

We set

$$y(t) = \int_0^t e^{-\alpha(t-s)}\, Au(s)\, ds.$$

Then

$$y'(t) + \alpha y(t) = Au(t), \quad \text{a.e.} \quad \text{on }]0, \infty[\tag{4.91}$$

and therefore

$$\psi(u(t)) + \frac{\varepsilon}{2}\,|y(t)|^2 + \alpha\varepsilon \int_0^t |y(s)|^2\,\mathrm{d}s \leqslant \psi(u(0)), \quad \text{on } [0, \infty[. \tag{4.92}$$

We define

$$g(t) = \int_0^t a(t-s)\,Au(s)\,\mathrm{d}s$$

and note that

$$g(t) = a(0)y(t) + \int_0^t (\alpha a(t-s) - a'(t-s))\,y(s)\mathrm{d}s.$$

If the second alternative of condition (f) holds (i.e., $a \in L^1(0, \infty)$) then inequality (4.92) clearly implies that $g(t)$ is bounded (in H) on $[0, \infty[$. Now suppose that $a(\infty) > 0$. We then define "the energy function"

$$E(t) = \psi(u(t)) + \frac{1}{2}\,a(t)\,|F(t)|^2 - \frac{1}{2}\int_0^t a'(t-s)\,|F(t) - F(s)|^2\,\mathrm{d}s,$$

where

$$F(t) = \int_0^t Au(s)\,\mathrm{d}s.$$

Assumption (e) implies just as in Section 4 that $E'(t) \leqslant 0$ a.e. $t > 0$. Therefore,

$$\psi(u(t)) + \frac{1}{2}\,a(t)\,|F(t)|^2 \leqslant \psi(u_0) \quad \text{for} \quad t \geqslant 0. \tag{4.93}$$

Since $g(t)$ can be written as

$$g(t) = a(0)\,F(t) + \int_0^t a'(t-s)\,F(s)\,\mathrm{d}s,$$

we conclude that

$$|g(t)| \leqslant M \text{ for } 0 \leqslant t < \infty, \tag{4.94}$$

where M is a positive constant.
Next, we take the inner product of Eq. (4.89) with $u'(t)$ to obtain

$$|u'(t)|^2 + \frac{\mathrm{d}}{\mathrm{d}t}\,\varphi(u(t)) \leqslant |g(t)|\,|u'(t)|, \quad \text{a.e.} \quad t > 0.$$

Let u_0 be arbitrary but fixed in $D(\varphi)$. By the definition of $\partial\varphi$ we have

$$\varphi(u(t)) \leqslant \varphi(u_0) - (u'(t) + g(t), \quad u(t) - u_0), \quad \text{a.e.} \quad t > 0.$$

Since A^{-1} is bounded from W' to W, A_0^{-1} is also bounded on H. Thus (4.93) and Proposition 2.6 in Chapter II imply that $|u(t)|$ is bounded on $[0, \infty[$. We may therefore conclude from the previous estimates that

$$(\varphi(u(t)))^2 + \frac{\mathrm{d}}{\mathrm{d}t}\varphi(u(t)) \leqslant C_1, \quad \text{a.e.} \quad t > 0$$

because φ is bounded from below by an affine function. The last inequality shows that $\varphi(u(t))$ is bounded on $[0, \infty[$. Thus once again invoking condition (c) we deduce that

$$\|u(t)\|_V \text{ is bounded on } [0, \infty[. \tag{4.95}$$

Let w be arbitrary in W. Since $u \in C([0, \infty]; W)$ and A is demicontinuous from W to W', we may conclude that the function $t \to (Au(t), w)$ is uniformly continuous on $[0, \infty[$. Since $(y(t), w)$ is bounded, it follows from (4.92) that $\frac{\mathrm{d}}{\mathrm{d}t}(y(t), w)$ is bounded and uniformly continuous on $[0, \infty[$. Furthermore, function $(y(t), w)$ belongs to $L^2(0, \infty)$. This implies (see e.g. A. Halanay [1], Chapter II, Lemma 4) that $(y(t), w) \to 0$ and $(y'(t), w) \to 0$ as $t \to \infty$. Thus, Eq. (4.91) gives

$$Au(t) \to 0 \text{ weakly in } W' \text{ as } t \to \infty.$$

The injection of V into W being compact, Eq. (4.91) and assumption (g) then imply that

$$\lim_{t\to\infty} u(t) = A^{-1}(0) \quad \text{strongly in } W,$$

which concludes the proof.

Remark. The condition $u \in C([0, \infty]; W)$ is quite natural and not very restrictive. Indeed, we have already mentioned that every solution $u \in W^{1,2}_{\mathrm{loc}}(0, \infty; H)$ of Eq. (4.63) is continuous from $[0, \infty[$ to W.

We conclude this Section with an asymptotic result for the solutions of Eq. (4.86).

THEOREM 4.7. *Let us consider* Eq. (4.86) *under the following conditions:*

1. $A = \partial\varphi$ *and* $m > 0$.

2. φ *is convex and lower-semicontinuous from* H to $]-\infty, +\infty]$. *Moreover, every level set* $\{u; \varphi(u) \leqslant \lambda\}$ *is compact in* H.

3. $A^{-1}(0)$ *consists of a single element.*

4. $a(t)$ *defines a strongly positive kernel.*

If $u \in W^{1,2}_{loc}(0, \infty; H)$ *is a solution* of Eq. (4.86) *with* $f(t) = 0$, *such that* $Au \in L^2_{loc}(0, \infty; H)$, *then*

$$|\lim_{t\to\infty} u(t)| = A^{-1}(0) \text{ strongly in } H. \tag{4.96}$$

Proof. Let $u \in W^{1,2}_{loc}(0, \infty; H)$ be any solution of Eq. (4.86) such that $Au \in L^2_{loc}(0, \infty; H)$. In other words, there exists $w \in L^2_{loc}(0, \infty; H)$ such that $w(t) \in Au(t)$ a.e. on $]0, \infty[$ and

$$u'(t) + mw(t) + \int_0^t a(t-s)w(s)ds = 0, \text{ a.e. on }]0, \infty[. \tag{4.97}$$

From Lemma 2.1, $t \to \varphi(u(t))$ is absolutely continuous on every finite interval $[0, T]$ and

$$\frac{d}{dt}\varphi(u(t)) = (w(t), u'(t)), \quad \text{a.e. on }]0, \infty[.$$

Taking into account condition 4, we obtain

$$\frac{d}{dt}\left(\varphi(u(t)) + m\int_0^t |w(s)|^2 ds + \frac{\varepsilon}{2}|y(t)|^2 + \alpha\varepsilon\int_0^t |y(s)|^2 ds\right) \leqslant 0, \quad \text{a.e. } t > 0 \tag{4.98}$$

where α, ε are positive and $y(t)$ is defined as in the proof of Theorem 4.6, i.e.,

$$y(t) = \int_0^t e^{-\alpha(t-s)} w(s)ds.$$

Noticing that

$$y'(t) + \alpha y(t) = w(t), \quad \text{a.e. on }]0, \infty[\tag{4.99}$$

it follows from (4.98) that $y' \in L^2(0, \infty; H)$. Since $y(t)$ also belongs to $L^2(0, \infty; H)$, this implies that $\lim_{t\to\infty} |y(t)| = 0$. Thus from (4.98) we deduce that $\lim_{t\to\infty} \varphi(u(t)) = \varphi_\infty$ exists. Now, by definition of the subdifferential $\partial\varphi$, we have

$$\varphi(u(t)) \leqslant \varphi(v) + (w(t), u(t) - v) \text{ for every } v \in H \tag{4.100}$$

and using once again inequality (4.98) we find that

$$\varphi(u(t)) + m\int_0^t |w(s)|^2 ds + \frac{\varepsilon}{2}|y(t)|^2 + \alpha\varepsilon\int_0^t |y(s)|^2 ds \leqslant \varphi(u(0)).$$

In view of condition 2, the latter inequality implies that $\{u(t); t \geqslant 0\}$ is a compact subset of H. Furthermore, inequality (4.100) implies that

$$\varphi_\infty \leqslant \varphi(v) \quad \text{for every } v \in H$$

because $w \in L^2(0, \infty; H)$. This simply says that

$$\varphi_\infty = \varphi(A^{-1}0) = \min \{\varphi(v);\ v \in H\}.$$

Since from condition 3, φ has a unique minimum point on H, the latter implies relation (4.96) as claimed.

Remark. We close our discussion here by observing that we can rewrite Eq. (4.1) as

$$u'(t) + a(0)\,Au(t) + \int_0^t a'(t-s)Au(s)\,\mathrm{d}s \ni f'(t), \quad \text{a.e. } t > 0$$

$$u(0) = f(0)$$

which has the form of Eq. (4.86) under investigation. However, it should be noticed that Theorem 4.7 does not contain Theorem 4.2 as a special case because conditions of a different nature are assumed.

Bibliographical notes

§ 1. Theorem 1.1 was first proved by Y. Kōmura [2] but the proof that we have given here is due to T. Kato [4]. Theorem 1.2 was obtained independently by J. Dorroh [1] for the case of semigroups defined on the whole space and by M. G. Crandall and A. Pazy [1] in the case of semigroups defined on a closed convex subset of a Hilbert space. Theorem 1.3 is due to Y. Kōmura [2]. A simpler proof of this result was derived by H. Brézis and A. Pazy [1] from a selection theorem for multivalued semigroups. Theorems 1.4, 1.5 and 1.6 together with their proofs are essentially contained in the paper by H. Brézis and A. Pazy [1]. A linear version of Theorem 1.4 for general Banach spaces has been given by P. R. Chernoff [1]. Other results of a similar nature were obtained by several authors, see e.g. S. Ôharu [1], J. Mermin [1], [2], I. Miyadera and S. Ôharu [1], I. Miyadera [1], [2]. Theorems 1.7 and 1.8 were obtained by H. Brézis and A. Pazy [1] (see also Brézis' book [8] for other results of a similar nature). Corollary 1.1 was first proved by J. Watanabe [1].

The results of this paragraph are due to H. Brézis [5], [7]. Under more restrictive assumptions a result similar to Theorem 2.1 was independently obtained by J. Watanabe [2]. For other regularity results for differential equations associated with subgradient mappings and several applications to nonlinear parabolic problems we refer the reader to H. Brézis [5], [6], [7].

Equations having the form (2.50) were studied by many authors. We mention in this connection the papers by A. Friedman [2], [3], D. G. Aronson [1], [2], J. A. Dubinski [1], [2], O. A. Oleinik, A. S. Kalashnikov and Chzou Yui-Lin [1], S. Ouchi [1], L. A. Peletier [1] and Y. Konishi [5]. Further results and references can be found in the book by J. L. Lions [1].

Recently, H. Brézis [11], J. Watanabe [3], J. C. Peralba [1], [2], (see also M. Biroli [3]) investigated by a method similar to that used in Section 2.1, the time dependent equations of the type

$$\frac{du}{dt}(t) + \partial\varphi(t, u(t)) \ni 0, \quad \text{a.e.} \quad \text{on }]0, T[$$

under suitable regularity conditions on the function $t \to \varphi(t,.)$. The main achievement of these papers consists in the fact the dependence on t is allowed for the domain $D(\varphi(t,.))$. A noteworthy contribution related to this problem, was given by J. J. Moreau [3], [4], [5], in the case in which $\varphi(t,.) = I_{C(t)}$ where $\{C(t); 0 \leqslant t \leqslant T\}$ is a family of closed convex subsets on a Hilbert space H.

§ 3. The results given in Sections 3.1 and 3.2 are sharpened forms of the earlier results of Lions due to H. Brézis [6]; see J. L. Lions [1] [5] for a discussion of the literature on the subject. For more details on the applications of the theory we refer the reader to the survey of J. L. Lions [4] and the book by G. Duvaut and J. L. Lions [1]. The approach used here is similar to that developed in the quoted paper by H. Brézis which contains many other important results in this field.

The results of Section 3.3 are due to the author [8]. These results may be compared most closely with those of C. Bardos and H. Brézis [1], and H. Gajewski and K. Zacharias [1], [2]. However, there is not a large overlap and the methods as well as the conditions used here are very different.

§ 4. The results of Section 4.2, 4.3 and 4.4 are due to the author [7], [9]. In a long series of papers, J. J. Levin and J. Nohel [1], [2], [3] have contributed remarkably to the problem of asymptotic stability of the solutions of Eqs. (4.1) and (4.63) in the special case $H = R^1$. Other remarkable results in this direction have been obtained by A. Halanay [2], K. B. Hannsgen [1], and S. O. Londen [1]. Various results and further references on this subject can be found in the book by C. Corduneanu [2] and also in R. K. Miller's book [1]. This paragraph attempts to extend the ideas of Levin and Nohel to functional-differential equations on Hilbert spaces. Among the first papers devoted to such topics and whose contents are similar to the spirit of this paragraph, we shall mention those by M. Artola [1] and C. Dafermos [1]. Some of Artola's results are generalized by those included in Section 4.4. Recently, R. C. MacCamy and J. S. Wong [1] developed some ideas of A. Halanay [2] into the general framework of Hilbert spaces to obtain stability results for the equations having the form investigated in Section 4.4. Recently, other results related to those given in Sections 4.2 and 4.3 were established by R. C. MacCamy [2], [3] and S. O. Londen [2].

However the conditions that these authors have formulated, while more general on a, involve stronger assumptions on A than ours (boundedness, continuity, ect.) and certainly the methods are quite different. In concluding these notes, we shall mention that for the most part these results are also applicable to functional-differential equations with operator kernels. Along these lines the papers quoted above and also the work of S. Aizicovici [1] can be cited.

Chapter V

Second order nonlinear differential equations

In this chapter we shall develop and apply the existence theory for first order differential equations to the study of certain second order differential equations associated with nonlinear maximal monotone operators in Banach spaces. We shall treat two types of problems. One is concerned with nonlinear differential equations of hyperbolic type. The other is concerned with boundary value problems for second order differential equations.

§ 1. Nonlinear differential equations of hyperbolic type

We are given two real Hilbert spaces V and H such that $V \subset H$ and the inclusion mapping of V into H is continuous and densely defined. V' denotes the dual space of V and H is identified with its own dual H'. Notations for inner products and norms will be the same as in Chapter IV, Section 3.1. In particular, (v_1, v_2) denotes the inner product between an element of V and an element of V'; if $v_1, v_2 \in H$, this is the ordinary inner product in H.

Throughout this paragraph A will be a linear continuous and symmetric operator from V to V' and M a nonlinear (multivalued) monotone operator from V to V'.

1.1. The equation $\dfrac{d^2u}{dt^2} + Au + M\left(\dfrac{du}{dt}\right) \ni f$

In this section we are interested in the problem

$$\begin{aligned} &\frac{d^2u}{dt^2} + Au + M\left(\frac{du}{dt}\right) \ni f(t), \quad \text{a.e. on }]0, T[\\ &u(0) = u_0, \quad \frac{du}{dt}(0) = v_0 \end{aligned} \tag{1.1}$$

where M is a maximal monotone set in $V \times V'$ and $A \in L(V, V')$ is assumed to satisfy the following coercivity condition

$$(Au, u) + \alpha|u|^2 \geqslant \omega\|u\|^2, \quad \forall u \in V. \tag{1.2}$$

where $\alpha \in R^1$ and $\omega > 0$.

The main result is

THEOREM 1.1. *Let f be given in $W^{1,1}(0, T; H)$. Let u_0, v_0 be given satisfying*

$$\begin{aligned} &u_0 \in V, \quad v_0 \in D(M) \\ &\{Au_0 + Mv_0\} \cap H \neq \emptyset. \end{aligned} \tag{1.3}$$

Then there exists a unique solution $u(t)$ of problem (1.1) *which satisfies:*

$$u \in W^{1,\infty}(0, T; V) \cap W^{2,\infty}(0, T; H), \tag{1.4}$$

$$\frac{\mathrm{d}^+}{\mathrm{d}t} u(t) \text{ and } \frac{\mathrm{d}^+}{\mathrm{d}t}\left(\frac{\mathrm{d}u}{\mathrm{d}t}(t)\right) \text{ exist in } V, \text{ respectively in } H, \text{ for all } t \in [0, T[, \tag{1.5}$$

$$\frac{\mathrm{d}^+}{\mathrm{d}t}\frac{\mathrm{d}u}{\mathrm{d}t}(t) + Au(t) + M\left(\frac{\mathrm{d}u}{\mathrm{d}t}(t)\right) \ni f(t), \quad 0 \leqslant t < T. \tag{1.6}$$

Proof. Let $X = V \times H$ be the Hilbert space endowed with the inner product

$$\langle U_1, U_2\rangle = (Au_1, u_2) + \alpha(u_1, u_2) + (v_1, v_2),$$

where $U_1 = [u_1, v_1]$, $U_2 = [u_2, v_2]$.

Let $\mathcal{M} \subset X \times X$ be the set defined by

$$D(\mathcal{M}) = \{[u, v] \in V \times H;\ \{Au + Mv\} \cap H \neq \emptyset\},$$

$$\mathcal{M}[u, v] = [-v, \{Au + Mv\} \cap H] + \sigma[u, v];\ [u, v] \in D(\mathcal{M}),$$

where

$$\sigma = \sup\left\{\frac{\alpha(u, v)}{(Au, u) + \alpha|u|^2 + |v|^2};\ u \in V;\ v \in H,\ \ \|u\| + |v| \neq 0\right\}. \tag{1.7}$$

Then problem (1.1) can be written as

$$\begin{cases} \dfrac{\mathrm{d}}{\mathrm{d}t} U(t) \in -\mathcal{M}U(t) + \sigma U(t) + F(t), \quad 0 \leqslant t < T, \\ U(0) = U_0, \end{cases} \tag{1.8}$$

while

$$U(t) = \left[u(t), \frac{du}{dt}(t) \right], \quad F(t) = [0, f(t)], \, U_0 = [u_0, v_0].$$

It is easily verified that operator $\mathfrak{M}$ is monotone in $X \times X$. We assert that $\mathfrak{M}$ is maximal monotone in $X \times X$, i.e. $R(1 + \mathfrak{M}) = V \times H$. Let $[g, h]$ be arbitrary given in $V \times H$. Equation $U + \mathfrak{M}U = [g, h]$ can be written as

$$u + (-v + \sigma u) = g$$

$$v + (Au + Mv + \sigma v) \ni h$$

or

$$\begin{aligned} &(1 + \sigma)\, u - v = g \\ &(1 + \sigma)\, v + Au + Mv \ni h. \end{aligned} \tag{1.9}$$

Substituting $u = (1 + \sigma)^{-1}(v + g)$ in the second equation of system (1.9), we obtain

$$(1 + \sigma)\, v + (1 + \sigma)^{-1} Av + Mv \ni h - (1 + \sigma)^{-1} Ag. \tag{1.10}$$

Note that our assumptions about A imply that operator $v \to (1 + \sigma)v + (1 + \sigma)^{-1}Av$ is continuous, positive and coercive from V to V'. Then it follows from Corollary 1.3 in Chapter II that the range of the mapping $v \to (1 + \sigma)\, v + (1 + \sigma)^{-1} Av + Mv$ is all of V'. In particular, this shows that Eq. (1.10) has a unique solution $v \in D(M)$, and *a fortiori* that $[h, g] \in R \in (1 + \mathfrak{M})$. Then the conclusion of Theorem 1.1 follows from Theorem 2.3 stated in Chapter III.

Suppose now that $M = \partial\varphi$ where φ is a lower-semicontinuous proper convex function from V to $]-\infty, +\infty]$. As an application of Theorem 1.1, we have the following existence result on *hyperbolic variational inequations*.

COROLLARY 1.1. *Let $f \in W^{1,1}(0, T; H)$, $u_0 \in V$ and $v_0 \in D(\partial\varphi)$ be such that*

$$\{Au_0 + \partial\varphi(v_0)\} \cap H \neq \emptyset. \tag{1.11}$$

Then there exists a unique function $u \in C(0, T; V)$ which satisfies

$$\frac{du}{dt} \in C(0, T; H) \cap L^\infty(0, T; V), \quad \frac{du}{dt}(t) \in D(\varphi), \quad \forall t \in]0, T[\tag{1.12}$$

$$\frac{d^2u}{dt^2} \in L^\infty(0, T; H), \tag{1.13}$$

$$\left(\frac{d^2u}{dt} + Au - f,\ v - \frac{du}{dt} \right) \geqslant \varphi\left(\frac{du}{dt} \right) - \varphi(v), \quad \text{a.e. } t \in]0, T[, \ \forall v \in V \tag{1.14}$$

$$u(0) = u_0; \quad \frac{du}{dt}(0) = v_0. \tag{1.15}$$

Remark. An analogue of Theorem 1.1 may also be proved for Eq. (1.1) in which the initial conditions $u(0) = u_0$, $\frac{du}{dt}(0) = v_0$ are replaced by the "periodicity" conditions $u(0) = u(T)$, $\frac{du}{dt}(0) = \frac{du}{dt}(T)$. To this purpose we may use Theorem 2.4 in Chapter III but we shall not give the details.

1.2. Further investigation of the preceding case

In the present section we shall consider the sharpening of Corollary 1.1 that can be obtained if one assumes that operator A satisfies certain additional conditions. More precisely, we shall assume that

(a) There is $h \in H$ such that for every $u \in D(\varphi)$ and $\varepsilon \in]0, 1[$ solution u_ε of the equation $u_\varepsilon + \varepsilon A u_\varepsilon = u + \varepsilon h$ lies in $D(\varphi)$ and satisfies

$$\varphi(u_\varepsilon) \leqslant \varphi(u). \tag{1.16}$$

We shall assume also that A satisfies condition (1.2) strengthened to

$$(Au, u) \geqslant \omega \|u\|^2 \quad \text{for all } u \in V, \tag{1.17}$$

where $\omega > 0$.

The results we shall give here may be compared most closely with those given in Section 3.1, Chapter IV.

THEOREM 1.2. *Let A and φ satisfy conditions* (a) *and* (1.17) *and let f, u_0, v_0 be given such that*

$$Au_0 \in H, \ v_0 \in \overline{D(\varphi)}^H \cap V, \ \ f \in W^{1,2}(0, T; H). \tag{1.18}$$

Then there exists a unique function $u \in C(0, T; V)$ which satisfies

$$\frac{du}{dt} \in C(0, T; H) \cap L^\infty(0, T; V), \ \frac{du}{dt} \in D(\varphi), \text{ a.e. } t \in]0, T[\tag{1.19}$$

$$\frac{du}{dt} \textit{ is a.e. differentiable on }]0, T[\textit{ and } \sqrt{t}\,\frac{d^2u}{dt^2} \in L^2(0, T; H) \tag{1.20}$$

$$Au \in L^\infty(0, T; H), \ \ t\frac{d^2u}{dt^2} \in L^\infty(0, T; H), \tag{1.21}$$

$$\left(\frac{d^2u}{dt^2} + Au - f, \ v - \frac{du}{dt}\right) \geqslant \varphi\left(\frac{du}{dt}\right) - \varphi(v), \ \forall v \in V, \ \text{ a.e. } on\ ]0, T[\tag{1.22}$$

$$u(0) = u_0; \ \frac{du}{dt}(0) = v_0. \tag{1.23}$$

Moreover, if $v_0 \in D(\varphi)$ then $\frac{d^2u}{dt^2} \in L^2(0, T; H)$.

We shall prove Theorem 1.2 in several steps, beginning with the following Lemma.

LEMMA 1.1. *Let the conditions of Theorem* 1.2 *be satisfied. We further assume that*

$$Au_0 \in H, \quad \partial\varphi(v_0) \cap H \neq \emptyset. \tag{1.24}$$

Then solution $u(t)$ of problem (1.22) $\sim$ (1.23) *satisfies the following estimate*

$$|Au(t)|^2 + \left\| \frac{\mathrm{d}u}{\mathrm{d}t}(t) \right\|^2 \leqslant C \Bigg(|Au_0|^2 + \|v_0\|^2 + |f(0)|^2 + \\ + \int_0^T \left| \frac{\mathrm{d}f}{\mathrm{d}t}(t) \right|^2 \mathrm{d}t + 1 \Bigg), \quad \text{for } t \in [0, T[. \tag{1.25}$$

where C is a positive constant independent of u_0 and v_0.

Proof. First, we note that if the assumptions of Corollary 1.1 are satisfied, the variational problem (1.22) has a unique solution $u \in W^{1,\infty}(0, T; V) \cap W^{2,\infty}(0, T; H)$. We set $D(A) = \{u \in V;\ Au \in H\}$ and denote again by A the operator $u \to Au$ with the domain $D(A)$. From the hypotheses of our Lemma, we know that this operator is maximal monotone in $H \times H$. We set

$$A_\varepsilon = \varepsilon^{-1}(1 - (1 + \varepsilon A)^{-1}) = A(1 + \varepsilon A)^{-1}, \quad \varepsilon > 0$$

and note that A_ε is monotone, continuous and self-adjoint from H to itself.

In inequality (1.22) we take $v = (1 + \varepsilon A)^{-1}\left(\frac{\mathrm{d}u}{\mathrm{d}t} + \varepsilon h\right)$ and apply condition (a). We obtain

$$-\left(\frac{\mathrm{d}^2u}{\mathrm{d}t^2},\ A_\varepsilon \frac{\mathrm{d}u}{\mathrm{d}t}\right) - \left(Au,\ A_\varepsilon \frac{\mathrm{d}u}{\mathrm{d}t}\right) \geqslant -\left(f,\ A_\varepsilon \frac{\mathrm{d}u}{\mathrm{d}t}\right) - \\ - \left(\frac{\mathrm{d}^2u}{\mathrm{d}t^2} + Au - f,\ (1 + \varepsilon A)^{-1}h\right), \quad \text{a.e. } t \in]0, T[.$$

Hence

$$2^{-1}\frac{\mathrm{d}}{\mathrm{d}t}\left[\left(A_\varepsilon \frac{\mathrm{d}u}{\mathrm{d}t},\ \frac{\mathrm{d}u}{\mathrm{d}t}\right) + (Au, A_\varepsilon u)\right] \leqslant \left(A_\varepsilon f,\ \frac{\mathrm{d}u}{\mathrm{d}t}\right) + \\ + (Au - f,\ (1 + \varepsilon A)^{-1}h) + \frac{\mathrm{d}}{\mathrm{d}t}\left(\frac{\mathrm{d}u}{\mathrm{d}t},\ (1 + \varepsilon A)^{-1}h\right). \tag{1.26}$$

Integrating over $]0, t[$, yields

$$\left(A_\varepsilon \frac{\mathrm{d}u}{\mathrm{d}t}, \frac{\mathrm{d}u}{\mathrm{d}t}\right) + (Au, A_\varepsilon u) \leqslant (A_\varepsilon v_0, v_0) + (Au_0, A_\varepsilon u_0) +$$

$$+ 2\left[\int_0^t \left(f(s), A_\varepsilon \frac{\mathrm{d}u}{\mathrm{d}s}\right) \mathrm{d}s + \int_0^t (Au - f(s), (1 + \varepsilon A)^{-1}h) \, \mathrm{d}s +\right.$$

$$\left. + \left(\frac{\mathrm{d}u}{\mathrm{d}t} - v_0, (1 + \varepsilon A)^{-1}h\right)\right], \quad 0 \leqslant t \leqslant T. \tag{1.27}$$

On the other hand we multiply equation

$$\frac{\mathrm{d}^2u}{\mathrm{d}t^2} + Au + \partial\varphi\left(\frac{\mathrm{d}u}{\mathrm{d}t}\right) \ni f$$

by $\dfrac{\mathrm{d}u}{\mathrm{d}t} - v_0$, to obtain that

$$2^{-1}\frac{\mathrm{d}}{\mathrm{d}t}\left[\left|\frac{\mathrm{d}u}{\mathrm{d}t} - v_0\right|^2 + (Au, u)\right] \leqslant \left(f, \frac{\mathrm{d}u}{\mathrm{d}t} - v_0\right) - \left(v_0^*, \frac{\mathrm{d}u}{\mathrm{d}t} - v_0\right) +$$

$$+ (Au, v_0), \quad \text{a.e. } t \in]0, T[$$

where v_0^* is some element in $\partial\varphi(v_0) \cap H$. Integrating this last inequality over $]0, t[$ and rearranging we find

$$\left|\frac{\mathrm{d}u}{\mathrm{d}t} - v_0\right|^2 + (Au, u) \leqslant (Au_0, u_0) + C_1\left(\int_0^t \left|\frac{\mathrm{d}u}{\mathrm{d}t} - v_0\right|^2 \mathrm{d}s +\right.$$

$$\left. + \int_0^t (|f(s)|^2 + \|u(s)\|^2) \, \mathrm{d}s + 1\right), \quad 0 \leqslant t < T, \tag{1.28}$$

where C_1 is some positive constant independent of u_0, v_0 and f.

Using condition (1.17) and Gronwal's inequality we obtain that

$$\left|\frac{\mathrm{d}u}{\mathrm{d}t}(t)\right| + \|u(t)\| \leqslant C_2(\|u_0\| + |v_0| + \|f\|_{L^2(0, T; H)} + 1). \tag{1.29}$$

We use the identity

$$(1 + \varepsilon A)^{-1}x + \varepsilon A (1 + \varepsilon A)^{-1}x = x, \quad \forall x \in H$$

which implies by condition (1.17) that

$$\omega\|(1+\varepsilon A)^{-1}x\|^2+\varepsilon|A_\varepsilon x|^2 \leqslant (A_\varepsilon x, x) \leqslant \|A\|\,\|(1+\varepsilon A)^{-1}x\|\,\|x\| \tag{1.30}$$

for all $x \in V$. Here $\|A\|$ denotes the norm of A in $L(V, V')$.

On the other hand we have

$$\int_0^t \left(f(s), A_\varepsilon \frac{du}{ds}(s)\right) ds = (f(t), A_\varepsilon u(t)) - (f(0), A_\varepsilon u_0)$$

$$-\int_0^t \left(\frac{df}{ds}(s), A_\varepsilon u(s)\right) ds.$$

Putting this relation in inequality (1.27) and using (1.30) we find that

$$\left(A_\varepsilon \frac{du}{dt}, \frac{du}{dt}\right) + (A_\varepsilon u, Au) \leqslant C_3\Bigg(\|v_0\|^2 + |A_\varepsilon u(t)|\,|f(t)| +$$

$$+\left|\frac{du}{dt} - v_0\right| + \int_0^t \left(\left|\frac{df}{ds}(s)\right|^2 + |A_\varepsilon u(s)|^2 + |f(s)|^2\right) ds +$$

$$+|f(0)|^2 + |Au_0|^2\Bigg).$$

Comparing this inequality with (1.29) and making use of the inequality

$$(A_\varepsilon u, Au) \geqslant |A_\varepsilon u|^2, \qquad \forall u \in V, \quad \varepsilon > 0.$$

we conclude that

$$\left(A_\varepsilon \frac{du}{dt}, \frac{du}{dt}\right) + |A_\varepsilon u|^2 \leqslant C_4\Bigg(|Au_0|^2 + \|v_0\|^2 + |f(0)|^2$$

$$+\int_0^t \left|\frac{df}{dt}(t)\right|^2 dt + \int_0^t |A_\varepsilon u(s)|^2 ds + 1\Bigg). \tag{1.31}$$

In particular, this inequality (see, for example Proposition 3.6 in Chapter II) shows that $\lim_{\varepsilon\to 0} A_\varepsilon u(t) = Au(t)$ in H for every $t \in [0, T]$. Therefore, letting $\varepsilon \to 0$, we obtain from (1.30) and (1.31) that

$$\omega\left\|\frac{du}{dt}(t)\right\|^2 + |Au(t)|^2 \leqslant C_4\Bigg(|Au_0|^2 + \|v_0\|^2 + |f(0)|^2 +$$

$$+\left\|\frac{df}{dt}\right\|^2_{L^2(0,T;\,H)} + \int_0^t |Au(s)|^2 ds + 1\Bigg)$$

which clearly implies the desired estimate (1.25).

LEMMA 1.2. *Let v_0 be given in $\overline{D(\varphi)}^H \cap V$. Then there exists a sequence $\{v_n\} \subset D(\partial\varphi)$ such that $\partial\varphi(v_n) \cap H \neq \emptyset$, $\lim_{n\to\infty} v_n = v_0$ in H and $\{v_n\}$ is bounded in V.*

Proof. Inasmuch as A is monotone, continuous and coercive from V to V' and $\partial\varphi$ is maximal monotone in $V \times V'$, for every positive integer n, the equation

$$v_n + n^{-1}(A + \partial\varphi)v_n \ni v_0 \tag{1.32}$$

has a unique solution $v_n \in D(\partial\varphi)$. By (a) we have

$$(A_\varepsilon v_n, v_n) + n^{-1}|A_\varepsilon v_n|^2 \leqslant (A_\varepsilon v_n, v_0) + $$
$$+ (v_n, (1 + \varepsilon A)^{-1}h) + n^{-1}(A_\varepsilon v_n, h). \tag{1.33}$$

Hence

$$(A_\varepsilon v_n, v_n) + n^{-1}|A_\varepsilon v_n|^2 \leqslant |A_\varepsilon v_n| (| v_0 | + n^{-1}|h|) + |v_n|\, |h|.$$

Therefore $Av_n \in H$ and $\lim_{\varepsilon\to 0} A_\varepsilon v_n = Av_n$ in H for every n. In particular, this implies that $\partial\varphi(v_n) \cap H \neq \emptyset$ for all n. Letting $\varepsilon \to 0$ in inequality (1.33), yields the following estimate for v_n:

$$(Av_n, v_n) + n^{-1}|Av_n|^2 \leqslant (Av_n, v_0) + (v_n, h) + n^{-1}(Av_n, h).$$

Finally

$$\omega\|v_n\|^2 + n^{-1}|Av_n|^2 \leqslant \|v_n\|\, \|Av_0\|_* + | v_n |\, |h| + h^{-1}|Av_n|\, |h|.$$

This implies that $\{v_n\}$ is a bounded set of V. Next we multiply Eq. (1.32) with $v_n - v_0$. If $v_0 \in D(\varphi)$ we obtain

$$|v_n - v_0|^2 + n^{-1}(Av_0, v_n - v_0) + n^{-1}(\varphi(v_n) - \varphi(v_0)) \leqslant 0.$$

Recalling that φ is bounded from below by an affine function on V (see Proposition 2.1 in Chapter II) we may conclude that

$$\lim_{n\to\infty} v_n = v_0 \quad \text{in } H.$$

Obviously, the last relation extends to all of $\overline{D(\varphi)}^H \cap V$.

LEMMA 1.3. *Under the assumptions of Lemma 1.3, solution $u(t)$ of problem (1.22) satisfies the following estimates*

$$\int_0^T t\left|\frac{d^2u}{dt^2}\right|^2 dt \leqslant C\left(|Au_0|^2 + \|v_0\|^2 + |f(0)|^2 + \int_0^T \left|\frac{df}{dt}\right|^2 dt + 1\right), \quad (1.34)$$

$$\int_0^T \left|\frac{d^2u}{dt^2}\right|^2 dt \leqslant C\left(\varphi(v_0) + |Au_0|^2 + \|v_0\|^2 + |f(0)|^2 + \int_0^T \left|\frac{df}{dt}\right|^2 dt + 1\right), \quad (1.35)$$

$$t\left|\frac{d^2u}{dt^2}\right| \leqslant C\left(|Au_0| + \|v_0\| + |f(0)| + \left(\int_0^T \left|\frac{df}{dt}\right|^2 dt\right)^{1/2} + 1\right), \quad (1.36)$$

where C is a positive constant independent of u_0, v_0 and f.

Proof. Let ψ be the extended real-valued function on H defined by

$$\psi(u) = \begin{cases} \varphi(u) & \text{if } u \in V \\ +\infty & \text{if } u \bar{\in} V \end{cases}$$

and denote by $\bar{\psi}$ the *closure* of ψ in H, i.e.,

$$\bar{\psi}(u) = \liminf_{v \to u} \psi(v) \quad \text{for all } u \in H$$

where "lim inf" is considered in the strong topology of H. Let $[u_1, v_1] \in \partial\varphi$ be such that $v_1 \in H$. By the definition of $\partial\varphi$ we have

$$\psi(v) \geqslant \psi(u_1) + (v_1, v - u_1) \quad \text{for all } v \text{ in } H,$$

and this implies that $\bar{\psi}(u) > -\infty$ for all $u \in H$. Moreover, $\bar{\psi}$ is convex and lower-semicontinuous on H. We note also that $\partial\varphi \subset \partial\bar{\psi}$. In other words

$$\partial\varphi(u) \cap H \subset \partial\bar{\psi}(u) \quad \text{for all } u \in V.$$

Since inequation (1.22) is equivalent to

$$\frac{d^2u}{dt^2} + Au + \partial\varphi\left(\frac{du}{dt}\right) \ni f,$$

we may conclude that

$$\frac{d^2u}{dt^2} + Au + \partial\bar{\psi}\left(\frac{du}{dt}\right) \ni f, \quad \text{a.e. on }]0, T[\quad (1.37)$$

Since $\dfrac{d^2u}{dt^2}$ and Au belong to $L^\infty(0, T; H)$, it follows from Lemma 2.1 in Chapter IV that

$$\frac{d}{dt}\bar{\psi}\left(\frac{du}{dt}\right) = \left(f - \frac{d^2u}{dt^2} - Au, \frac{d^2u}{dt^2}\right), \quad \text{a.e. on }]0, T[.$$

Thus multiplying Eq. (1.37) by $t\dfrac{d^2u}{dt^2}$ and integrating the result, yields

$$\int_0^T t\left|\frac{d^2u}{dt^2}\right|^2 dt + T\bar{\psi}\left(\frac{du}{dt}(T)\right) = \int_0^T \bar{\psi}\left(\frac{du}{dt}\right) dt + \\ + \int_0^T t\left(f - Au, \frac{d^2u}{dt^2}\right) dt. \tag{1.38}$$

On the other hand, using once again Eq. (1.37), we find that

$$\bar{\psi}\left(\frac{du}{dt}\right) - \bar{\psi}(v) \leqslant \left(f - \frac{d^2u}{dt^2} - Au, \frac{du}{dt} - v\right) \quad \text{for all } v \in H$$

and therefore

$$\int_0^T \bar{\psi}\left(\frac{du}{dt}\right) dt \leqslant T\bar{\psi}(v) + \frac{1}{2}|v_0 - v|^2 + \frac{1}{2}\int_0^T |f - Au|^2 \, dt \\ + \frac{1}{2}\int_0^T \left|\frac{du}{dt} - v\right|^2 dt \quad \text{for all } v \in H.$$

Estimate (1.25) then implies that

$$\int_0^T \bar{\psi}\left(\frac{du}{dt}\right) dt \leqslant C\left(\|v_0\|^2 + |Au_0|^2 + |f(0)|^2 + \int_0^T \left|\frac{df}{dt}\right|^2 dt + 1\right).$$

On the other hand, $\bar{\psi}$ being lower-semicontinuous, there exist $u_1 \in H$ and $\alpha \in R^1$ such that

$$\bar{\psi}\left(\frac{du}{dt}(T)\right) \geqslant \left(u_1, \frac{du}{dt}(T)\right) + \alpha.$$

Putting all these estimates into Eq. (1.38) and using once again (1.25), one obtains (1.34). Estimate (1.35) may be obtained in a similar manner.

In order to prove estimate (1.36), we note first that from estimate (1.37) we have

$$\frac{1}{2}\frac{\mathrm{d}}{\mathrm{d}t}\left(\left|\frac{\mathrm{d}u}{\mathrm{d}t}(t+h)-\frac{\mathrm{d}u}{\mathrm{d}t}(t)\right|^2+(Au(t+h)-Au(t),u(t+h)-u(t))\right)\leqslant$$

$$\left(f(t+h)-f(t),\frac{\mathrm{d}u}{\mathrm{d}t}(t+h)-\frac{\mathrm{d}u}{\mathrm{d}t}(t)\right),\quad \text{a.e. } t\in\,]0,T[.$$

Hence

$$\left|\frac{\mathrm{d}^2u}{\mathrm{d}t^2}(t)\right|\leqslant\left|\frac{\mathrm{d}^2u}{\mathrm{d}t^2}(s)\right|+C\int_s^t\left|\frac{\mathrm{d}f}{\mathrm{d}\tau}(\tau)\right|\mathrm{d}\tau,\quad 0<s<t<T. \tag{1.39}$$

This is exactly the same result as in the proof of Theorem 2.2 in Chapter IV; hence we may conclude the argument just as there.

Proof of Theorem 1.2. Let f, u_0 and v_0 be given, satisfying condition (1.18). Let v_n be chosen as in Lemma 1.2. Now denote by $u_n(t)$ the solution of the initial value problem

$$\frac{\mathrm{d}^2u_n}{\mathrm{d}t^2}+Au_n+\partial\varphi\left(\frac{\mathrm{d}u_n}{\mathrm{d}t}\right)\ni f_n,\quad \text{a.e. } t\in\,]0\,,T[;\quad u_n(0)=u_0, \tag{1.40}$$

$$\frac{\mathrm{d}u_n}{\mathrm{d}t}(0)=v_n.$$

Using the fact that $\partial\varphi$ is monotone in $V\times V'$, we conclude that

$$\frac{\mathrm{d}}{\mathrm{d}t}\left[\left|\frac{\mathrm{d}u_n}{\mathrm{d}t}-\frac{\mathrm{d}u_m}{\mathrm{d}t}\right|^2+(A(u_n-u_m),\,u_n-u_m)\right]\leqslant$$

$$\leqslant 2\left(f_n-f_m,\frac{\mathrm{d}u_n}{\mathrm{d}t}-\frac{\mathrm{d}u_m}{\mathrm{d}t}\right),\quad \text{a.e. on }]0,T[.$$

Hence

$$\left|\frac{\mathrm{d}u_n}{\mathrm{d}t}-\frac{\mathrm{d}u_m}{\mathrm{d}t}\right|^2+\omega\|u_n-u_m\|^2\leqslant|v_n-v_m|^2+\int_0^t|f_n-f_m|^2\,\mathrm{d}t+$$

$$+\int_0^t\left|\frac{\mathrm{d}u_n}{\mathrm{d}t}-\frac{\mathrm{d}u_m}{\mathrm{d}t}\right|^2\mathrm{d}t,\quad 0\leqslant t\leqslant T.$$

so that

$$\lim_{n\to\infty} u_n(t) = u(t) \qquad \text{uniformly on } [0, T] \text{ in } V$$

$$\lim_{n\to\infty} \frac{\mathrm{d}u_n}{\mathrm{d}t}(t) = \frac{\mathrm{d}u}{\mathrm{d}t}(t) \qquad \text{uniformly on } [0, T] \text{ in } H.$$

But $\sqrt{t}\,\dfrac{\mathrm{d}^2u_n}{\mathrm{d}t^2}$ is bounded in $L^2(0, T; H)$, $\dfrac{\mathrm{d}u_n}{\mathrm{d}t}$ is bounded in $L^\infty(0, T; V)$ and $t\,\dfrac{\mathrm{d}^2u_n}{\mathrm{d}t^2}$, Au_n are both bounded in $L^\infty(0, T; H)$ by the previous lemmas. Taking a weakly convergent subsequence we may therefore infer that

$$\sqrt{t}\,\frac{\mathrm{d}^2u}{\mathrm{d}t^2} \in L^2(0, T; H), \; Au \in L^\infty(0, T; H), \tag{1.41}$$

and

$$t\,\frac{\mathrm{d}^2u}{\mathrm{d}t^2} \in L^\infty(0, T; H), \; \frac{\mathrm{d}u}{\mathrm{d}t} \in L^\infty(0, T; V). \tag{1.42}$$

Furthermore, if $u_0 \in D(\varphi)$ then again from Lemmas 1.2 and 1.3 we infer that $\dfrac{\mathrm{d}^2u}{\mathrm{d}t} \in L^2(0, T; H)$. It remains to be proven that $u(t)$ satisfies Eq. (1.22) almost everywhere on $]0, T[$.

Let $t_0 \in]0, T[$ be such that $\dfrac{\mathrm{d}^2u}{\mathrm{d}t}(t_0)$ exists and let $[x, y]$ be arbitrarily given in $\partial\varphi$. We have

$$\frac{1}{2}\frac{\mathrm{d}}{\mathrm{d}t}\left|\frac{\mathrm{d}u_n}{\mathrm{d}t} - x\right|^2 + \left(Au_n, \frac{\mathrm{d}u_n}{\mathrm{d}t} - x\right) \geqslant \left(f_n - y, \frac{\mathrm{d}u_n}{\mathrm{d}t} - x\right)$$

so that

$$\left|\frac{\mathrm{d}u_n}{\mathrm{d}t}(t) - x\right|^2 - \left|\frac{\mathrm{d}u_n}{\mathrm{d}t}(t_0) - x\right|^2 + (Au_n(t), u_n(t)) - (Au_n(t_0), u_n(t_0)) \geqslant$$

$$2\int_{t_0}^{t}\left(f_n(s) - y, \frac{\mathrm{d}u_n}{\mathrm{d}t}(s) - x\right)\mathrm{d}s + 2\int_{t_0}^{t}(Au_n(s), x)\,\mathrm{d}s.$$

Therefore, letting $n \to \infty$, we obtain

$$\left|\frac{\mathrm{d}u}{\mathrm{d}t}(t) - x\right|^2 - \left|\frac{\mathrm{d}u}{\mathrm{d}t}(t_0) - x\right|^2 + (Au(t), u(t)) - (Au(t_0), u(t_0)) \geqslant$$

$$\geqslant 2\int_{t_0}^{t}\left(f(s) - y, \frac{\mathrm{d}u}{\mathrm{d}t}(s) - x\right)\mathrm{d}s + 2\int_{t_0}^{t}(Au(s), x)\,\mathrm{d}s. \tag{1.43}$$

(Extracting further subsequence if necessary, we may assume that $Au_n \to Au$ in the weak-star topology of $L^\infty(0, T; H)$).

We note that

$$(Au(t), u(t)) - (Au(t_0), u(t_0)) = 2\int_{t_0}^{t} \left(Au(s), \frac{du}{dt}(s)\right) ds$$

and

$$2\left(\frac{du}{dt}(t) - \frac{du}{dt}(t_0), \frac{du}{dt}(t) - x\right) \geqslant \left|\frac{du}{dt}(t) - x\right|^2 - \left|\frac{du}{dt}(t_0) - x\right|^2.$$

We further suppose that $t = t_0$ is a Lebesgue point for the functions $Au \in L^\infty(0, T; H)$ and $\dfrac{du}{dt} \in L^\infty(0, T; V)$. Putting all these inequalities in (1.43), dividing by $t - t_0$, and letting $t \to t_0$, we obtain

$$\left(\frac{d^2u}{dt^2}(t_0) + Au(t_0) - f(t_0), \frac{du}{dt}(t_0) - x\right) \geqslant -\left(y, \frac{du}{dt}(t_0) - x\right).$$

Since $\partial\varphi$ is maximal monotone in $V \times V'$ we conclude that

$$\frac{d^2u}{dt^2}(t_0) + Au(t_0) - f(t_0) \in -\partial\varphi\left(\frac{du}{dt}(t_0)\right),$$

as claimed. The uniqueness follows by a standard procedure. This completes the proof of Theorem 1.2.

Remark. If the condition on f is replaced by

$$f \in L^2(0, T; V),$$

then we can prove the existence and uniqueness of a solution $u(t)$ of the above initial value problem satisfying

$$\frac{du}{dt} \in C(0, T; H) \cap L^\infty(0, T; V), \quad \sqrt{t}\,\frac{d^2u}{dt^2} \in L^2(0, T; H).$$

We omit the proof which is a variant of that given before for Theorem 1.2.

1.3. Examples

Example 1. Let β be a maximal monotone graph in $R^1 \times R^1$ such that $0 \in D(\beta)$. Let $j: R^1 \to]-\infty, +\infty]$ be a lower-semicontinuous convex function such that $\beta = \partial j$.

Consider the hyperbolic equation

$$\frac{\partial^2 u}{\partial t^2} - \Delta u + \beta\left(\frac{\partial u}{\partial t}\right) \ni f, \quad \text{a.e. on }]0, T[\times\Omega \tag{1.44}$$

subject to the initial conditions

$$u(0, x) = u_0(x), \quad \frac{\partial u}{\partial t}(0, x) = v_0(x), \quad \text{a.e. on } \Omega \tag{1.45}$$

and to the boundary condition

$$u(t, x) = 0, \quad \text{a.e. on }]0, T[\times\Gamma. \tag{1.46}$$

Here Ω is an open and bounded subset of R^n and Γ denotes its boundary.

Various examples of equations of type (1.44) where β is a nonlinear scalar function on R^1 arise in physics. For instance if $\beta(r) = |r|\, r$ the equation represents a classical vibrating membrane with the resistance proportional to the velocity.

PROPOSITION 1.1. *Let f, u_0, v_0 be given satisfying*

$$f, \frac{\partial f}{\partial t} \in L^2(0, T; L^2(\Omega))$$

$$u_0 \in H_0^1(\Omega) \cap H^2(\Omega),$$

$$v_0 \in H_0^1(\Omega), \ v_0(x) \in \overline{D(\beta)} \quad \text{a.e. } on\ \Omega.$$

Then there exists a unique solution $u(t, x)$ for problems (1.44) ~ (1.46) *which satisfies:*

$$u \in C(0, T; H_0^1(\Omega)) \cap L^\infty(0, T; H^2(\Omega)) \tag{1.47}$$

$$\frac{\partial u}{\partial t} \in C(0, T; L^2(\Omega)) \cap L^\infty(0, T; H_0^1(\Omega)) \tag{1.48}$$

$$\sqrt{t}\, \frac{\partial^2 u}{\partial t^2} \in L^2(0, T; L^2(\Omega)), \quad t\, \frac{\partial^2 u}{\partial t^2} \in L^\infty(0, T; L^2(\Omega)). \tag{1.49}$$

Furthermore, if $j(v_0) \in L^1(\Omega)$ then

$$\frac{\partial^2 u}{\partial t^2} \in L^2(0, T; L^2(\Omega)).$$

Suppose further that there exists $g \in L^2(\Omega)$ *such that* $g_0(x) \in \beta(v_0(x))$ *a.e. on* Ω. *Then* $\dfrac{\partial^2 u}{\partial t^2} \in L^\infty(0, T; L^2(\Omega))$ *and*

$$\frac{\partial^+}{\partial t}\left(\frac{\partial u}{\partial t}(t)\right) - \Delta u(t) + \beta(u(t)) \ni f(t), \quad \forall t \in [0, T[, \quad \text{a.e. } on\ \Omega.$$

Proof. We shall apply Theorem 1.1 and Theorem 1.2 where $V = H_0^1(\Omega)$, $H = L^2(\Omega)$, $Au = -\Delta u$ for $u \in H_0^1(\Omega)$, i.e.,

$$(Au, v) = \int_\Omega \operatorname{grad} u \operatorname{grad} v \, \mathrm{d}x, \quad \text{for all } v \in H_0^1(\Omega)$$

and

$$\varphi(u) = \int_\Omega j(u(x)) \, \mathrm{d}x \quad \text{for} \quad u \in H_0^1(\Omega).$$

Conditions (1.12) are clearly implied by our present hypotheses on f, u_0 and v_0 (We note that $\{u \in L^2(\Omega);\ Au \in L^2(\Omega)\} = H_0^1(\Omega) \cap H^2(\Omega)$). To conclude the proof it suffices to verify hypothesis (a) of Theorem 1.2.

To this end we denote by $u_\varepsilon \in H_0^1(\Omega) \cap H^2(\Omega)$ the solution of the Dirichlet problem

$$u_\varepsilon - \varepsilon \Delta u_\varepsilon = u \quad \text{on } \Omega$$

$$u_\varepsilon = 0 \quad \text{on } \Gamma$$

where u is some element of $L^2(\Omega)$ and $\varepsilon > 0$. In other words, $u_\varepsilon = (1 + \varepsilon A)^{-1} u$. Let β_λ be Yosida approximation of β and let $j_\lambda \colon R^1 \to R^1$ be such that $\beta_\lambda = \partial j_\lambda$ (see Theorem 2.2 in Chapter II). Without loss of generality we may assume that $0 \in \beta(0)$. In particular, this implies that $\beta_\lambda(0) = 0$ for all $\lambda > 0$. By definition of ∂j_λ we have

$$j_\lambda(u_\varepsilon) - j_\lambda(u) \leqslant \beta_\lambda(u_\varepsilon)(u_\varepsilon - u), \quad \forall u \in L^2(\Omega).$$

Hence

$$\int_\Omega j_\lambda(u_\varepsilon) \, \mathrm{d}x \leqslant \int_\Omega j_\lambda(u) \, \mathrm{d}x + \varepsilon \int_\Omega \beta_\lambda(u_\varepsilon) \Delta u_\varepsilon \, \mathrm{d}x.$$

The function β_λ being monotone, we have

$$\int_\Omega \beta_\lambda(u_\varepsilon) \Delta u_\varepsilon \, \mathrm{d}x = -\int_\Omega \operatorname{grad} \beta_\lambda(u_\varepsilon) \operatorname{grad} u_\varepsilon \, \mathrm{d}x \leqslant 0,$$

and therefore,

$$\int_\Omega j_\lambda(u_\varepsilon)\,\mathrm{d}x \leqslant \int_\Omega j_\lambda(u)\,\mathrm{d}x, \quad \forall u \in L^2(\Omega).$$

Letting $\lambda \to 0$ gives

$$\int_\Omega j(u_\varepsilon(x))\,\mathrm{d}x \leqslant \int_\Omega j(u(x))\,\mathrm{d}x, \quad \text{for all } \varepsilon > 0.$$

Thus we have proved that

$$\varphi((1 + \varepsilon A)^{-1}u) \leqslant \varphi(u), \quad \forall u \in L^2(\Omega),$$

which is just condition (a) with $h = 0$.

Example 2. Let $\psi \in H^2(\Omega)$ be such that $\psi(x) \leqslant 0$ a.e. on Γ. Consider the following hyperbolic unilateral problem:

$$\begin{aligned}
&\frac{\partial u}{\partial t} \geqslant \psi && \text{a.e. on }]0, T[\times\Omega,\\
&\frac{\partial^2 u}{\partial t^2} = \Delta u + f && \text{a.e. on } \left\{(t, x);\ \frac{\partial u}{\partial t} > \psi\right\},\\
&\frac{\partial^2 u}{\partial t^2} \geqslant \Delta u + f && \text{a.e. on } \left\{(t, x);\ \frac{\partial u}{\partial t} = \psi\right\},\\
&u(t, x) = 0 && \text{a.e. on }]0, T[\times\Gamma,\\
&u(0, x) = u_0(x); && \frac{\partial u}{\partial t}(0, x) = v_0(x) \quad \text{a.e. on } \Omega.
\end{aligned} \tag{1.50}$$

PROPOSITION 1.2. *Let u_0, v_0 and f be given satisfying:*

$$u_0 \in H^1_0(\Omega) \cap H^2(\Omega),\ v_0 \in H^1_0(\Omega),\ v_0 \geqslant \psi \quad \text{a.e. on } \Omega. \tag{1.51}$$

$$f,\ \frac{\partial f}{\partial t} \in L^2(0, T; L^2(\Omega)). \tag{1.52}$$

Then problem (1.50) *has a unique solution* $u(t, x)$ *which satisfies:*

$$u \in C(0, T; H_0^1(\Omega)) \cap L^\infty(0, T; H^2(\Omega)), \tag{1.53}$$

$$\frac{\partial u}{\partial t} \in L^\infty(0, T; H_0^1(\Omega)) \cap C(0, T; L^2(\Omega)), \tag{1.54}$$

$$\frac{\partial^2 u}{\partial t^2} \in L^\infty(0, T; L^2(\Omega)). \tag{1.55}$$

Proof. We take $V = H_0^1(\Omega)$, $H = L^2(\Omega)$, $A = -\Delta$ and $\varphi = I_K$ where I_K is the indicator function of the closed convex subset K of $H_0^1(\Omega)$ defined by

$$K = \{u \in H_0^1(\Omega);\ u \geqslant \psi \text{ a.e. on } \Omega\}.$$

Our present assumptions clearly imply that

$$Au_0 \in H \quad \text{and} \quad \partial I_K(v_0) \cap H \neq \emptyset. \tag{1.56}$$

Now hypothesis (a) easily follows from the maximum principle, taking $h = -\Delta\psi$. The general theory can therefore be applied to the present situation to conclude that there exists a unique function u which satisfies conditions (1.53) $\sim$ (1.55) and inequality

$$\int_\Omega \left(\frac{\partial^2 u}{\partial t^2} - \Delta u - f \right)\left(v - \frac{\partial u}{\partial t} \right) \mathrm{d}x \geqslant 0, \quad \text{a.e. } t \in]0, T[, \quad \forall v \in K. \tag{1.57}$$

The last argument in the proof of Corollary 3.4 in Chapter IV can now be repeated here to deduce that $u(t)$ is just the solution of problem (1.50).

1.4. Singular perturbations and hyperbolic variational inequations

Let A be a linear, continuous and symmetric operator from V to V' such that

$$(Au, u) \geqslant \omega\ \|u\|^2, \quad \forall u \in V,\ \omega > 0.$$

Let K be a closed convex subset of V such that $0 \in K$. We are also given u_0, v_0 and f which are assumed to satisfy

$$f \in W^{1,2}(0, T; H), \tag{1.58}$$

$$Au_0 \in H;\ v_0 \in K. \tag{1.59}$$

Applying Theorem 1.1 where $Mu = \frac{1}{\varepsilon}(u + \partial I_K(u))$ we conclude that for every $\varepsilon > 0$ there exists a unique function $u_\varepsilon(t) \in C(0, T; V)$ which satisfies

$$\frac{\mathrm{d}u_\varepsilon}{\mathrm{d}t} \in C(0, T; H) \cap L^\infty(0, T; V), \frac{\mathrm{d}u_\varepsilon}{\mathrm{d}t} \in K, \quad \text{a.e.} \quad t \in]0, T[, \tag{1.60}$$

$$\frac{\mathrm{d}^2u_\varepsilon}{\mathrm{d}t^2} \in L^\infty(0, T; H), \tag{1.61}$$

$$\left(\varepsilon \frac{\mathrm{d}^2u_\varepsilon}{\mathrm{d}t^2}(t) + \frac{\mathrm{d}u_\varepsilon}{\mathrm{d}t}(t) + Au_\varepsilon(t) - f(t),\ v - \frac{\mathrm{d}u_\varepsilon}{\mathrm{d}t}(t)\right) \geqslant 0, \quad \text{a.e. on }]0, T[\tag{1.62}$$

for every $v \in K$, and

$$u_\varepsilon(0) = u_0, \quad \frac{\mathrm{d}u_\varepsilon}{\mathrm{d}t}(0) = v_0. \tag{1.63}$$

Secondly we consider the variational evolution inequation

$$\left(\frac{\mathrm{d}u}{\mathrm{d}t}(t) + Au(t) - f(t),\ v - \frac{\mathrm{d}u}{\mathrm{d}t}(t)\right) \geqslant 0, \quad \text{a.e. } t \in]0, T[\ \forall v \in K, \tag{1.64}$$

$$u(0) = u_0; \quad \frac{\mathrm{d}u}{\mathrm{d}t}(t) \in K \quad \text{a.e.} \quad t \in]0, T[, \tag{1.65}$$

already studied in Chapter IV. We have proved there (see Theorem 3.3) that if

$$f(0) - Au_0 \in K \tag{1.66}$$

then problems (1.64) $\sim$ (1.65) have a unique solution $u \in C(0, T; V)$ satisfying

$$\frac{\mathrm{d}u}{\mathrm{d}t} \in L^2(0, T; V), \quad \frac{\mathrm{d}^2u}{\mathrm{d}t^2} \in L^2(0, T; H). \tag{1.67}$$

One expects that $u_\varepsilon(t)$ will converge to $u(t)$ as $\varepsilon \to 0$. Our next results show that this is indeed the case under certain conditions.

THEOREM 1.3. *Let u_0, v_0 and f satisfy conditions* (1.58) *and* (1.59). *Then*

$$\|u_\varepsilon(t) - u(t)\| \leqslant C\sqrt{\varepsilon}, \quad \text{for all } t \in [0, T]. \tag{1.68}$$

$$\left(\int_0^T \left|\frac{\mathrm{d}u_\varepsilon}{\mathrm{d}t} - \frac{\mathrm{d}u}{\mathrm{d}t}\right|^2 \mathrm{d}t\right)^{1/2} \leqslant C\sqrt{\varepsilon}. \tag{1.69}$$

Proof. Let $v_\varepsilon = u_\varepsilon - u$. We take $v = \dfrac{\mathrm{d}u}{\mathrm{d}t}$ in relation (1.62) and $v_\varepsilon = \dfrac{\mathrm{d}u_\varepsilon}{\mathrm{d}t}$ in (1.64). Subtraction of the corresponding equations yields

$$\left(\varepsilon\frac{\mathrm{d}^2u_\varepsilon}{\mathrm{d}t^2} + \frac{\mathrm{d}v_\varepsilon}{\mathrm{d}t} + Av_\varepsilon, \frac{\mathrm{d}v_\varepsilon}{\mathrm{d}t}\right) \leqslant 0, \quad \text{a.e. on }]0, T[.$$

Hence

$$\left|\frac{\mathrm{d}v_\varepsilon}{\mathrm{d}t}(t)\right|^2 + \frac{1}{2}\frac{\mathrm{d}}{\mathrm{d}t}(Av_\varepsilon(t), v_\varepsilon(t)) \leqslant -\varepsilon\left(\frac{\mathrm{d}^2u_\varepsilon}{\mathrm{d}t^2}(t), \frac{\mathrm{d}v_\varepsilon}{\mathrm{d}t^2}(t)\right) \leqslant$$

$$\leqslant \frac{1}{2}\left|\frac{\mathrm{d}v_\varepsilon}{\mathrm{d}t}(t)\right|^2 + \frac{\varepsilon^2}{2}\left|\frac{\mathrm{d}^2u_\varepsilon}{\mathrm{d}t^2}(t)\right|^2, \quad \text{a.e. } t \in]0, T[,$$

and therefore,

$$\left|\frac{\mathrm{d}v_\varepsilon}{\mathrm{d}t}(t)\right|^2 + \frac{\mathrm{d}}{\mathrm{d}t}(Av_\varepsilon(t), v_\varepsilon(t)) \leqslant \varepsilon^2\left|\frac{\mathrm{d}^2u_\varepsilon}{\mathrm{d}t^2}(t)\right|^2, \quad \text{a.e. } t \in]0, T[.$$

Finally,

$$\int_0^t \left|\frac{\mathrm{d}v_\varepsilon}{\mathrm{d}t}\right|^2 \mathrm{d}t + \omega\|v_\varepsilon(t)\|^2 \leqslant \varepsilon^2 \int_0^T \left|\frac{\mathrm{d}^2u_\varepsilon}{\mathrm{d}t^2}\right|^2 \mathrm{d}t, \quad 0 \leqslant t \leqslant T, \ \varepsilon > 0. \tag{1.70}$$

From Theorem 1.1, we know that $\dfrac{\mathrm{d}^+}{\mathrm{d}t}\left(\dfrac{\mathrm{d}u_\varepsilon}{\mathrm{d}t}(t)\right)$ exists for all $t \in [0, T]$ so that Eq. (1.62) can be written as

$$\varepsilon\frac{\mathrm{d}^+}{\mathrm{d}t}\frac{\mathrm{d}u_\varepsilon}{\mathrm{d}t}(t) + \frac{\mathrm{d}u_\varepsilon}{\mathrm{d}t}(t) + Au_\varepsilon(t) + \partial I_K\left(\frac{\mathrm{d}u_\varepsilon}{\mathrm{d}t}(t)\right) \ni f(t).$$

Then

$$\varepsilon\frac{\mathrm{d}^+}{\mathrm{d}t}\frac{\mathrm{d}}{\mathrm{d}t}u_\varepsilon(0) \in f(0) - v_0 - Au_0 - \partial I_K(v_0).$$

More precisely, it follows from the proof of Theorem 1.1 that $\varepsilon \frac{d^+}{dt}\left(\frac{d^+}{dt}u_\varepsilon(0)\right) = (f(0) - v_0 - Au_0 - \partial I_K(v_0))^0$; so that we may infer that

$$\varepsilon \frac{d^+}{dt}\left(\frac{d}{dt}u_\varepsilon(0)\right) \text{ remains in a bounded set of } H. \tag{1.71}$$

Since ∂I_K is monotone, we have

$$\frac{\varepsilon}{2}\frac{d}{dt}\left|\frac{du_\varepsilon}{dt}(t+h) - \frac{du_\varepsilon}{dt}(t)\right|^2 + \left|\frac{du_\varepsilon}{dt}(t+h) - \frac{du_\varepsilon}{dt}(t)\right|^2 +$$

$$+ \left(Au_\varepsilon(t+h) - Au_\varepsilon(t), \frac{du_\varepsilon}{dt}(t+h) - \frac{du_\varepsilon}{dt}(t)\right) \leqslant$$

$$\leqslant \left(f(t+h) - f(t), \frac{du_\varepsilon}{dt}(t+h) - \frac{du_\varepsilon}{dt}(t)\right).$$

Integrating from 0 to t, we find that

$$\int_0^t \left|\frac{d}{dt}(u_\varepsilon(t+h) - u_\varepsilon(t))\right|^2 dt \leqslant \frac{\varepsilon}{2}\left|\frac{du_\varepsilon}{dt}(h) - v_0\right|^2 + C\|u_\varepsilon(h) - u_0\|^2 +$$

$$+ \int_0^T |f(t+h) - f(t)| \left|\frac{d}{dt}(u_\varepsilon(t+h) - u_\varepsilon(t))\right| dt.$$

But once again from Theorem 1.1, the right derivative $\frac{d^+}{dt}u(t)$ exists in V for all $t \in [0, T[$; so we may infer that

$$\int_0^T \left|\frac{d^2u_\varepsilon}{dt^2}\right|^2 dt \leqslant \frac{\varepsilon}{2}\left|\frac{d^+}{dt}\frac{d}{dt}u_\varepsilon(0)\right|^2 + C\|v_0\|^2 + \int_0^T \left|\frac{df}{dt}\right| \left|\frac{d^2u_\varepsilon}{dt^2}\right| dt.$$

Consequently, we conclude from relation (1.71) that

$$\varepsilon \int_0^T \left|\frac{d^2u_\varepsilon}{dt^2}\right|^2 dt \leqslant C, \quad \varepsilon > 0. \tag{1.72}$$

and the proof is complete.

The question which arises naturally is whether $\lim_{\varepsilon\to 0}\frac{du_\varepsilon}{dt} = \frac{du}{dt}$ uniformly in t in the strong topology of H.

As mentioned in the proof of Theorem 3.3 from Chapter IV, problems (1.64) and (1.65) can be written under the following equivalent form

$$\frac{\mathrm{d}v}{\mathrm{d}t} + \Lambda v = \frac{\mathrm{d}g}{\mathrm{d}t}$$

$$v(0) = A^{-1}f(0) - u_0,$$

where $\Lambda = (1 + \partial I_k)^{-1}A$, $v = g - u$ and $g = A^{-1}f$. Our condition (1.66) implies that $v(0) \in D(\Lambda)$, so that the general theory can be applied to the present situation to deduce that $\frac{\mathrm{d}^+}{\mathrm{d}t} v(0) = \frac{\mathrm{d}g}{\mathrm{d}t}(0) - \Lambda v(0)$. But explicit calculation shows that $\Lambda v(0) = f(0) - Au_0$. In other words

$$\frac{\mathrm{d}^+}{\mathrm{d}t} u(0) = f(0) - Au_0.$$

This shows that the answer to the question raised before is negative unless $v_0 = f(0) - Au_0$. However, we shall see that under certain conditions on A and K, $\frac{\mathrm{d}u_\varepsilon}{\mathrm{d}t} \to \frac{\mathrm{d}u}{\mathrm{d}t}$ "almost uniformly" on $[0, T]$.

THEOREM 1.4. *In Theorem* 1.3 *we further suppose that*

$$A(v_0 + Au_0 - f(0)) \in H, \tag{1.73}$$

$$\theta(v_0 + Au_0 - f(0)) + v \in K, \quad \forall v \in K, \quad 0 \leqslant \theta \leqslant 1. \tag{1.74}$$

Then

$$\left| \frac{\mathrm{d}}{\mathrm{d}t} (u_\varepsilon(t) - u(t) - \psi_\varepsilon(t)) \right| \leqslant C\sqrt{\varepsilon}, \quad t \in [0, T], \tag{1.75}$$

where

$$\psi_\varepsilon(t) = (v_0 + Au_0 - f(0)) \int_0^t \mathrm{e}^{-s/\varepsilon}\, \mathrm{d}s.$$

Proof. We set

$$\psi = v_0 - f(0) + Au_0; \quad v_\varepsilon = u_\varepsilon - u - \psi_\varepsilon.$$

Let $v = \frac{\mathrm{d}u}{\mathrm{d}t} + \frac{\mathrm{d}\psi_\varepsilon}{\mathrm{d}t} = \frac{\mathrm{d}u}{\mathrm{d}t} + \psi \mathrm{e}^{-t/\varepsilon}$ in (1.62) and let $v = \frac{\mathrm{d}u_\varepsilon}{\mathrm{d}t} - \frac{\mathrm{d}\psi_\varepsilon}{\mathrm{d}t}$ in (1.64) to obtain

$$-\left(\varepsilon \frac{\mathrm{d}^2 u_\varepsilon}{\mathrm{d}t^2} + \frac{\mathrm{d}u_\varepsilon}{\mathrm{d}t} + Au_\varepsilon - f - \left(\frac{\mathrm{d}u}{\mathrm{d}t} + Au - f\right), \frac{\mathrm{d}v_\varepsilon}{\mathrm{d}t}\right) \geqslant 0$$

or

$$-\left(\varepsilon \frac{d^2 v_\varepsilon}{dt^2} + \frac{dv_\varepsilon}{dt} + Av_\varepsilon + A\psi_\varepsilon + \varepsilon \frac{d^2 u}{dt^2},\ \frac{dv_\varepsilon}{dt}\right) \geqslant 0.$$

Hence

$$\frac{\varepsilon}{2} \frac{d}{dt} \left|\frac{dv_\varepsilon}{dt}\right|^2 + \left|\frac{dv_\varepsilon}{dt}\right|^2 + \frac{1}{2} \frac{d}{dt} (Av_\varepsilon,\ v_\varepsilon) \leqslant -\left(\varepsilon \frac{d^2 u}{dt^2} + A\psi_\varepsilon,\ \frac{dv_\varepsilon}{dt}\right).$$

We observe that $A\psi_\varepsilon = \varepsilon(1 - e^{-t/\varepsilon})\, A\psi$; so that $|A\psi_\varepsilon| \leqslant \varepsilon|A\psi|$. Therefore, the right-hand term of the last inequality is bounded by

$$\varepsilon \left|\frac{d^2 u}{dt^2}\right| \left|\frac{dv_\varepsilon}{dt}\right| + \varepsilon|A\psi| \left|\frac{dv_\varepsilon}{dt}\right| \leqslant \left|\frac{dv_\varepsilon}{dt}\right|^2 + \frac{\varepsilon^2}{2} |A\psi|^2 + \frac{\varepsilon^2}{2} \left|\frac{d^2 u}{dt^2}\right|^2.$$

Hence

$$\frac{\varepsilon}{2} \frac{d}{dt} \left|\frac{dv_\varepsilon}{dt}\right|^2 + \frac{1}{2} \frac{d}{dt} (Av_\varepsilon, v_\varepsilon) \leqslant \frac{\varepsilon^2}{2}\left(|A\psi|^2 + \left|\frac{d^2 u}{dt^2}\right|^2\right), \quad \text{a.e. } t \in]0, T[. \tag{1.76}$$

Integrating inequality (1.76) over $]0, T[$, yields

$$\varepsilon \left|\frac{dv_\varepsilon}{dt}(t)\right|^2 + \omega \int_0^t \|v_\varepsilon(t)\|^2\, dt \leqslant \varepsilon^2 \left(\int_0^t \left|\frac{d^2 u}{dt^2}\right|^2 dt + t|A\psi|^2\right), \quad 0 \leqslant t \leqslant T \tag{1.77}$$

since $v_\varepsilon(0) = 0$ and $\dfrac{dv_\varepsilon}{dt}(0) = \dfrac{du_\varepsilon}{dt}(0) - \dfrac{du}{dt}(0) - \psi = 0$. This implies the desired estimate. The proof is complete.

1.5. Nonlinear wave equation

In this section we shall consider the differential equation

$$\frac{d^2 u}{dt^2} + Au + Mu \ni f, \quad 0 < t < T \tag{1.78}$$

subject to the initial conditions

$$u(0) = u_0, \quad \frac{du}{dt}(0) = v_0 \tag{1.79}$$

where M is a monotone set in $H \times H$ (or in $V \times V'$) whereas A is a linear continuous operator from V to V' which satisfies

$$(Au, v) = (u, Av) \quad \text{for} \quad \text{all } u, v \text{ in } V. \tag{1.80}$$

$$(Au, u) \geqslant \omega \|u\|^2 \quad \text{for} \quad \text{all } u \in V; \ \omega > 0. \tag{1.81}$$

First we shall consider the case in which M is Lipschitzian on H, i.e.,

$$|Mu - Mv| \leqslant L|u - v| \quad \text{for} \quad \text{all } u, v \text{ in } H, \tag{1.82}$$

where L is a positive constant independent of u and v.

THEOREM 1.5. *Let f, u_0 and v_0 be given satisfying*

$$f \in W^{1,2}(0, T; H), \quad u_0 \in V; \ v_0 \in H. \tag{1.83}$$

Then the initial value problem (1.78) ~ (1.79) *has a unique solution $u(t)$ which satisfies:*

$$u \in C(0, T; H) \cap L^\infty(0, T; V) \tag{1.84}$$

$$\frac{du}{dt} \in L^\infty(0, T; H) \cap C(0, T; V') \tag{1.85}$$

$$\frac{d^2u}{dt^2} \in L^\infty(0, T; V'). \tag{1.86}$$

Moreover, if $Au_0 \in H$ and $v_0 \in V$, then

$$Au \in L^\infty(0, T; H), \tag{1.87}$$

$$\frac{du}{dt} \in L^\infty(0, T; V), \quad \frac{d^2u}{dt^2} \in L^\infty(0, T; H). \tag{1.88}$$

Proof. Let A_H denote the self-adjoint operator in H with domain $D(A) = \{u \in V; Au \in H\}$ which is given by $A_H u = Au$ for $u \in D(A)$. Let $A^{1/2}$ denote the square root of A_H. We note that $D(A^{1/2}) = V$ and

$$(Au, v) = (A^{1/2}u, A^{1/2}v) \text{ for all } u, v \text{ in } V.$$

Let $\mathscr{H} = H \times H$ be the product space endowed with the inner product

$$\langle U, V \rangle = (u_1, v_1) + (u_2, v_2); \quad U = [u_1, u_2]; \ V = [v_1, v_2].$$

Let $\mathscr{L} : \mathscr{H} \to \mathscr{H}$ be the operator defined by

$$D(\mathscr{L}) = V \times V,$$

$$\mathscr{L}U = [-A^{1/2}v, \quad A^{1/2}u + A^{-1/2}Mu], \quad U = [u, v].$$

We observe that in view of condition (1.81) we have

$$(Au, u) = |A^{1/2}u|^2 \geqslant \omega\|u\|^2 \geqslant C|u|^2, \quad \forall u \in V,$$

which implies that operator $A^{-1/2}$ is well defined and continuous from H to itself. Inasmuch as M is Lipschitzian on H, there is a real α such that $\alpha + \mathscr{L}$ is monotone in $\mathscr{H} \times \mathscr{H}$. We assert that if α is large enough then $\alpha + \mathscr{L}$ is maximal monotone in $\mathscr{H} \times \mathscr{H}$. To show this we shall consider the equation

$$(\lambda + \alpha + \mathscr{L})U = F, \quad \lambda > 0 \tag{1.89}$$

where $F = [f, g] \in H \times V$ and $U = [u, v]$.

Equivalently

$$(\alpha + \lambda)\, u - A^{1/2}v = f$$

$$(\alpha + \lambda)^2 u + Au + Mu = (\alpha + \lambda) f + A^{1/2}g$$

or

$$(\alpha + \lambda)u - A^{1/2}v = f$$

$$u = ((\alpha + \lambda)^2 + A)^{-1}(-Mu + (\alpha + \lambda)\, f + A^{1/2}g). \tag{1.90}$$

Because operator $(1+(\alpha+\lambda)^{-2}A)^{-1}$ is nonexpansive on H, we can deduce that the second equation in (1.90) has a unique solution $u \in D(A)$ whenever $L(\alpha + \lambda)^{-2} < 1$. Thus we have shown that $R(\lambda + \alpha + \mathscr{L}) \supset H \times V$ for all $\lambda > 0$ and $\alpha > L^{1/2}$. Since $H \times V$ is dense in $\mathscr{H}$ and $\mathscr{L}$ is closed, this clearly implies that

$$R(\lambda + \alpha + \mathscr{L}) \supset \mathscr{H}, \quad \forall \lambda > 0,\ \alpha > L^{1/2}. \tag{1.91}$$

Now we refer to Theorem 2.3 in Chapter III to obtain a solution $U(t) \in C(0, T; \mathscr{H})$ for

$$\frac{\mathrm{d}u}{\mathrm{d}t}(t) + \mathscr{L}U(t) = [0, A^{-1/2}f(t)], \quad \text{a.e. on }]0, T[$$

with the initial condition

$$U(0) = [u_0,\ A^{-1/2}v_0] \in V \times V$$

which satisfies

$$\frac{dU}{dt} \in L^\infty(0, T; \mathscr{H}).$$

Note that our present hypotheses on f imply $F = [0, A^{-1/2}f] \in W^{1,2}(0, T; \mathscr{H})$. In other words, the differential system

$$\frac{du}{dt} = A^{1/2}v,$$
$$\frac{dv}{dt} = -A^{1/2}u - A^{-1/2}Mu + A^{-1/2}f(t), \tag{1.92}$$

with initial condition

$$u(0) = u_0, \ v(0) = A^{-1/2}v_0. \tag{1.93}$$

has a unique solution $[u, v] \in C(0, T; \mathscr{H})$ such that $\frac{du}{dt} \in L^\infty(0, T; H)$ and $\frac{dv}{dt} \in L^\infty(0, T; H)$.

It follows that $A^{1/2}u \in L^\infty(0, T; H)$, so that $u \in L^\infty(0, T; V)$ and $A^{1/2}\frac{dv}{dt} \in L^\infty(0, T; V')$. In particular this implies that the function $t \to \frac{du}{dt}(t) = A^{1/2}v(t)$ is V'-valued absolutely continuous on $[0, T]$ and $\frac{d^2u}{dt^2} = A^{1/2}\frac{dv}{dt} \in L^\infty(0, T; V')$. Therefore, we may differentiate the first equation of system (1.92) and replace the result in the second equation to obtain

$$\frac{d^2u}{dt^2} + Au + Mu = f, \quad \text{a.e. on }]0, T[.$$

Furthermore, $u(0) = u_0$ and $\frac{du}{dt}(0) = v_0$ by (1.93). Hence $u(t)$ is the required solution.

Now suppose that $Au_0 \in H$ and $v_0 \in V$.

Setting

$$g(t) = f(t) - Mu(t)$$

we can regard $u(t)$ as the solution of linear hyperbolic equation

$$\frac{d^2u}{dt^2} + Au = g, \quad \text{a.e. on }]0, T[, \tag{1.94}$$

subject to the initial conditions

$$u(0) = u_0, \quad \frac{du}{dt}(0) = v_0 \tag{1.95}$$

Since M is Lipschitzian on H and $u \in W^{1,\infty}(0, T; H)$, so is Mu. Thus $g \in W^{1,1}(0, T; H)$. Theorem 1.1 can therefore be applied to the present situation to conclude that $u(t)$ satisfies conditions (1.87) and (1.88). This completes the proof of Theorem 1.5.

Let W be a real reflexive Banach space which satisfies

$$W \subset H$$

with inclusion mapping continuous and densely defined in H. As usual, we denote by (,) the pairing between W and its dual space W'. The norm in W and W' will be denoted by $\|.\|_W$ and respectively, $\|.\|_{W'}$.

We shall again consider Eq. (1.78), essentially under the same assumptions on A, V and H as in the preceding theorem. More precisely, we maintain conditions (1.80) and (1.81) of Theorem 1.5 and further suppose that

(i) *V is compact in H.*

(ii) *$M : W \to W'$ is single-valued and cyclically maximal monotone in $W \times W'$. Moreover, M is bounded on the bounded subsets of W and coercive,* i.e.,

$$\lim_{\|u\|_W \to \infty} (Mu, u)/_{\|u\|_W} = \infty. \tag{1.96}$$

(iii) *Suppose we are given a sequence of functions*

$\{u_n\} \subset W^{1,\infty}(0, T; H) \cap L^\infty(0, T; W)$ such that

$u_n \to u$ in the strong topology of $L^2(0, T; H)$

$u_n \to u$ in the weak-star topology of $L^\infty(0, T; W)$.

Then a subsequence $\{Mu_{n_k}\}$ can be extracted such that

$$Mu_{nk} \to Mu \text{ weakly-star in } L^\infty(0, T; W'). \tag{1.97}$$

Let M_H denote the monotone operator on H with domain $D(M_H) = \{u \in W;\ Mu \in H\}$ which is given by $M_H u = Mu$ for $u \in D(M_H)$. Inasmuch as M is maximal monotone and coercive in $W \times W'$ we may conclude that M_H is maximal monotone in $H \times H$. This implies that for every $\varepsilon > 0$ operator $(1 + \varepsilon M_H)^{-1}$ is well defined and nonexpansive on all of H.

Our last hypothesis may be stated as follows:

(iv) *$(1 + \varepsilon M_H)^{-1} V \subset V$ and*

$$[(1 + \varepsilon M_H)^{-1} u]^2 \leqslant [u]^2 \quad \text{for all } u \in V \text{ and } \varepsilon > 0 \tag{1.98}$$

where $[u]^2 = (Au, u)$.

THEOREM 1.6. *Let V, H, W, A, M be given, satisfying the above hypotheses. Let u_0, v_0, f be given, satisfying*

$$u_0 \in V \cap W, \quad v_0 \in H, \quad f \in L^1(0, T; H). \tag{1.99}$$

Then there exists at least one function u which satisfies

$$u \in L^\infty(0, T; V \cap W) \tag{1.100}$$

$$\frac{\mathrm{d}u}{\mathrm{d}t} \in L^\infty(0, T; H) \tag{1.101}$$

$$\frac{\mathrm{d}^2u}{\mathrm{d}t^2} + Au + Mu = f(t) \quad \text{on }]0, T[\tag{1.102}$$

$$u(0) = u_0, \quad \frac{\mathrm{d}u}{\mathrm{d}t}(0) = v_0. \tag{1.103}$$

Remark. Equation (1.102) is taken in the sense of vectorial distributions on $]0, T[$. However, since $\frac{\mathrm{d}^2u}{\mathrm{d}t^2} \in L^1(0, T; H) + L^\infty(0, T; V') + L^\infty(0, T; W')$ we conclude that $\frac{\mathrm{d}^2u}{\mathrm{d}t^2} \in L^1(0, T; (V \cap W)')$ so that this equation is valid almost everywhere on $]0, T[$ (see Theorem 2.2. in Chapter I). In particular this implies that $\frac{\mathrm{d}u}{\mathrm{d}t}$ is continuous with values in $(V \cap W)'$ and $\frac{\mathrm{d}u}{\mathrm{d}t}(0)$ makes sense.

Proof of Theorem 1.6. Note that hypothesis (ii) implies (see Theorem 2.3 in Chapter II) the existence of a lower-semicontinuous convex function $\psi : W \to]-\infty, +\infty]$ such that

$$Mu = \partial\psi(u), \quad \forall\, u \in W = D(\partial\psi). \tag{1.104}$$

Moreover, the coercivity condition (1.96) implies that A^{-1} is bounded on W'. Proposition 2.6 from Chapter II then implies that

$$\lim_{\|u\|_W \to \infty} \psi(u)/\|u\|_W = \infty. \tag{1.105}$$

Let $\psi_H : H \to]-\infty, +\infty]$ be the convex function defined by

$$\psi_H(u) = \begin{cases} \psi(u) & \text{if } u \in W \\ +\infty & \text{if } u \in H \setminus W. \end{cases}$$

Since W is reflexive, it follows from assumption (1.105) that every level set $\{u \in H; \psi_H(u) \leqslant \lambda\}$ is weakly (equivalently, strongly) closed in H. This shows that ψ_H is lower-semicontinuous on H and $\partial\psi_H$ is maximal monotone in $H \times H$.

Therefore,

$$M_H = \partial\psi_H.$$

In particular, it follows that $D(M_H)$ is a dense subset of $W = D(\psi_H)$ in the topology of H. We set

$$\psi_\lambda(u) = \inf\left\{\frac{|u-v|^2}{2\lambda} + \psi_H(v);\ v \in H\right\}$$

and recall (see Theorem 2.2 in Chapter II) that $(M_H)_\lambda = \partial\psi_\lambda$ for all $\lambda > 0$.

For the time being, let us assume

LEMMA 1.4. *Let $D(A) = \{u \in V;\ Au \in H\}$. Then $D(A) \cap W$ is a dense subset of $V \cap W$.*

Now denote by $u_\lambda(t)$ the solution of the initial value problem

$$\frac{d^2u_\lambda}{dt^2} + Au_\lambda + \partial\psi_\lambda(u_\lambda) = f_\lambda, \quad \text{a.e. on }]0, T[\tag{1.106}$$

$$u_\lambda(0) = u_\lambda^0, \quad \frac{du_\lambda}{dt}(0) = v_\lambda^0 \tag{1.107}$$

where u_λ^0, v_λ^0 and f_λ satisfy

$$Au_\lambda^0 \in H, \quad v_\lambda^0 \in V, \quad u_\lambda^0 \in W \tag{1.108}$$

$$f_\lambda \in L^2(0, T; H), \quad \frac{df_\lambda}{dt} \in L^2(0, T; H) \tag{1.109}$$

$$\lim_{\lambda\to 0} u_\lambda^0 = u_0 \text{ in } V \cap W, \quad \lim_{\lambda\to 0} v_\lambda^0 = v_0 \text{ in } H, \tag{1.110}$$

and

$$\lim_{\lambda\to 0} f_\lambda = f \quad \text{in } L^1(0, T; H).$$

Since the operator $\partial\psi_\lambda$ is Lipschitz on H it follows from Theorem 1.5 that there is a unique solution u_λ for problems (1.106) and (1.107), which satisfies

$$\frac{du_\lambda}{dt} \in L^\infty(0, T; V)$$

$$\frac{d^2u_\lambda}{dt^2} \in L^\infty(0, T; H), \ Au_\lambda \in L^\infty(0, T; H).$$

Taking the inner products of Eq. (1.106) with $\frac{du_\lambda}{dt}$, we obtain

$$\frac{d}{dt}\left(\frac{1}{2}\left|\frac{du_\lambda}{dt}\right|^2 + \frac{1}{2}(Au_\lambda, u_\lambda) + \psi_\lambda(u_\lambda)\right) = \left(f_\lambda, \frac{du_\lambda}{dt}\right).$$

Integrating this equation from 0 to t and using condition (1.81) gives:

$$\frac{1}{2}\left|\frac{du_\lambda}{dt}(t)\right|^2 + \frac{1}{2}\omega\,\|u_\lambda(t)\|^2 + \psi_\lambda(u_\lambda(t)) \leqslant \frac{1}{2}(|v_\lambda^0|^2 + C\|u_\lambda^0\|^2) +$$

$$+ \psi_\lambda(u_\lambda^0) + \int_0^t |f_\lambda(s)|\left|\frac{du_\lambda}{ds}(s)\right| ds; \quad 0 \leqslant t \leqslant T. \tag{1.111}$$

We choose $v_1 \in W$ such that $Mv_1 = 0$ (This choice is possible because $R(M) = W'$.) Recalling that $|(\partial\psi_\lambda)^0 v_1| \leqslant |M_H v_1| = 0$, we deduce that

$$\psi_\lambda(u_\lambda) \geqslant \psi_\lambda(v_1)$$

so that the last integral inequality gives

$$\left|\frac{du_\lambda}{dt}(t)\right| \leqslant \sqrt{|v_\lambda^0|^2 + 2C\|u_\lambda^0\|^2 + 2\psi_\lambda(u_\lambda^0) - 2\,|\psi_\lambda(v_1)|} +$$

$$+ \int_0^t |f_\lambda(s)|\, ds, \quad 0 \leqslant t \leqslant T. \tag{1.112}$$

Hence

$$\|u_\lambda(t)\|^2 \leqslant C(|v_\lambda^0|^2 + \|u_\lambda^0\|^2 + \psi_\lambda(u_\lambda^0) + \|f_\lambda\|^2_{L^1(0,\,T;\,H)}), \quad 0 \leqslant t \leqslant T. \tag{1.113}$$

Noting that $\psi_\lambda(u_\lambda^0) \leqslant \psi(u_\lambda^0) \leqslant (Mu_\lambda^0, u_\lambda^0 - v_1) + \psi_\lambda(v_1)$, we have that $\psi_\lambda(u_\lambda^0)$ is bounded as $\lambda \to 0$. Comparing this with (1.112) and (1.113) we obtain

$$\left|\frac{du_\lambda}{dt}(t)\right| + \|u_\lambda(t)\| \leqslant C, \quad \lambda > 0, \quad 0 \leqslant t \leqslant T. \tag{1.114}$$

Next from inequality (1.111)

$$\psi_\lambda(u_\lambda(t)) \leqslant C \quad \text{for } \lambda > 0, \quad t \in [0, T], \tag{1.115}$$

where C is some positive constant independent of λ.

Now set

$$w_\lambda(t) = (1 + \lambda M_H)^{-1} u_\lambda(t).$$

Since $(1 + \lambda M_H)^{-1}$ is nonexpansive on H, we have

$$\left|\frac{\mathrm{d}w_\lambda}{\mathrm{d}t}(t)\right| \leqslant \left|\frac{\mathrm{d}u_\lambda}{\mathrm{d}t}(t)\right|, \quad \text{a.e. on }]0, T[,$$

and therefore using condition (iv) and estimate (1.114),

$$\left|\frac{\mathrm{d}w_\lambda}{\mathrm{d}t}(t)\right| + \|w_\lambda(t)\| \leqslant C, \quad \text{a.e., on }]0, T[. \tag{1.116}$$

Furthermore, since $\psi_\lambda(u_\lambda) \geqslant \psi((1 + \lambda M_H)^{-1}u_\lambda)$, it follows from (1.105) that

$$\|w_\lambda(t)\|_W \leqslant C \quad \text{on } [0, T]. \tag{1.117}$$

Since the injection mapping of V into H is compact, it follows from (1.116), and a well-known compactness theorem, that there exist subsequences $\{u_{\lambda_n}\} \subset \{u_\lambda\}$, $\{w_{\lambda_n}\} \subset \{w_\lambda\}$ such that $\lim_{n\to\infty} \lambda_n = 0$ and

$$\begin{aligned} u_{\lambda_n} &\to u \quad \text{strongly in } L^2(0, T; H) \\ w_{\lambda_n} &\to w \quad \text{strongly in } L^2(0, T; H). \end{aligned} \tag{1.118}$$

Since $\lim_{\lambda\to 0} (1 + \lambda M_H)^{-1}u = u$ for every $u \in \overline{D(M_H)} = H$, we may conclude that $w = u$. Extracting further subsequences we may assume that

$$u_{\lambda_n} \to u \text{ in the weak-star topology of } L^\infty(0, T; V)$$

$$w_{\lambda_n} \to u \text{ in the weak-star topology of } L^\infty(0, T; W)$$

$$\frac{\mathrm{d}u_{\lambda_n}}{\mathrm{d}t} \to \frac{\mathrm{d}u}{\mathrm{d}t} \text{ in the weak-star topology of } L^\infty(0, T; H)$$

(It is easy to show that these limits have the relationship stated).

Furthermore, since M is bounded on every bounded subset of W it follows by (1.17) that $Mw_\lambda = \partial\psi_\lambda(u_\lambda)$ are bounded in $L^\infty(0, T; W')$. This implies that $\dfrac{\mathrm{d}^2u_\lambda}{\mathrm{d}t^2}$ are bounded in $L^\infty(0, T; W') + L^\infty(0, T; V')$. Consequently, we may assume that

$$\frac{\mathrm{d}^2u_{\lambda_n}}{\mathrm{d}t^2} \to \frac{\mathrm{d}^2u}{\mathrm{d}t^2} \text{ weakly-star in } L^\infty(0, T; (V \cap W)')$$

$$Mw_{\lambda_n} \to g \quad \text{weakly-star in } L^\infty(0, T; W')$$

$$Au_{\lambda_n} \to Au \text{ weakly-star in } L^\infty(0, T; V').$$

The previous relations then imply by virtue of hypothesis (iii) that $g(t) \in Mu(t)$ a.e. on $]0, T[$ and

$$\frac{d^2u}{dt^2} + Au + Mu = f.$$

It also follows that

$$u(0) = u_0 \quad \text{and} \quad \frac{du}{dt}(0) = v_0. \tag{1.119}$$

The former relation is true since $u_{\lambda_n}(0) \to u(0)$ weakly in H while $u^0_{\lambda_n} \to u_0$ strongly in H. To prove the latter relation in (1.119) let $\varphi \in C^1(0, T; V \cap W)$ be such that $\varphi(T) = 0$. We have

$$\int_0^T \left(\frac{d^2u_{\lambda_n}}{dt^2}, \varphi\right) dt = -\left(\frac{du_{\lambda_n}}{dt}(0), \varphi(0)\right) - \int_0^T \left(\frac{du_{\lambda_n}}{dt}, \frac{d\varphi}{dt}\right) dt$$

which converges to

$$\int_0^T \left(\frac{d^2u}{dt^2}, \varphi\right) dt = -(v_0, \varphi(0)) - \int_0^T \left(\frac{du}{dt}, \frac{d\varphi}{dt}\right) dt$$

as $n \to \infty$. Therefore

$$\left(v_0 - \frac{du}{dt}(0), \varphi(0)\right) = 0.$$

Since $\varphi(0)$ is in fact arbitrary in $V \cap W$, this implies that $\dfrac{du}{dt}(0) = v_0$ as claimed.

Proof of Lemma 1.4. Let A_H be the self-adjoint operator on H with domain $D(A)$ which is given by $A_H u = Au$ for $u \in D(A)$. We have already seen that $A_H = \partial\varphi$ where $\varphi : H \to]-\infty, +\infty]$ is given by

$$\varphi(u) = \begin{cases} \dfrac{1}{2}(Au, u) & \text{if } u \in V \\ +\infty & \text{if } u \in H \setminus V. \end{cases}$$

Then assumption (iv) can be written as

$$\varphi((1 + \varepsilon M_H)^{-1}u) \leqslant \varphi(u) \quad \text{for all } u \in V \text{ and } \varepsilon > 0. \tag{1.120}$$

We apply Theorem 3.7 in Chapter II to obtain that $A_H + M_H$ is maximal monotone in $H \times H$ and

$$\overline{D(A) \cap D(M_H)} = H.$$

Let u be arbitrarily given in $V \cap W$ and let $u_\varepsilon \in D(A) \cap D(M_H)$ be the solution of the equation

$$u_\varepsilon + \varepsilon A u_\varepsilon + \varepsilon M_H u_\varepsilon = u, \quad \varepsilon > 0. \tag{1.121}$$

We know that $\lim_{\varepsilon \to 0} u_\varepsilon = u$ in the strong topology of H. Taking inner products of this equation with Au_ε, we obtain

$$\omega \|u_\varepsilon\|^2 + \varepsilon |Au_\varepsilon|^2 \leqslant (Au_\varepsilon, u) \leqslant C \|u_\varepsilon\| \; \|u\|$$

because $(Au_\varepsilon, M_H u_\varepsilon) \geqslant 0$ in virtue of assumption (1.120). We may therefore infer that

$$\{u_\varepsilon\} \text{ is bounded in } V. \tag{1.122}$$

Taking the inner product of (1.121) with $M_H u_\varepsilon$, we conclude, just as above, that

$$(M_H u_\varepsilon, u_\varepsilon - u) + \varepsilon |M_H u_\varepsilon|^2 \leqslant 0.$$

In other words

$$\psi(u_\varepsilon) \leqslant \psi(u) \quad \text{for all } \varepsilon > 0.$$

Then it follows from condition (1.105) that

$$\{u_\varepsilon\} \text{ is bounded in } W. \tag{1.123}$$

Since V and W are reflexive, we may take weakly convergent subsequences. There exists a subsequence $\{u_{\varepsilon_n}\}$ such that

$$u_{\varepsilon_n} \to u \quad \text{weakly in } V$$

$$u_{\varepsilon_n} \to u \quad \text{weakly in } W.$$

Thus we have proved that u belongs to the weak (equivalently, strong) closure of $D(A) \cap D(M_H)$ in the space $V \cap W$. This completes the proof of Lemma 1.4.

As an example, we take $H = L^2(\Omega)$, $V = H_0^1(\Omega)$, $W = L^p(\Omega)$ where $2 \leqslant p < \infty$ and $A = -\Delta$. Here Ω is a bounded and open subset of R^n, with smooth boundary.

We define

$$(Mu)(x) = \beta(u(x)), \quad \text{a.e. on } \Omega, \ u \in L^p(\Omega)$$

where β is a monotone (nondecreasing) continuous function on real axis which satisfies

$$|\beta(r)| \leqslant C_1 |r|^{p-1} + C_2, \quad \forall\, r \in R;\ C_1, C_2 > 0 \tag{1.124}$$

$$r\beta(r) \geqslant C_3 |r|^p - C_4, \quad \forall\, r \in R,\ C_3, C_4 > 0 \tag{1.125}$$

We also assume that $\beta(0) = 0$.

Hypotheses (i) and (ii) are easily verified (We note that $H_0^1(\Omega)$ is compact in $L^2(\Omega)$). In order to verify hypothesis (iii), we take a sequence of functions $\{u_n\}$ satisfying:

$$u_n \to u \text{ strongly in } L^2(0, T;\ L^2(\Omega))$$

$$u_n \to u \text{ weakly-star in } L^\infty(0, T;\ L^p(\Omega)).$$

Therefore we may extract a subsequence $\{u_{n_k}\}$ such that

$$u_{n_k}(t, x) \to u(t, x), \quad \text{a.e. on }]0, T[\times\Omega$$

and moreover such that

$$\beta(u_{n_k}(t, x)) \to \beta(u(t, x)), \quad \text{a.e. on }]0, T[\times\Omega. \tag{1.126}$$

It follows from condition (1.124) that $u_{n_k}(t, x)$ are bounded in $L^{p'}(Q)$ where $1/p + 1/p' = 1$ and $Q =]0, T[\times\Omega$. Then (1.126) implies (see e.g. J. L. Lions [1], Lemma 1.3, Chapter I) that

$$\beta(u_{n_k}) \to \beta(u) \text{ weakly in } L^{p'}(Q).$$

Now, extracting a further subsequence if necessary, we find that $\beta(u_{n_k}) \to \beta(u)$ weakly star in $L^\infty(0, T; L^{p'}(\Omega))$. This proves hypothesis (iii). As for hypothesis (iv) it is clear that $(1 + \varepsilon M_H)^{-1} H_0^1(\Omega) \subset H_0^1(\Omega)$ and

$$[(1 + \varepsilon\beta)^{-1} u]^2 = \int_\Omega |\text{grad}\,(1 + \varepsilon\beta)^{-1} u|^2\, dx \leqslant \int_\Omega |\text{grad}\, u|^2\, dx, \quad \forall\, u \in V$$

because $\beta(0) = 0$ and $(1 + \varepsilon\beta)^{-1}$ is nonexpansive on R^1.

Theorem 1.6 may therefore be applied to yield the following result.

THEOREM 1.7. *Let f, u_0, v_0 be given satisfying:*

$$f \in L^1(0, T;\ L^2(\Omega))$$

$$u_0 \in H_0^1(\Omega) \cap L^p(\Omega), \quad v_0 \in L^2(\Omega).$$

Then there exists a function $u(t, x)$ satisfying:

$$u \in L^\infty(0, T;\ H_0^1(\Omega) \cap L^p(\Omega)) \tag{1.127}$$

$$\frac{\partial u}{\partial t} \in L^\infty(0, T;\ L^2(\Omega)) \tag{1.128}$$

$$\frac{\partial^2 u}{\partial t^2} - \Delta u + \beta(u) = f \qquad \textit{on }]0, T[\times\Omega \tag{1.129}$$

$$u(0, x) = u_0(x), \quad \frac{\partial u}{\partial t}(0, x) = v_0(x) \quad \textit{on } \Omega. \tag{1.130}$$

Remark. If $\beta(u) = ku^3$, Eq. (1.129) reduces to the Klein-Gordon equation:

$$\square u + m^2 u + ku^3 = f, \quad \square = -\frac{\partial^2}{\partial t^2} + \Delta, \; k > 0$$

which arises in the quantum field theory (see K. Jörgens [1], I. Segal [1]). In this connection, it should be noted that if $\beta(u) = |u|^{p-2}\, u$ where $2 \leqslant p \leqslant 2\,\dfrac{n-1}{n-2}$, then we can prove the uniqueness of the solution u given by Theorem 1.6. For the proof we refer to J. L. Lions [1].

§ 2. Boundary value problems for second order nonlinear differential equations

2.1. A class of two-point boundary value problems

This section deals with the boundary problems

$$-\frac{\mathrm{d}}{\mathrm{d}t}\left(\partial\varphi\left(\frac{\mathrm{d}u}{\mathrm{d}t}(t)\right)\right) + Au(t) \ni f(t), \quad \text{a.e. on }]0, T[\tag{2.1}$$

$$\left[\partial\varphi\left(\frac{\mathrm{d}u}{\mathrm{d}t}(0)\right), -\partial\varphi\left(\frac{\mathrm{d}u}{\mathrm{d}t}(T)\right)\right] - \partial l(u(0)),\, u(T)) \ni 0 \tag{2.2}$$

in a real Hilbert space H, where $0 < T < \infty$ and A is a (multivalued) maximal monotone operator from H into itself. Here φ (respectively l) are lower semicontinuous convex functions from H (respectively, $H \times H$) to $]-\infty, +\infty]$ and $\partial\varphi$ (respectively, ∂l) denote their subdifferentials.

As usual, we shall denote by $|.|$ (respectively $(,)$) the norm (respectively the inner product) in H.

Problem (1.1), (1.2) includes as special cases many other types of nonlinear two-point boundary problems. For instance, if $l(u_1, u_2) = l_1(u) + l_2(u)$ then the boundary conditions (2.2) can be written as

$$\partial\varphi\left(\frac{\mathrm{d}u}{\mathrm{d}t}(0)\right) - \partial l_1(u(0)) \ni 0, \; \partial\varphi\left(\frac{\mathrm{d}u}{\mathrm{d}t}(T)\right) + \partial l_2(u(T)) \ni 0.$$

Other types of boundary conditions for equations (2.1) are implicitly incorporated into the problem through the conditions $l(u(0), u(T)) < \infty$ and will be discussed below.

We now state the hypotheses that will be referred to constantly in the sequel.

(A) *Functions* φ *and* l *are proper, convex and lower-semicontinuous from* H *(respectively* $H \times H$*) into* $]-\infty, +\infty]$*. Furthermore, there exists* $\alpha > 1$ *such that*

$$\varphi(u)/|u|^{\alpha} \to \infty \text{ as } |u| \to \infty \tag{2.3}$$

and

$$l(u, v)/(|u| + |v|)^{\alpha} \to +\infty \text{ as } |u| \text{ and } |v| \to +\infty. \tag{2.4}$$

(B) *There exists* $d > 1$ *such that*

$$\varphi(\lambda u) = \lambda^{d} \varphi(u) \text{ for all } \lambda > 0 \text{ and } u \in D(\varphi). \tag{2.5}$$

Let D_{φ} denote the set of all end points pairs $[u(0), u(T)]$ arising from functions $u \in W^{1,1}(0, T; H)$ such that $\varphi\left(\frac{du}{dt}\right) \in L^{1}(0, T)$.

(C) *There exists a pair* $[u_1, u_2] \in D(l) \cap D_{\varphi}$ *such that one of the following two conditions holds*

$$u_1 \in \operatorname{int}\{u \in H;\ [u, u_2] \in D(l)\}. \tag{2.6}$$

$$u_1 \in \operatorname{int}\{u \in H;\ [u, u_2] \in D_{\varphi}\}. \tag{2.7}$$

(D) A *is a maximal monotone set in* $H \times H$*. Moreover,* A *is* $\partial\varphi$*-monotone*, i.e.,

$$\varphi((1 + \lambda A)^{-1} x - (1 + \lambda A)^{-1} y) \leqslant \varphi(x - y) \text{ for all } \lambda > 0;\ x, y \in H \tag{2.8}$$

and there exist h_1 *and* h_2 *in* D_{φ} *such that for all* $\lambda > 0$ *and* x, y *in* H*, the following inequality holds*

$$l((1 + \lambda A)^{-1}(x + \lambda h_1), (1 + \lambda A)^{-1}(x + \lambda h_2)) \leqslant l(x, y). \tag{2.9}$$

The main existence result for problems (2.1) and (2.2) may now be stated.

THEOREM 2.1. *Assume that hypotheses A, B, C and D hold. Let* $f \in L^{2}(0, T; H)$*. Then there exist the functions* $u \in W^{1,\infty}(0, T; H)$ *and* $p \in W^{1,2}(0, T; H)$ *satisfying:*

$$-\frac{dp}{dt}(t) + Au(t) \ni f(t), \quad \text{a.e. on }]0, T[\tag{2.10}$$

$$p(t) - \partial\varphi\left(\frac{du}{dt}(t)\right) \ni 0, \quad \text{a.e. on }]0, T[\tag{2.11}$$

$$[p(0), -p(T)] \in \partial l(u(0), u(T)). \tag{2.12}$$

Proof. Define a function $F : L^2(0, T; H) \to]-\infty, +\infty]$ by the following relation:

$$F(u) = \begin{cases} \int_0^T \varphi\left(\frac{\mathrm{d}u}{\mathrm{d}t}\right) \mathrm{d}t + l(u(0), u(T)) \text{ if } u \in W^{1,1}(0, T; H) \text{ and } \varphi\left(\frac{\mathrm{d}u}{\mathrm{d}t}\right) \in \\ \qquad\qquad\qquad\qquad\qquad\qquad\qquad\qquad \in L^1(0, T) \\ +\infty \qquad\qquad\qquad\qquad\qquad\qquad\qquad\qquad \text{otherwise.} \end{cases}$$

We observe that F is convex and nonidentically $+\infty$ on $L^2(0, T; H)$. In particular, hypothesis (A) implies that $D(F) \subset W^{1,\alpha}(0, T; H)$. On the other hand, a standard device involving Fatou's lemma (see the proof of Proposition 2.8 in Chapter II) shows that F is lower-semicontinuous on $W^{1,\alpha}(0, T; H)$ endowed with the norm

$$\|u\|_{1,\alpha}^{\alpha} = |u(0)|^{\alpha} + \int_0^T \left|\frac{\mathrm{d}u}{\mathrm{d}t}(t)\right|^{\alpha} \mathrm{d}t. \tag{2.13}$$

We note that Hypothesis (A) implies that every level set $\{u \in L^2(0, T; H);\ F(u) \leqslant \lambda\}$ is bounded and therefore weakly compact in $W^{1,\alpha}(0, T; H)$. Since F is convex this clearly implies that F is lower-semicontinuous on $L^2(0, T; H)$.

Let $\mathscr{A} : L^2(0, T; H) \to L^2(0, T; H)$ be the multivalued operator defined by

$$\mathscr{A}\, u = -\frac{\mathrm{d}p}{\mathrm{d}t} \quad \text{for } u \in W^{1,\infty}(0, T; H) \tag{2.14}$$

where $p \in W^{1,2}(0, T; H)$ is such that

$$p(t) \in \partial\varphi\left(\frac{\mathrm{d}u}{\mathrm{d}t}(t)\right), \quad \text{a.e. on }]0, T[\tag{2.15}$$

$$[p(0), -p(T)] \in \partial l(u(0), u(T)).$$

The main step of this proof is given by the following lemma which is quite interesting in itself.

LEMMA 2.1. *Let hypotheses* (A) *and* (C) *hold. Then* $\mathscr{A} = \partial F$.

Proof. An explicit calculation involving Eqs. (2.14) and (2.15) shows that $\mathscr{A}u \subset \partial F(u)$ for all $u \in D(\mathscr{A})$. To prove that $\mathscr{A} = \partial F$ it suffices to show that $\mathscr{A}$ is maximal monotone in $L^2(0, T; H) \times L^2(0, T; H)$. Let f be given in $L^2(0,T;H)$. Equation

$$u + \mathscr{A}u \ni f \tag{2.16}$$

can be written as

$$\begin{aligned} & p(t) - \partial\varphi(u'(t)) \ni 0, \quad \text{a.e. on }]0, T[\\ & -p'(t) + u(t) = f(t), \quad \text{a.e. on }]0, T[\\ & [p(0), -p(T)] \in \partial l(u(0), u(T)), \end{aligned} \tag{2.17}$$

where $u \in W^{1,\infty}(0, T; H)$ and $p \in W^{1,2}(0, T; H)$ (for the sake of simplicity we shall write $'$ instead of $\dfrac{\mathrm{d}}{\mathrm{d}t}$).

For any $\lambda > 0$, we define

$$\begin{aligned} F_\lambda(u) = & \int_0^T \varphi_\lambda(u'(t))\,\mathrm{d}t + \int_0^T \left(\frac{1}{2}|u(t)|^2 - (f(t), u(t))\right)\mathrm{d}t + \\ & + l_\lambda(u(0), u(T)) \end{aligned} \tag{2.18}$$

where $\varphi_\lambda : H \to]-\infty, +\infty[$ and $l_\lambda : H \times H \to]-\infty, +\infty[$ are defined as in Chapter II, Section 2.2 (see Theorem 2.2). Recalling that

$$\varphi_\lambda(u) = \frac{\lambda}{2}|(\partial\varphi)_\lambda u|^2 + \varphi((1 + \lambda\partial\varphi)^{-1}u)$$

(and

$$u - (1 + \lambda\partial\varphi)^{-1}u = \lambda(\partial\varphi)_\lambda u$$

it follows from condition (2.3) that

$$\varphi_\lambda(u)/|u|^{1+\beta} \to \infty \text{ as } |u| \to \infty \tag{2.19}$$

where β is some positive constant.

Similarly, condition (2.4) implies that

$$l_\lambda(u_1, u_2) \to \infty \text{ as } |u_1| \to \infty \text{ and } |u_2| \to \infty. \tag{2.20}$$

Thus by the same argument as that used for F, we may conclude that for every λ, F_λ is lower-semicontinuous from $L^2(0, T; H)$ to $]-\infty, +\infty]$. Furthermore, we have shown that

$$F_\lambda(u) \to \infty \text{ as } \|u\| \xrightarrow[L^2(0,T;H)]{} \infty.$$

Thus there is $u_\lambda \in W^{1,1}(0, T; H)$ such that

$$F_\lambda(u_\lambda) \leqslant F_\lambda(v) \quad \text{for all } v \in W^{1,1}(0, T; H). \tag{2.21}$$

Since φ_λ and l_λ are Frechet differentiable, relation (2.21) implies that

$$\int_0^T (\partial\varphi_\lambda(u'_\lambda),\ v')\,dt + \int_0^T (u_\lambda - f,\ v)\,dt + \tag{2.22}$$

$$+\langle \partial l_\lambda(u_\lambda(0),\ u_\lambda(T)),\ [v(0),\ v(T)]\rangle = 0 \text{ for all } v \in W^{1,\infty}(0, T; H)$$

where $\langle , \rangle$ denotes the usual inner product in $H \times H$. In particular (2.22) yields the following equation

$$-(\partial\varphi_\lambda(u'_\lambda))' + u_\lambda - f = 0$$

taken in the sense of distributions over $]0, T[$. However, since $(\partial\varphi_\lambda(u'_\lambda))' = u_\lambda - f$ belongs to $L^2(0, T; H)$ it follows that

$$-(\partial\varphi_\lambda(u'_\lambda))' + u_\lambda = f, \quad \text{a.e. on }]0, T[. \tag{2.23}$$

Now we take inner products of both sides of Eqs. (2.23) with $v \in W^{1,2}(0, T; H)$. Integrating by parts over $]0, T[$ and using (2.22) we obtain

$$(\partial\varphi_\lambda(u'_\lambda(0)),\ v(0)) - (\partial\varphi_\lambda(u'_\lambda(T)),\ v(T)) =$$

$$= \langle \partial l_\lambda(u_\lambda(0),\ u_\lambda(T)),\ [v(0),\ v(T)]\rangle.$$

Since v is arbitrary in $W^{1,2}(0, T; H)$, we conclude that

$$[\partial\varphi_\lambda(u'_\lambda(0)),\ -\partial\varphi_\lambda(u'_\lambda(T))] = \partial l_\lambda(u_\lambda(0),\ u_\lambda(T)). \tag{2.24}$$

Note that (2.19) implies that $(\partial\varphi_\lambda)^{-1}$ is defined on all of H and is bounded on every bounded subset (see Proposition 2.6 in Chapter II). Since $\partial\varphi_\lambda(u'_\lambda) \in W^{1,2}(0, T; H)$ this clearly implies that $u'_\lambda \in L^\infty(0, T; H)$, i.e., $u_\lambda \in W^{1,\infty}(0, T; H)$. Moreover, using once again relation (2.21) we find that

$$\int_0^T |u_\lambda(t)|^2\,dt + |u_\lambda(0)|^2 + |u_\lambda(T)|^2 \leqslant C \quad \text{for all } \lambda > 0. \tag{2.25}$$

Therefore,

$$\{(\partial\varphi_\lambda(u'_\lambda))'\} \text{ remain in a bounded set of } L^2(0, T; H). \tag{2.26}$$

Now we need to show that

$$\{|\partial\varphi_\lambda(u'_\lambda(0))| + |\partial\varphi_\lambda(u'_\lambda(T))|\} \text{ is bounded} \tag{2.27}$$

To this end there are two cases to be considered separately. We set $p_\lambda(t) = \partial\varphi_\lambda(u'_\lambda(t))$ and suppose first that condition (2.6) in hypothesis (C) holds. Thus there exists a function $x \in W^{1,1}(0, T; H)$ such that $\varphi(x') \in L^1(0, T)$ and

$$x(0) \in \operatorname{int}\{y \in H;\ [y, x(T)] \in D(l)\}. \tag{2.28}$$

In other words, $x(0)$ belongs to the interior of the effective domain of function $y \to l(y, x(T))$. But this implies (see Chapter II, Proposition 2.2) that $l(., x(T))$ is locally bounded at $y = x(0)$. Consequently, there are some positive constants ρ and C such that

$$l(x(0) + \rho w, x(T)) \leqslant C \quad \text{for all } w \in H,\ |w| = 1. \tag{2.29}$$

On the other hand, it follows from Eq. (2.24) that

$$l_\lambda(u_\lambda(0),\ u_\lambda(T)) \leqslant l_\lambda(x(0) + \rho w, x(T)) + (p_\lambda(0), u_\lambda(0) - x(0) - \rho w) - (p_\lambda(T), u_\lambda(T) - x(T)). \tag{2.30}$$

We now take the inner product of Eq. (2.23) by $u_\lambda - x$ and integrate by parts on $]0, T[$. We obtain

$$-\int_0^T (\partial\varphi_\lambda(u'_\lambda),\ u'_\lambda - x')\,dt + \int_0^T (f - u_\lambda,\ u_\lambda - x)\,dt = (p_\lambda(0),\ u_\lambda(0) - x(0)) - (p_\lambda(T), u_\lambda(T) - x(T)).$$

We observe that condition (2.3) implies that $\inf\{\varphi(u);\ u \in H\} = M > -\infty$. Hence $\varphi_\lambda(u) \geqslant M$ for all $\lambda > 0$ and $u \in H$ so that the above inequality combined with relation (2.25) shows that

$$(p_\lambda(0), u_\lambda(0) - x(0)) - (p_\lambda(T), u_\lambda(T) - x(T)) \leqslant C$$

for all $\lambda > 0$.

We now take $w = p_\lambda(0)/|p_\lambda(0)|$ in relation (2.30) and use (2.29) to obtain a bound independent of λ for $|p_\lambda(0)|$. Combined with (2.26) this implies (2.27).

Here we have used in particular condition (2.4) which implies that $\inf\{l_\lambda(u_1, u_2); [u_1, u_2] \in H \times H\} = \inf\{l(u_1, u_2); [u_1, u_2] \in H \times H\} = M > -\infty$.

Now we suppose that condition (2.7) holds. Let $x \in W^{1,1}(0, T; H)$ be such that $l(x(0), x(T)) < \infty$ and

$$x(0) \in \operatorname{int}\{y \in H;\ [y, x(T)] \in D_\varphi\}. \tag{2.31}$$

Let $\Phi(y_1, y_2)$ denote the infimum of

$$\int_0^T \left(\varphi(y') + \frac{1}{2}|y|^2 - (f, y)\right) dt$$

over all arcs $y \in W^{1,1}(0, T; H)$ such that $y(0) = y_1$, and $y(T) = y_2$. It is easily seen that Φ is convex and lower-semicontinuous on $H \times H$. Moreover, $D(\Phi) = D_\varphi$ and the infimum defining Φ is attained. Thus by the interiority condition (2.31) it follows that there exists $\rho > 0$ such that

$$\Phi(x(0) + \rho w, x(T)) \leqslant M \quad \text{for all } w \in H,\ |w| = 1. \tag{2.32}$$

Using once again Eq. (2.23), we find that

$$- (p_\lambda(0), u_\lambda(0) - x(0) - \rho w) + (p_\lambda(T), u_\lambda(T) - x(T)) \geqslant$$
$$\geqslant \int_0^T (\varphi_\lambda(u'_\lambda) + \frac{1}{2} |u_\lambda|^2 - (f, u_\lambda))\, dt - \Phi(x(0) + \rho w,\ x(T)).$$

Next, from conditions (2.30) and (2.32) we get

$$- \rho(p_\lambda(0),\ w) \leqslant C + l_\lambda(x(0),\ x(T)) \leqslant C + l(x(0,\ x(T)). \tag{2.33}$$

In relation (2.33) we may take $w = -p_\lambda(0)/|p_\lambda(0)|$ to deduce that $|p_\lambda(0))|$ is bounded. Thus we have proved (2.27) in both cases of hypothesis (C). Combined with (2.26), this implies that

$$\{\partial\varphi_\lambda(u'_\lambda)\} \text{ is bounded in } L^\infty(0,\ T;\ H) \tag{2.34}$$

and therefore

$$\{u'_\lambda\} \text{ is bounded in } L^\infty(0,\ T;\ H) \tag{2.35}$$

because $\varphi(1 + \lambda\partial\varphi)^{-1}u'_\lambda) \leqslant \varphi_\lambda(u'_\lambda) \leqslant \varphi_\lambda(u_0) + (\partial\varphi_\lambda(u'_\lambda),\ u'_\lambda - u_0)$ and $\varphi(u)/ /|u| \to \infty$ as $|u| \to \infty$.

Now we pass to the limit. The estimates (2.25), (2.26), (2.34) and (2.35) imply that a subsequence (denoted again u_λ) can be extracted from $\{u_\lambda\}$ such that:

$$u_\lambda \to u \text{ weakly-star in } L^\infty(0,\ T;\ H) \tag{2.36}$$

$$u'_\lambda \to u' \text{ weakly-star in } L^\infty(0,\ T;\ H)$$

$$\partial\varphi_\lambda(u'_\lambda) \to p \text{ weakly-star in } L^\infty(0,\ T;\ H)$$

$$(\partial\varphi_\lambda(u'_\lambda))' \to p' \text{ weakly in } L^2(0,\ T;\ H)$$

as $\lambda \to 0$. Extracting further subsequences if necessary, we may also assume that

$$[u_\lambda(0),\ u_\lambda(T)] \to [u_0, u_T] \text{ weakly in } H \times H \tag{2.37}$$

$$[\partial\varphi_\lambda(u'_\lambda(0)),\ \partial\varphi_\lambda(u'_\lambda(T))] \to [p_0, p_T] \text{ weakly in } H \times H.$$

We assert that $u_0 = u(0)$, $u_T = u(T)$, $p_0 = p(0)$ and $p_T = p(T)$. To show the limits are as stated, note that (2.36) implies that

$$\int_0^T (u', \varphi)\,dt = (u_T, \varphi(T)) - (u_0, \varphi(0)) - \int_0^T (u, \varphi')\,dt$$

for every $\varphi \in C^1(0, T; H)$. Since $u \in W^{1,1}(0, T; H)$ this implies that

$$(u_T - u(T), \varphi(T)) - (u_0 - u(0), \varphi(0)) = 0, \ \forall \varphi \in C^1(0, T; H)$$

which reduce to the desired equalities. A similar argument is applied for $\partial\varphi_\lambda(u'_\lambda)$.

Let u_λ and u_μ be two solutions of Eqs. (2.23) and (2.24). Subtracting the corresponding equations, multiplying by $u_\lambda - u_\mu$ and integrating the result, yields:

$$\int_0^T (\partial\varphi_\lambda(u'_\lambda) - \partial\varphi_\mu(u'_\mu),\ u'_\lambda - u'_\mu)\,dt + \int_0^T |u_\lambda - u_\mu|^2\,dt +$$

$$+\langle \partial l_\lambda(u_\lambda(0), u_\lambda(T)) - \partial l_\mu(u_\mu(0), u_\mu(T)),\ [u_\lambda(0) - u_\mu(0),$$

$$u_\lambda(T) - u_\mu(T)]\rangle = 0.$$

Since $\partial\varphi_\lambda(u'_\lambda)$ and $\partial l_\lambda(u_\lambda(0), u_\lambda(T))$ are bounded in $L^2(0, T; H)$ and respectively, in $H \times H$, this implies that

$$u_\lambda \to u \text{ strongly in } L^2(0, T; H)$$

$$\limsup_{\lambda,\mu\to 0} \int_0^T (\partial\varphi_\lambda(u'_\lambda) - \partial\varphi_\mu(u'_\mu),\ u'_\lambda - u'_\mu)\,dt \leqslant 0$$

$$\limsup_{\lambda,\mu\to 0} \langle \partial l_\lambda(u_\lambda(0), u_\lambda(T)) - \partial l_\mu(u_\mu(0), u_\mu(T)),$$

$$[u_\lambda(0) - u_\mu(0),\ u_\lambda(T) - u_\mu(T)]\rangle \leqslant 0$$

because $\partial\varphi$ and ∂l are monotone.

Therefore, referring to Proposition 1.1 in Chapter II we conclude that

$$p(t) \in \partial\varphi(u'(t)), \quad \text{a.e. on }]0, T[$$

and

$$[p(0), -p(T)] \in \partial l(u(0), u(T)).$$

In other words, we have shown that Eq. (2.16) has a solution $u \in D(\mathscr{A})$. The arbitrariness of f in $L^2(0, T; H)$ then implies that $\mathscr{A}$ is maximal monotone in $L^2(0, T; H) \times L^2(0, T; H)$. The proof of the lemma is complete.

We return to the proof of the theorem. Let $\tilde{A}$ denote the realization of A in $L^2(0, T; H)$, i.e.,

$$\tilde{A} = \{[u, v] \in L^2(0, T; H) \times L^2(0, T; H);\ u(t) \in D(A) \text{ and}$$
$$v(t) \in Au(t), \quad \text{a.e. on }]0, T[\}.$$

Then according to Lemma 2.1, the boundary value problem (2.10) – (2.12) can be expressed as

$$\tilde{A}u + \partial F(u) \ni f \tag{2.38}$$

Let $[h_1, h_2] \in D_\varphi$ be chosen as in hypothesis (D) and let $h \in W^{1,1}(0, T; H)$ be such that $\varphi\left(\frac{\mathrm{d}h}{\mathrm{d}t}\right) \in L^1(0, T)$ and $h(0) = h_1$, $h(T) = h_2$. Then by homogeneousness condition (2.5) and hypothesis (D),

$$\varphi_\mu\left(\frac{\mathrm{d}}{\mathrm{d}t}(1 + \lambda A)^{-1}(u(t) + \lambda h(t))\right) \leqslant \varphi_\mu\left(\frac{\mathrm{d}u}{\mathrm{d}t}(t) +\right.$$
$$\left. + \lambda \frac{\mathrm{d}h}{\mathrm{d}t}(t)\right), \quad \text{a.e. on }]0, T[\tag{2.39}$$

for every $u \in W^{1,1}(0, T; H)$ and $\mu > 0$.

We note that the convexity of φ combined with the homogeneousness condition (2.5), gives

$$\varphi(u + \lambda v) \leqslant (1 + \lambda)^{d-1}\varphi(u) + \lambda(1 + \lambda)^{d-1}\varphi(v),$$

for all $\lambda > 0$ and u, v in H. Putting this into relation (2.39), we obtain

$$\varphi\left(\frac{\mathrm{d}}{\mathrm{d}t}(1 + \lambda A)^{-1}(u(t) + \lambda h(t))\right) \leqslant (1 + \omega(\lambda))\,\varphi\left(\frac{\mathrm{d}u}{\mathrm{d}t}(t)\right) +$$
$$+ \lambda(1 + \lambda)^{d-1}\varphi\left(\frac{\mathrm{d}h}{\mathrm{d}t}(t)\right), \quad \text{a.e. on }]0, T[$$

where

$$\lim_{\lambda \to 0} \omega(\lambda)/\lambda = d - 1.$$

Then by hypothesis (D),

$$F((1 + \lambda\tilde{A})^{-1}(u + \lambda h)) \leqslant (1 + \omega(\lambda))F(u) + \lambda(1 + \lambda)^{d-1}\int_0^T \varphi\left(\frac{\mathrm{d}h}{\mathrm{d}t}\right)\mathrm{d}t \tag{2.40}$$

for all $u \in W^{1,1}(0, T; H) \supset D(F)$ and all $\lambda > 0$.

Theorem 3.7 from Chapter II can therefore be applied to the present situation to deduce that $\tilde{A} + \partial F$ is maximal monotone in $L^2(0, T; H) \times$ $\times L^2(0, T; H)$. On the other hand, as mentioned above, hypothesis (*A*) implies that

$$F(u)/\|u\|_{L^2(0,T;\,H)} \to \infty \text{ as } \|u\|_{L^2(0,T;\,H)} \to \infty.$$

Since ∂F is coercive, so is $\tilde{A} + \partial F$, so that $R(\tilde{A} + \partial F)$ is all of $L^2(0, T; H)$. In particular, for every $f \in L^2(0, T; H)$ Eq. (2.38) has at least a solution $u \in D(\tilde{A}) \cap D(\partial F)$, q.e.d.

Remarks 1°. The preceding proof shows that hypothesis (C) is symmetric with respect to u_1 and u_2. Also, hypotheses (B) and (D) are unnecessary if one of the following two conditions holds

$$(\text{int } D(F)) \cap D(A) \neq \emptyset$$

$$D(F) \cap \text{int } D(A) \neq \emptyset.$$

This is a consequence of Theorem 1.7 given in Chapter II.

2°. Lemma 2.1 and its proof remain valid if function F is replaced by

$$\int_0^T \gamma(t)\, \varphi\left(\frac{du}{dt}\right) dt + l(u(0), u(T))$$

where $\gamma(t)$ is a scalar function of class C^1 on $[0, T]$ such that $\gamma(t) \neq 0$ everywhere on $[0, T]$. In this case ∂F coincides with the operator $u \to$ $\to -\frac{d}{dt}\left(\gamma\, \partial\varphi\left(\frac{du}{dt}\right)\right)$ defined on all $u \in W^{1,\infty}(0, T; H)$ with the property that there exists at least a function $p \in W^{1,2}(0, T; H)$ satisfying

$$p(t) \in \partial\varphi\left(\frac{du}{dt}(t)\right), \quad \text{a.e. on }]0, T[$$

and

$$[\gamma(0)p(0), -\gamma(T)p(T)] \in \partial l(u(0), u(T)).$$

3°. If A is coercive, i.e.,$(Au, u) \geqslant \omega\,|u|^2$ for all $u \in H$, then condition (2.4) in hypothesis (A) is unnecessary.

2.2. Examples

Example 1. We shall start with the following immediate consequence of Theorem 2.1.

THEOREM 2.2. *Let A be a maximal monotone set in $H \times H$ and let l be a lower-semicontinuous, proper convex function from $H \times H$ to $]-\infty, +\infty]$ satisfying conditions* (2.4) *and* (2.9). *Let $f \in L^2(0, T; H)$ be given.*

Then there exists a function $u \in W^{2,2}(0, T; H)$ *which satisfies*

$$\frac{d^2u}{dt^2}(t) \in Au(t) + f(t), \quad \text{a.e.} \quad on \ ]0, T[\tag{2.41}$$

$$\left[\frac{du}{dt}(0), -\frac{du}{dt}(T)\right] \in \partial l(u(0), u(T)). \tag{2.42}$$

Moreover, the solution $u(t)$ *is unique up to an additive constant.*

Proof. For existence we shall apply Theorem 2.1 where $\varphi(u) = \frac{1}{2}|u|^2$. We note that in this case $D_\varphi = H \times H$ so that hypothesis (C) is trivially satisfied. Hypotheses (A), (B) and (D) follow from the corresponding properties of φ and l.

Let u and v be two solutions of problems (2.41) and (2.42) and let $w = u - v$. Subtracting the corresponding equations, and multiplying by w, yields

$$\frac{1}{2}\frac{d^2}{dt^2}|w(t)|^2 \geqslant \left|\frac{dw}{dt}(t)\right|^2, \quad \text{a.e.} \quad \text{on }]0, T[.$$

Here we have used the monotonicity of A. Integrating this inequality over $]0, T[$ and using the boundary conditions (2.42) we obtain

$$\int_0^T \left|\frac{dw}{dt}(t)\right|^2 dt = 0$$

because ∂l is monotone in $(H \times H) \times (H \times H)$. Hence $w(t) = u(t) - v(t)$ is constant over $[0, T]$.

COROLLARY 2.1. *Let* A *be a maximal monotone set in* $H \times H$. *Let* f, x, y *be given such that*

$$f \in L^2(0, T; H), \quad x \in D(A), \quad y \in D(A). \tag{2.43}$$

Then there exists a unique function $u \in W^{2,2}(0, T; H)$ *which satisfies:*

$$\frac{d^2u}{dt^2}(t) \in Au(t) + f(t), \quad \text{a.e.} \quad on \ ]0, T[\tag{2.44}$$

$$u(0) = x, \ u(T) = y. \tag{2.45}$$

Proof. We apply Theorem 2.2 where

$$l(u_1, u_2) = \begin{cases} 0 & \text{if } u_1 = x \quad \text{and} \quad u_2 = y \\ +\infty & \text{otherwise.} \end{cases}$$

We observe that in the present situation, condition (2.9) holds with $h_1 \in Ax$ and $h_2 \in Ay$.

COROLLARY 2.2. *Let A be maximal monotone in $H \times H$ and let ψ be a lower-semicontinuous proper convex function from H into $]-\infty, +\infty]$ such that*

$$\psi((1 + \lambda A)^{-1}(u + \lambda h)) \leqslant \psi(u) \quad \text{for all } \lambda > 0 \quad \text{and} \quad u \in H \tag{2.46}$$

where h is some element of H.

Let f and x be given, satisfying

$$f \in L^2(0, T; H), \quad x \in D(A). \tag{2.47}$$

Then there exists a unique function $u \in W^{2,2}(0, T; H)$ which satisfies

$$\frac{d^2u}{dt^2} \in Au + f, \quad \text{a.e. on }]0, T[\tag{2.48}$$

$$u(0) = x, \quad \frac{du}{dt}(T) + \partial\psi\,(u(T)) \ni 0. \tag{2.49}$$

Proof. There is nothing left to do, except to apply Theorem 2.2 where

$$l(u_1, u_2) = \begin{cases} \psi\,(u_2) & \text{if } u_1 = x \\ +\infty & \text{otherwise.} \end{cases}$$

Example 2. Let β and γ be two maximal monotone graphs in $R^1 \times R^1$. Then there exists a lower-semicontinuous convex function $j\colon R^1 \to]-\infty, +\infty]$ such that $\beta = \partial j$.
We shall assume that the following conditions hold

$$j(r)/|r|^{1+\alpha} \to +\infty \qquad \text{as } |r| \to \infty, \tag{2.50}$$

$$j(\lambda r) = \lambda^d\, j(r) \qquad \text{for all } \lambda > 0 \text{ and } r \in R^1, \tag{2.51}$$

$$j((1+\lambda\gamma)^{-1}r_1 - (1+\lambda\gamma)^{-1}r_2) \leqslant j(r_1 - r_2) \quad \text{for all } \lambda > 0, \tag{2.52}$$

and all r_1, r_2 in R^1. Here $\alpha > 0$ and $d > 1$.

Besides these graphs we are given a lower-semicontinuous proper convex function $l\colon R^1 \times R^1 \to]-\infty, +\infty]$, which is assumed to satisfy:

$$l(r_1, r_2)/(|r_1| + |r_2|)^{1+\alpha} \to +\infty \quad \text{as } |r_1| \text{ and } |r_2| \to \infty. \tag{2.53}$$

Suppose further that there exists $[h_1, h_2] \in R^1 \times R^1$ such that

$$l((1 + \lambda\gamma)^{-1}(r_1 + \lambda h_1), (1 + \lambda\gamma)^{-1}(r_2 + \lambda h_2)) \leqslant l(r_1, r_2), \tag{2.54}$$

for all $\lambda > 0$ and all r_1, r_2 in R^1.

We finally assume that there exists $[u_0, v_0] \in D(l)$ such that

$$u_0 - v_0 \in \operatorname{int} D(j). \tag{2.55}$$

Let us verify the hypotheses of Theorem 2.1 where $H = R^1$, $\varphi = j$ and $A = \gamma$. Hypotheses (A), (B) and (D) follow easily from the corresponding properties of β, γ and l. As for hypothesis (C) we note that condition (2.55) implies that $j\left(\frac{\mathrm{d}}{\mathrm{d}t}\left(\frac{t}{T}v_0 + \left(1 - \frac{t}{T}\right)u\right)\right) \in L^1(0, T)$ for all u in a sufficiently small neighbourhood of u_0. The latter clearly implies condition (2.7) in hypothesis (C).

Theorem 2.1 may therefore be applied to yield an existence result for the nonlinear boundary value problem

$$-\frac{\mathrm{d}}{\mathrm{d}t}\left(\beta\left(\frac{\mathrm{d}u}{\mathrm{d}t}(t)\right)\right) + \gamma(u(t)) \ni f(t), \quad \text{a.e. on }]0, T[\tag{2.56}$$

$$\left[\beta\left(\frac{\mathrm{d}u}{\mathrm{d}t}(0)\right), -\beta\left(\frac{\mathrm{d}u}{\mathrm{d}t}(T)\right)\right] - \partial l(u(0), u(T)) \ni 0. \tag{2.57}$$

Namely,

COROLLARY 2.3. *Let conditions* (2.50) $\sim$ (2.55) *be satisfied. Let* $f \in L^2(0, T)$ *be given. Then problems* (2.56) and (2.57) *have at least a solution* $u(t)$ *which satisfies*

$$u \in W^{1,\infty}(0, T), \quad \beta\left(\frac{\mathrm{d}u}{\mathrm{d}t}\right) \in W^{1,2}(0, T). \tag{2.58}$$

Remarks 1°. Let $g: R^1 \to]-\infty, +\infty]$ be such that $\partial g = \gamma$. Then Eq.(2.56) can be interpreted as a generalized Euler-Lagrange equation associated with the *convex problem of Bolza*,

$$\min\left\{\int_0^T\left(j\left(\frac{\mathrm{d}u}{\mathrm{d}t}\right) + g(u)\right)\mathrm{d}t + l(u(0), u(T));\ u \in W^{1,1}(0, T)\right\}. \tag{2.59}$$

In this context, the boundary conditions (2.57) can be viewed as *transversality conditions* corresponding to the optimization problem (2.59). Comments on these aspects of the problems, along with a number of results and techniques involved in the question of necessary and sufficient conditions for optimality for convex problem of Bolza, can be found in R.T. Rockafellar's papers [7], [8] (see also author's papers [10], [11]) and we do not develop them further here.

2°. If l is taken as in Corollary 2.1., i.e.,

$$l(u_1, u_2) = \begin{cases} 0 & \text{if } u_1 = x \text{ and } u_2 = y \\ +\infty & \text{otherwise} \end{cases}$$

where $x, y \in D(\gamma)$ then conditions (2.53), (2.54) are trivially satisfied and the boundary conditions (2.57) are equivalent to

$$u(0) = x, \quad u(T) = y.$$

Condition (2.55) is satisfied by assuming that $x - y \in \operatorname{int} D(j)$. We finally observe that conditions (2.50), (2.51), (2.52) and (2.57) are satisfied for $\beta(r) = |r|^{\alpha-2}r$ where $\alpha > 1$.

Example 3. Let Ω be a bounded and open subset of R^n with a sufficiently smooth boundary Γ. We set $Q =]0, T[\times \Omega$ and $\Sigma' =]0, T[\times \Gamma$.

Let β be a maximal monotone graph in $R^1 \times R^1$ and let $j: R^1 \to]-\infty, +\infty]$ be such that $\partial j = \beta$. We shall assume that $0 \in D(\beta)$ and

$$j(r)/r^2 \geqslant M \quad \text{for } |r| \text{ sufficiently large} \tag{2.60}$$

and

$$j(\lambda r) = \lambda^d j(r) \quad \text{for all } \lambda > 0 \quad \text{and } r \in R^1 \tag{2.61}$$

where $d > 1$.

Let $A = -\sum_{i=1}^{n} \frac{\partial}{\partial x_i}\left(\left|\frac{\partial u}{\partial x_i}\right|^{p-2}\frac{\partial u}{\partial x_i}\right)$, $2 \leqslant p < \infty$ be the differential operator defined from $W_0^{1,p}(\Omega)$ to $(W_0^{1,p}(\Omega))' = W^{-1,p'}(\Omega)$, by the following relation

$$(Au, v) = \sum_{i=1}^{n} \int_\Omega \left|\frac{\partial u}{\partial x_i}\right|^{p-2} \frac{\partial u}{\partial x_i}\frac{\partial v}{\partial x_i}\, dx \text{ for } u, v \in W_0^{1,p}(\Omega).$$

We shall denote again A as the restriction of this operator to $D(A)=\{u \in L^2(\Omega);\ Au \in L^2(\Omega)\}$ and recall that $A = \partial\psi$ where $\psi: L^2(\Omega) \to]-\infty, +\infty]$ is given by

$$\psi(u) = \begin{cases} \dfrac{1}{p}\sum_{i=1}^{n}\displaystyle\int_\Omega \left|\frac{\partial u}{\partial x_i}\right|^p dx & \text{if } u \in W_0^{1,p}(\Omega) \\ +\infty & \text{otherwise.} \end{cases}$$

Define

$$\varphi(u) = \begin{cases} \displaystyle\int_\Omega j(u(x))\, dx & \text{if } j(u) \in L^1(\Omega) \\ +\infty & \text{otherwise.} \end{cases}$$

By Proposition 2.8 from Chapter II, φ is lower-semicontinuous on $L^2(\Omega)$ and its subdifferential $\partial\varphi \subset L^2(\Omega) \times L^2(\Omega)$ is given by

$$(\partial\varphi(u))(x) = \beta(u(x)), \quad \text{a.e.} \quad \text{on } \Omega.$$

Finally, let β_1 and β_2 be two maximal monotone graphs in $R^1 \times R^1$ such that $0 \in D(\beta_1)$ and

$$j_1(r) / r^2 \geqslant M \text{ for all sufficiently large } |r| \tag{2.62}$$

$$j_2(u) \in L^1(\Omega) \text{ for all } u \in L^2(\Omega), \tag{2.63}$$

where

$$\partial j_i = \beta_i, \quad i = 1, 2.$$

We shall verify the hypotheses of Theorem 2.1 with $H = L^2(\Omega)$, φ and A defined as before and

$$l(u, v) = \int_\Omega (j_1(u(x)) + j_2(u(x)))\, dx, \quad \text{for } u, v \in L^2(\Omega).$$

Hypotheses (A) and (B) are easily verified. Let us consider hypothesis (D). For simplicity, we shall assume that $0 \in \beta(0)$. This implies that $\beta_\lambda(0) = 0$ ($\beta_\lambda = \lambda^{-1}(1 - (1 + \lambda\beta)^{-1})$) and using the monotonicity of β_λ we obtain

$$\int_\Omega \beta_\lambda(u(x) - v(x))\,(Au(x) - Av(x))\, dx \geqslant 0 \quad \text{for all } u, v \in W_0^{1,p}(\Omega).$$

Hence

$$(\partial\varphi_\lambda(u - v),\ Au - Av) \geqslant 0 \text{ for all } \lambda > 0 \text{ and } u, v \in D(A).$$

In virtue of Proposition 1.2 (Chapter IV) this implies that A is $\partial\varphi$-monotone. Similarly, one verifies condition (2.9).

For hypothesis (C), it suffices to require the following condition: There exist u_0 and v_0 in $L^2(\Omega)$ and $u \in W^{1,1}(0, T; L^2(\Omega))$ such that

$$u(0, x) = u_0(x),\ u(T, x) = v_0(x), \quad \text{a.e.} \quad \text{on } \Omega \tag{2.64}$$

$$j\left(\frac{\partial u}{\partial t}\right) \in L^1(Q), \quad j_1(u_0) \in L^1(\Omega). \tag{2.65}$$

Theorem 2.1 therefore yields the following result.

COROLLARY 2.4. *Let β, β_1 and β_2 be given satisfying the above hypotheses. Let $f \in L^2(Q)$ be given. Then the boundary value problem*

$$\begin{aligned}
&\frac{\partial}{\partial t}\left(\beta\left(\frac{\partial u}{\partial t}(t, x)\right)\right) + \\
&+ \sum_{i=1}^{n} \frac{\partial}{\partial x_i}\left(\left|\frac{\partial u}{\partial x_i}(t, x)\right|^{p-2} \frac{\partial u}{\partial x_i}(t, x)\right) \ni f(t, x), \quad \text{a.e. } on\ Q \qquad (2.66)\\
&\beta\left(\frac{\partial u}{\partial t}(0, x)\right) - \beta_1(u(0, x)) \ni 0, \qquad \text{a.e. } on\ \Omega\\
&\beta\left(\frac{\partial u}{\partial t}(T, x)\right) + \beta_2(u(T, x)) \ni 0,\\
&u(t, x) = 0 \qquad on\ \Sigma
\end{aligned}$$

has at least a solution $u(t, x)$ satisfying

$$\frac{\partial u}{\partial t} \in L^\infty(0, T; L^2(\Omega)), \quad Au \in L^2(Q) \tag{2.67}$$

$$u \in L^p(0, T; W_0^{1,p}(\Omega)). \tag{2.68}$$

2.3. A boundary value problem on half-axis

Let H be a real Hilbert space and let $(,)$ (respectively $|\cdot|$) denote the inner product (respectively the norm) in H.

In this section we shall consider the boundary value problem

$$\frac{d^2u}{dt^2}(t) \in Au(t), \quad \text{a.e. on }]0, \infty[\tag{2.69}$$

$$u(0) = x, \quad \sup_{t \geq 0} |u(t)| < \infty.$$

The main result is

THEOREM 2.3. *Let A be a maximal monotone set in $H \times H$ such that $0 \in R(A)$. Then for every $x \in \overline{D(A)}$ problem (2.69) has a unique solution $u \in W^{2,2}_{\text{loc}}(0, \infty; H)$ satisfying*

$$u \in C(0, \infty; H), \quad t\frac{du}{dt} \in L^\infty(0, \infty; H). \tag{2.70}$$

$$t^{1/2}\frac{du}{dt} \in L^2(0, \infty; H), \quad t^{3/2}\frac{d^2u}{dt^2} \in L^2(0, \infty; H). \tag{2.71}$$

Moreover, if $x \in D(A)$, then $u \in C^1(0, \infty; H)$ and

$$\frac{d^2u}{dt^2} \in L^2(0, \infty; H). \tag{2.72}$$

Remark. If $A = \partial\varphi$, φ being a lower-semicontinuous, proper convex function on H, then solution $u(t)$ for problem (2.69) is just the solution of the optimization problem.

Minimize

$$F(v) = \frac{1}{2}\int_0^\infty \left|\frac{dv}{dt}(t)\right|^2 dt + \int_0^\infty \varphi(v(t))\,dt$$

in the class of functions $v(t)$ defined on $[0, \infty[$ that possess the following properties: (i) v is in $W^{1,2}_{loc}(0, \infty; H)$ (ii) $\dfrac{dv}{dt} \in L^2(0, \infty; H)$, (iii) $v(0) = x$. The verification of this assertion is left to the reader (cf. also Example 2, Section 2.2).

Proof of Theorem 2.3. *Uniqueness.* Let u_x and u_y be two solutions for Eq. (2.69) satisfying the respective conditions

$$u_x(0) = x, \quad u_y(0) = y$$

$$\sup_{t \geqslant 0} |u_x(t)| < \infty, \quad \sup_{t \geqslant 0} |u_y(t)| < \infty.$$

Subtracting the corresponding differential equations and multiplying by $u_x - u_y$, yields:

$$\left(\frac{d^2}{dt^2}(u_x(t) - u_y(t)), u_x(t) - u_y(t)\right) \geqslant 0, \quad \text{a.e.} \quad t > 0,$$

because A is monotone. Therefore we obtain

$$\frac{1}{2}\frac{d^2}{dt^2}\left|u_x(t) - u_y(t)\right|^2 \geqslant \left|\frac{d}{dt}(u_x(t) - u_y(t))\right|^2 \geqslant 0, \quad \text{a.e.} \quad t > 0.$$

This implies that $|u_x(t) - u_y(t)|^2$ is a convex function of t. Since this function is bounded on $[0, \infty[$ we may conclude that it is nonincreasing on $[0, \infty[$. Hence

$$|u_x(t) - u_y(t)| \leqslant |u_x(0) - u_y(0)| = |x - y| \quad t \geqslant 0. \tag{2.73}$$

This proves the uniqueness.

2°. *Existence.* Let $u_0 \in D(A)$ be such that $0 \in Au_0$. Translating the domain of A by u_0, we can assume that

$$0 \in A0. \tag{2.74}$$

We first suppose that $x \in D(A)$.

Now denote by u_T the solution of the two-point boundary value problem

$$\frac{d^2u_T}{dt^2} \in Au_T, \quad \text{a.e.} \quad 0 < t < T \tag{2.75}$$

$$u_T(0) = u_T(T) = x.$$

By Corollary 2.1, $u_T \in W^{2,2}(0, T; H)$ exists and is unique for every $T > 0$. We shall prove that as $T \to \infty$, $u_T(t)$ converges uniformly on every compact interval of $[0, \infty[$ to a function $u(t)$ satisfying problem (2.69). To this purpose some *a priori* estimates on the solutions of the approximate problems (2.75) are necessary.

LEMMA 2.2. *There exists a positive constant C independent of T such that*

$$\int_0^{T/2} \left| \frac{d^2u_T}{dt^2} \right|^2 dt \leqslant C \quad \textit{for all } T > 0. \tag{2.76}$$

Proof. Let u_n be the solution of the boundary value problem

$$\frac{d^2u_n}{dt^2} = A_n u_n, \quad 0 \leqslant t \leqslant T,$$
$$u_n(0) = u_n(T) = x, \tag{2.77}$$

where $A_n = n\,(1 - (1 + n^{-1}A)^{-1})$, $n = 1,2, \ldots$

Multiplying Eq. (2.77) by $\dfrac{d^2u_n}{dt^2}$, we obtain

$$\left| \frac{d^2u_n}{dt^2} \right|^2 = \frac{d}{dt}\left(A_n u_n, \frac{du_n}{dt} \right) - \left(\frac{d}{dt} A_n u_n, \frac{du_n}{dt} \right). \tag{2.78}$$

By the monotonicity of A_n we have

$$\left(\frac{d}{dt} A_n u_n, \frac{du_n}{dt} \right) \geqslant 0, \quad \text{a.e. } t \in \,]0, T[.$$

We note that the function $t \to A_n u_n(t)$ is almost everywhere differentiable on $]0, T[$, because A_n is Lipschitz on H. Putting this inequality in Eq. (2.78) and integrating over $]0, T[$, we find

$$\int_0^T \left| \frac{d^2 u_n}{dt^2} \right|^2 dt \leqslant \int_0^T \left(A_n x, \frac{d^2 u_n}{dt^2} \right) dt.$$

Recalling that $|A_n x| \leqslant |A^0 x| = \inf \{|y|; y \in Ax\}$ we conclude that

$$\int_0^T \left| \frac{d^2 u_n}{dt^2} \right|^2 dt \leqslant T|A^0 x|^2 \quad \text{for all } n. \tag{2.79}$$

Now we multiply (2.77) by $u_n(t)$ and use the monotonicity of A_n. We obtain

$$\frac{d}{dt}\left(\frac{du_n}{dt}(t), u_n(t) \right) \geqslant \left| \frac{du_n}{dt}(t) \right|^2, \quad \text{a.e. on }]0, T[.$$

Integrating this inequality over $]0, T[$, gives

$$\int_0^T \left| \frac{du_n}{dt} \right|^2 dt \leqslant \int_0^T \left(\frac{d^2 u_n}{dt^2}, x \right) dt$$

which implies from relation (2.79) that

$$\int_0^T \left| \frac{du_n}{dt} \right|^2 dt \leqslant T|A^0 x|\,|x|. \tag{2.80}$$

It follows from relations (2.79) and (2.80) that the Lebesgue measure of $\{t \in]0, T[; \left| \frac{du_n}{dt}(t) \right|^2 + \left| \frac{d^2 u_n}{dt^2}(t) \right|^2 \geqslant 2|A^0 x|\,(|A^0 x| + |x|)\}$ is smaller than T/2. Let $\tau \in [0, T/2]$ be arbitrary but fixed. The latter inequality then shows that for every n there is $t_n \in [\tau, T]$ such that

$$\left| \frac{d^2}{dt^2}(u_n(t_n)) \right|^2 + \left| \frac{du_n}{dt}(t_n) \right|^2 \leqslant 2|A^0 x|\,(|A^0 x| + |x|). \tag{2.81}$$

Observe that Eq. (2.78) implies that

$$\left| \frac{d^2 u_n}{dt^2}(t) \right|^2 \leqslant \frac{d}{dt}\left(A_n u_n(t), \frac{du_n}{dt}(t) \right). \tag{2.82}$$

Therefore

$$\int_0^t \left| \frac{d^2 u_n}{ds^2}(s) \right|^2 ds \leqslant \left(\frac{d^2 u_n}{dt^2}(t), \frac{du_n}{dt}(t) \right) - \left(A_n x, \frac{du_n}{dt}(0) \right). \tag{2.83}$$

Note that relation (2.82) implies that $\left(\frac{d^2u_n}{dt^2}, \frac{du_n}{dt}\right)$ is nondecreasing as a function of t on $[0, T]$. Consequently, we conclude from inequalities (2.81) and (2.83) that

$$\int_0^\tau \left|\frac{d^2u_n}{dt^2}(t)\right|^2 dt \leqslant |A^0x| \left(|A^0x| + |x| + \left|\frac{du_n}{dt}(0)\right|\right). \tag{2.84}$$

Finally, we use the identity

$$\frac{du_n}{dt}(0) = \varepsilon^{-1}\left(u_n(\varepsilon) - x - \int_0^\varepsilon dt \int_0^t \frac{d^2u_n}{ds^2}(s)\, ds\right), \quad 0 < \varepsilon < T$$

which combined with relation (2.84) implies that

$$\int_0^\tau \left|\frac{d^2u_n}{dt^2}(t)\right|^2 dt \leqslant C, \quad n = 1,2, \ldots,$$

where C is a nonnegative constant independent of n and τ. (We note that inequality (2.73) and the end condition $u_n(0) = x$ imply that $|u_n(t)| \leqslant |x|$ on $[0, T]$). Since τ is arbitrary in $[0, T/2]$ we may conclude that

$$\int_0^{T/2} \left|\frac{d^2u_n}{dt^2}\right|^2 dt \leqslant C \quad \text{for all } n. \tag{2.85}$$

To conclude the proof of the lemma, we shall show that $\lim\limits_{n\to\infty} u_n(t) = u(t)$ uniformly on $[0, T]$. Let u_n and u_m be two solutions. Subtracting the corresponding differential equations and multiplying by $u_n - u_m$, yields:

$$\frac{d^2}{dt^2}|u_n(t) - u_n(t)|^2 = 2\left|\frac{d}{dt}(u_n(t) - u_m(t))\right|^2$$
$$+ 2(A_nu_n(t) - A_mu_m(t), u_n(t) - u_m(t))$$

and therefore

$$\int_0^T \left|\frac{d}{dt}(u_n(t) - u_m(t))\right|^2 dt \leqslant \int_0^T |A_nu_n(t) - A_mu_m(t)|\,|n^{-1}A_nu_n(t) -$$
$$m^{-1}A_mu_m(t)|\, dt$$

because $A_nu \in A(1 + n^{-1}A)^{-1}u$ and A is monotone. Using the estimate (2.80) and the end conditions $u_n(0) = u_m(0) = x$ it follows that

$$|u_n(t) - u_m(t)| \leqslant (CT(n^{-1} + m^{-1}))^{1/2}, \quad t \in [0, T]$$

where C is independent of n and m. Thus $u(t) = \lim_{n\to\infty} u_n(t)$ exists uniformly on $[0, T]$. Furthermore, since $\left\{\frac{d^2u_n}{dt^2}\right\}$ is bounded in $L^2(0, T; H)$ and converges to $\frac{d^2u}{dt^2}$ in $\mathscr{D}'(0, T; H)$ we may conclude that

$$\frac{d^2u_n}{dt^2} \to \frac{d^2u}{dt^2} \quad \text{weakly in } L^2(0, T; H). \tag{2.86}$$

Let $\mathscr{A}$ be the realization of A in $L^2(0, T; H)$, i.e.,

$$\mathscr{A} = \{[u, v] \in L^2(0, T; H) \times L^2(0, T; H);\, v(t) \in Au(t) \quad \text{a.e. on }]0, T[\}$$

Then $(\mathscr{A}_n u)(t) = A_n u(t)$ a.e. on $]0, T[$ so that (2.86) implies that

$$\mathscr{A}_n u_n \rightharpoonup \frac{d^2u}{dt^2} \quad \text{in } L^2(0, T; H).$$

Since $u_n \to u$ strongly in $L^2(0, T; H)$ and $\mathscr{A}$ is maximal monotone in $L^2(0, T; H) \times L^2(0, T; H)$ this implies (see Proposition 3.5 in Chapter II) $u \in D(\mathscr{A})$ and $\frac{d^2u}{dt^2} \in \mathscr{A}u$. In other words, $u(t)$ is a solution of problem (2.75). From the uniqueness of the solution it follows that $u = u_T$. Then (2.84) and (2.85) imply that

$$\int_0^{T/2} \left| \frac{d^2u_T}{dt^2} \right|^2 dt \leqslant C \quad \text{for all } T > 0$$

as claimed. This completes the proof of Lemma 2.2.

Now we return to the proof of the theorem. Let $[0, L]$ be an arbitrary finite interval of $[0, \infty[$. If $T > T' > L$, then by the monotonicity of A we have

$$\frac{d}{dt}\left(\frac{d}{dt}(u_T(t) - u_{T'}(t)),\, u_T(t) - u_{T'}(t)\right) \geqslant$$

$$\geqslant \left| \frac{d}{dt}(u_T(t) - u_{T'}(t)) \right|^2, \quad \text{a.e. on }]0, T'[.$$

Consequently

$$|u_T(t) - u_{T'}(t)|^2 \leqslant 2^{-1}\, t \frac{d}{dt} |u_T(t) - u_{T'}(t)|^2, \quad \text{a.e. } 0 < t < T'.$$

Integrating this inequality from L to T', we obtain

$$2\int_L^{T'} \frac{1}{t} |u_T(t) - u_{T'}(t)|^2 \, dt \leqslant |u_T(T') - x|^2.$$

Therefore

$$|u_T(L) - u_{T'}(L)| \leqslant |u_T(T') - x|/(\ln T'/L)^{1/2}. \tag{2.87}$$

The inequality

$$\frac{d^2}{dt^2}|u_T(t) - u_{T'}(t)|^2 \geqslant 0 \quad \text{a.e. } t \in]0, T'[$$

then implies by virtue of the end conditions that

$$|u_T(t) - u_{T'}(t)| \leqslant |u_T(L) - u_{T'}(L)| \text{ for } 0 < t < L. \tag{2.88}$$

Next, by (2.73) it follows that

$$|u_T(t)| \leqslant |x| \quad \text{for all } t \in [0, T].$$

(We note that $u=0$ satisfies Eq. (2.69). Putting this estimate in (2.87) and using relation (2.88) it follows that $\lim_{T\to\infty} u_T(t) = u(t)$ exists uniformly on every bounded interval $[0, L]$. Clearly, function $u(t)$ is bounded and continuous on $[0, \infty[$. Moreover, it follows from Lemma 2.2 that $\frac{d^2u}{dt^2} \in L^2(0, \infty; H)$ and

$$\frac{d^2u_T}{dt^2} \to \frac{d^2u}{dt^2} \quad \text{weakly in every } L^2(0, T; H).$$

To conclude the proof in the case in which $x \in D(A)$ it remains to be shown that $\frac{d^2u}{dt^2}(t) \in Au(t)$ a.e. on $]0, \infty[$. For this latter purpose an argument similar to that in the proof of Lemma 2.2 will be used.

For an arbitrary $L > 0$ define

$$\mathscr{A} = \{[u, v], u, v \in L^2(0, L; H), v(t) \in Au(t) \quad \text{a.e.} \quad t \in]0, L[\}.$$

Note that our hypotheses imply that $\mathscr{A}$ is a maximal monotone subset of $L^2(0, L; H) \times L^2(0, L; H)$. For T sufficiently large we have

$$\frac{d^2u_T}{dt^2} \in \mathscr{A}u_T.$$

$\mathscr{A}$ being maximal monotone, is demiclosed (see Proposition 3.5 quoted above). Thus since $u_T \to u$ strongly and $\dfrac{d^2u_T}{dt^2} \rightharpoonup \dfrac{d^2u}{dt^2}$ weakly in $L^2(0, L; H)$, we may infer that the restriction u_L of u on $]0, L[$ belongs to $D(\mathscr{A})$ and

$$\frac{d^2u_L}{dt^2} \in \mathscr{A}u_L .$$

Since L is arbitrary we conclude that

$$\frac{d^2u}{dt^2} \in Au(t), \quad \text{a.e. } t > 0.$$

Now we shall consider the case in which $x \in \overline{D(A)}$. The conclusions of the theorem will follow by a standard argument from the following lemma.

LEMMA 2.3. *Let $x \in D(A)$ and let $u(t)$ be the solution of problem* (2.69). *Then the following estimates hold:*

$$\int_0^\infty t^3 \left| \frac{d^2u}{dt^2}(t) \right|^2 dt \leqslant \frac{3}{2}|x|^2. \tag{2.89}$$

$$\int_0^\infty t \left| \frac{du}{dt}(t) \right|^2 dt \leqslant \frac{1}{2}|x|^2, \tag{2.90}$$

$$t \left| \frac{du}{dt}(t) \right| \leqslant |x| \quad \textit{for} \quad t \geqslant 0. \tag{2.91}$$

Proof. We multiply Eq. (2.69) by $tu(t)$, obtaining

$$t\left(\frac{d^2u}{dt^2}(t), u(t) \right) \geqslant 0, \quad \text{a.e. } t > 0,$$

because $0 \in A0$ and A is monotone in $H \times H$. Hence

$$t \left| \frac{du}{dt}(t) \right|^2 \leqslant \frac{d}{dt}\left(t\frac{du}{dt}(t), u(t) \right) - \frac{1}{2}\frac{d}{dt}|u(t)|^2, \quad \text{a.e. } t > 0. \tag{2.92}$$

Since $\dfrac{d^2}{dt^2}|u|^2 \geqslant 0$ a.e. on $]0, \infty[$ and $|u(t)|$ is bounded on $[0, \infty[$, it follows that $|u(t)|$ is nonincreasing as a function of t. Therefore, from condition (2.92) we obtain

$$\int_0^t s \left| \frac{du}{ds}(s) \right|^2 ds \leqslant \frac{1}{2}|x|^2 + t\left(\frac{du}{dt}(t), u(t) \right) \leqslant \frac{1}{2}|x|^2.$$

The desired inequality (2.90) therefore follows by letting $t \to \infty$.

To prove estimate (2.89) we set $v(t) = u(t+h) - u(t)$ where $h > 0$. Using once again the monotonicity of A, it follows from problem (2.69) that

$$\left(\frac{d^2v}{dt^2}(t), v(t)\right) \geqslant 0, \quad \text{a.e. on }]0, \infty[.$$

In particular, this implies that the function $t \to |v(t)|^2$ is convex and non-increasing on $[0, \infty[$.

The inequality

$$\int_0^t s^3 \left(\frac{d^2}{ds^2} v(s), v(s)\right) ds \geqslant 0, \quad \forall t > 0,$$

implies that

$$t^3 \left(\frac{dv}{dt}(t), v(t)\right) - \int_0^t \left(\frac{dv}{ds}(s), s^3 \frac{dv}{ds}(s) + 3s^2 v(s)\right) ds \geqslant 0.$$

Hence

$$\int_0^t s^3 \left|\frac{dv}{ds}(s)\right|^2 ds + \frac{3}{2}\int_0^t s^2 \frac{d}{ds} |v(s)|^2 \, ds \leqslant t^3 \left(\frac{dv}{dt}(t), v(t)\right) \leqslant 0.$$

Finally

$$\int_0^t s^3 \left|\frac{dv}{ds}(s)\right|^2 ds + \frac{3}{2} t^2 |v(t)|^2 \leqslant 3 \int_0^t s\, |v(s)|^2 \, ds \quad \text{for } t \geqslant 0.$$

Now we divide both sides of this inequality by h^2 and let $h \to 0$, obtaining:

$$\int_0^t s^3 \left|\frac{d^2u}{ds^2}(s)\right|^2 ds + \frac{3}{2} t^2 \left|\frac{du}{dt}(t)\right|^2 \leqslant 3 \int_0^t s \left|\frac{du}{ds}(s)\right|^2 ds, \quad \text{for } t \geqslant 0.$$

Combining this inequality with estimate (2.90) we obtain the estimates (2.89) and (2.91). This completes the proof of the lemma.

Let x be arbitrary in $\overline{D(A)}$ and let $\{x_n\} \subset D(A)$ be such that $x_n \to x$ as $n \to \infty$. We denote by $u_n(t)$ the solution of problem (2.69) with the initial condition $u_n(0) = x_n$. From Eq. (2.73) it follows that

$$|u_n(t) - u_m(t)| \leqslant |x_n - x_m| \quad \text{on } [0, \infty[.$$

Thus $\{u_n\}$ is a Cauchy sequence in $C(0, \infty; H)$. Let $u_n \to u$ in $C(0, \infty; H)$. Then the estimates given in Lemma 2.3 imply that

$$t^{1/2} \frac{du_n}{dt} \to t^{1/2} \frac{du}{dt} \quad \text{weakly in } L^2(0, \infty; H)$$

$$t \frac{du_n}{dt} \to t \frac{du}{dt} \quad \text{weakly-star in } L^\infty(0, \infty; H)$$

$$t^{3/2} \frac{d^2u_n}{dt^2} \to t^{3/2} \frac{d^2u}{dt^2} \quad \text{weakly in } L^2(0, \infty; H).$$

This shows in particular that $u \in W^{2,2}_{\mathrm{loc}}(0, \infty; H)$ and it satisfies relations (2.70) and (2.71) of the theorem. More precisely, $u(t)$ will satisfy estimates (2.89), (2.90) and (2.91) given in Lemma 2.3.

It remains to be shown that $\dfrac{\mathrm{d}^2 u}{\mathrm{d}t^2}(t) \in Au(t)$ a.e. on $[0, \infty[$. Let $[\alpha, \beta]$ be a bounded and closed interval of $]0, \infty[$ and let $y_n = \dfrac{\mathrm{d}^2 u_n}{\mathrm{d}t^2}\cdot$ We have

$$y_n \to \frac{\mathrm{d}^2 u}{\mathrm{d}t^2} \quad \text{weakly in } L^2(\alpha, \beta; H)$$

$$u_n \to u \quad \text{strongly in } L^2(\alpha, \beta; H)$$

$$y_n(t) \in Au_n(t), \quad \text{a.e. on }]\alpha, \beta[.$$

This is just the same result as in the previous case, so we may conclude the argument just as before. The proof of Theorem 2.3 is thereby complete.

Remark. More generally, we could consider a boundary value problem of the following type

$$\frac{\mathrm{d}^2 u}{\mathrm{d}t^2} \in Au, \quad \text{a.e. on }]0, \infty[\tag{2.93}$$

$$\frac{\mathrm{d}u}{\mathrm{d}t}(0) \in \partial\psi\,(u(0) - x), \ \sup_{t \geq 0} |u(t)| < \infty,$$

where A is maximal monotone in $H \times H$ and ψ is a lower-semicontinuous proper convex function from H into $[0, \infty]$ satisfying:

$$\psi(0) = 0 \qquad \lim_{|u| \to \infty} \psi(u)/|u| = \infty.$$

We observe that problem (2.69) can be written under this form taking

$$\psi(u) = \begin{cases} 0 & \text{if } u = 0 \\ +\infty & \text{otherwise.} \end{cases}$$

Suppose that A is $\partial\psi$-monotone and $0 \in R(A)$. If $x \in D(A)$ we can then prove the existence of a solution $u(t)$ for the above boundary value problem with

$$u \in C^1(0, \infty; H), \qquad \frac{\mathrm{d}^2 u}{\mathrm{d}t^2} \in L^2(0, \infty; H).$$

We have omitted the proof which is a variant of that given before for Theorem 2.3 and uses Corollary 2.3. For a different approach we refer the reader to H. Brézis [9].

Incidentally we have proved that the solution $u(t)$ for problem (2.69) satisfies the estimates (2.89) $\sim$ (2.91) given in Lemma 2.3.
In this connection, the following theorem is a quantitative refinement of Theorem 2.3 in the case in which $x \in D(A)$.

THEOREM 2.4. *Let* $u(t)$ *be the solution for problem* (2.69) *with* $x \in D(A)$. *Suppose that* $0 \in A0$. *Then*

$$\int_0^\infty \left| \frac{du}{dt}(t) \right|^2 dt \leqslant 2\,|A^0x|^{1/2}\,|x|^{3/2} \tag{2.94}$$

$$\int_0^\infty \left| \frac{d^2u}{dt^2}(t) \right|^2 dt \leqslant 2\,|A^0x|^{3/2}\,|x|^{1/2} \tag{2.95}$$

$$\left| \frac{du}{dt}(t) \right| \leqslant 2|A^0x|^{1/2}\,|x|^{1/2} \quad \text{on } [0, \infty[. \tag{2.96}$$

Proof. According to Theorem 2.3, the problem

$$\frac{d^2u_n}{dt^2} = A_n u_n, \quad 0 \leqslant t < \infty \tag{2.97}$$

$$u_n(0) = x; \quad \sup_{t \geqslant 0} |u_n(t)| < \infty$$

has a unique solution $u_n \in C^1(0, \infty; H)$ satisfying

$$\frac{d^2u_n}{dt^2} \in L^2(0, \infty; H).$$

Multiplying Eq. (2.97) by u_n and using the monotonicity of A_n we get

$$\left(\frac{d^2u_n}{dt^2}(t), u_n(t) \right) \geqslant 0,$$

and therefore

$$\left| \frac{du_n}{dt}(t) \right|^2 \leqslant \frac{d}{dt} \left(\frac{du_n}{dt}(t), u_n(t) \right).$$

But $|u_n(t)|^2$ is bounded on $[0, \infty[$ and by the previous inequality is convex. We may therefore infer that $|u_n(t)|$ is non-increasing as a function of t. Integrating over $]0, \infty[$ we obtain

$$\int_0^\infty \left| \frac{du_n}{dt}(t) \right|^2 dt \leqslant - \left(\frac{du_n}{dt}(0), x \right). \tag{2.98}$$

Next we integrate over $]0, \infty[$ the identity

$$\frac{1}{2}\frac{\mathrm{d}}{\mathrm{d}t}\left|\frac{\mathrm{d}u_n}{\mathrm{d}t}(t)\right|^2 = \left(\frac{\mathrm{d}^2u_n}{\mathrm{d}t^2}(t), \frac{\mathrm{d}u_n}{\mathrm{d}t}(t)\right)$$

to find that

$$\frac{1}{2}\left|\frac{\mathrm{d}u_n}{\mathrm{d}t}(0)\right|^2 = -\int_0^\infty \left(\frac{\mathrm{d}^2u_n}{\mathrm{d}t^2}(t), \frac{\mathrm{d}u_n}{\mathrm{d}t}(t)\right)\mathrm{d}t.$$

Here we have used, in particular, estimate (2.98) which implies that $\dfrac{\mathrm{d}u_n}{\mathrm{d}t} \in L^2(0, \infty; H)$. Using once again estimate (2.98), we conclude that

$$\frac{1}{2}\left|\frac{\mathrm{d}u_n}{\mathrm{d}t}(0)\right|^2 \leqslant \left(\int_0^\infty \left|\frac{\mathrm{d}^2u_n}{\mathrm{d}t^2}\right|^2 \mathrm{d}t\right)^{1/2} \left|\left(\frac{\mathrm{d}u_n}{\mathrm{d}t}(0), x\right)\right|^{1/2}. \tag{2.99}$$

On the other hand, taking inner products of Eq. (2.97) with $\dfrac{\mathrm{d}^2u_n}{\mathrm{d}t^2}$ and noting that

$$\left(\frac{\mathrm{d}}{\mathrm{d}t}A_nu_n(t), \frac{\mathrm{d}u_n}{\mathrm{d}t}(t)\right) \geqslant 0, \quad \text{a.e. on }]0, \infty[$$

yields

$$\left|\frac{\mathrm{d}^2u_n}{\mathrm{d}t^2}(t)\right|^2 \leqslant \frac{\mathrm{d}}{\mathrm{d}t}\left(A_nu_n(t), \frac{\mathrm{d}u_n}{\mathrm{d}t}(t)\right), \quad \text{a.e. on }]0, \infty[. \tag{2.100}$$

Hence

$$\int_0^\infty \left|\frac{\mathrm{d}^2u_n}{\mathrm{d}t^2}\right|^2 \mathrm{d}t \leqslant -\left(A_nx, \frac{\mathrm{d}u_n}{\mathrm{d}t}(0)\right) \leqslant |A^0x| \left|\frac{\mathrm{d}u_n}{\mathrm{d}t}(0)\right| \tag{2.101}$$

because A_nu_n and $\dfrac{\mathrm{d}u_n}{\mathrm{d}t}$ belong to $L^2(0, \infty; H)$.

Putting this estimate into relation (2.99) yields

$$\left|\frac{\mathrm{d}u_n}{\mathrm{d}t}(0)\right| \leqslant 2\,|A^0x|^{1/2}\,|x|^{1/2} \tag{2.102}$$

so that estimates (2.98) and (2.101) imply that

$$\int_0^\infty \left|\frac{\mathrm{d}u_n}{\mathrm{d}t}\right|^2 \mathrm{d}t \leqslant 2\,|A^0x|^{1/2}|x|^{3/2} \tag{2.103}$$

respectively

$$\int_0^\infty \left|\frac{d^2u_n}{dt^2}(t)\right|^2 dt \leqslant 2\,|A^0x|^{3/2}\,|x|^{1/2}. \tag{2.104}$$

Now we observe that inequality (2.100) implies that $\frac{d^2}{dt^2}\left|\frac{du_n}{dt}(t)\right|^2 \geqslant 0$ a.e. on $]0, \infty[$. Since $\left|\frac{du_n}{dt}\right| \in L^2(0, \infty)$ this implies that $\left|\frac{du_n}{dt}(t)\right|$ is a non-increasing function of t on $[0, \infty[$. Then estimate (2.102) yields

$$\left|\frac{du_n}{dt}(t)\right| \leqslant 2\,|A^0x|^{1/2}|x|^{1/2} \quad \text{on} \quad [0, \infty[. \tag{2.105}$$

Let u_n and u_m be two solutions of the respective equations

$$\frac{d^2u_n}{dt^2} = A_nu_n\,, \quad \frac{d^2u_m}{dt^2} = A_mu_m\,.$$

We subtract the defining equations for u_n and u_m and then multiply the result by $u_n - u_m$, thus obtaining

$$\frac{d}{dt}\left(\frac{d}{dt}(u_m(t) - u_n(t)),\, u_m(t) - u_n(t)\right) = \left|\frac{d}{dt}(u_n(t) - u_m(t))\right|^2 +$$
$$+ (A_mu_m(t) - A_nu_n(t),\, u_m(t) - u_n(t)).$$

Integrating from 0 to ∞ and recalling that $A_nu \in A(1 + n^{-1}A)^{-1}u$, we find that

$$\int_0^\infty \left|\frac{du_n}{dt} - \frac{du_m}{dt}\right|^2 dt \leqslant \int_0^\infty |A_nu_n - A_mu_m|\,|n^{-1}A_nu_n - m^{-1}A_mu_m|dt$$

so that relation (2.104) implies that

$$\int_0^\infty \left|\frac{du_n}{dt} - \frac{du_m}{dt}\right|^2 dt \leqslant C(n^{-1} + m^{-1})$$

where $C > 0$ is independent of m and n. Since $u_n(0) = u_m(0) = x$ we may therefore infer that there exists a function $v \in C(0, \infty; H)$ such that $\lim_{n\to\infty} u_n(t) = v(t)$ uniformly on every compact interval of $[0, \infty[$ and

$$\frac{du_n}{dt} \to \frac{dv}{dt} \quad \text{strongly in } L^2(0, \infty; H)$$

$$\frac{d^2u_n}{dt^2} \to \frac{d^2v}{dt^2} \quad \text{weakly in } L^2(0, \infty; H).$$

Moreover, it follows from estimates (2.102), (2.103) and (2.105) that

$$\int_0^\infty \left|\frac{\mathrm{d}v}{\mathrm{d}t}\right|^2 \mathrm{d}t \leqslant 2|A^0x|^{1/2}\,|x|^{3/2}$$

$$\int_0^\infty \left|\frac{\mathrm{d}^2v}{\mathrm{d}t^2}\right|^2 \mathrm{d}t \leqslant 2\,|A^0x|^{3/2}|x|^{1/2}$$

respectively

$$\left|\frac{\mathrm{d}v}{\mathrm{d}t}(t)\right| \leqslant 2|A^0x|^{1/2}|x|^{1/2}, \quad \text{a.e. on }]0, \infty[.$$

Since $\frac{\mathrm{d}v}{\mathrm{d}t}$ is continuous, the last inequality holds everywhere on $[0, \infty[$.

Finally, the last argument in the proof of Lemma 2.2 can now be repeated here to deduce that $\frac{\mathrm{d}^2v}{\mathrm{d}t^2} \in Av$, a.e. on $]0, \infty[$. This shows that $v = u$ and the proof is complete.

We now ask under what conditions on A does the solution $u(t)$ for problem (2.69) have compact support.

This problem is by no means trivial and has its own interest, for example in connection with the optimization theory (Cf. L. Berkovitz and H. Pollard [1], H. Brézis [12] for related problems). For simplicity we restrict ourselves to dealing with the case in which $A = \partial\varphi$ where φ is a lower-semicontinuous, proper convex function from H into $]-\infty, +\infty]$.

PROPOSITION 2.1. *Let* $A = \partial\varphi$ *be such that* $0 \in \operatorname{int} A0$. *Let* $x \in \overline{D(A)}$. *Then the solution* $u(t)$ *for problem* (2.69) *has compact support on* $[0, \infty[$.

Proof. Note that condition $0 \in A0$ implies that $\varphi(0) = \min\{\varphi(u); u \in H\}$. Taking, if necessary, $u \to \varphi(u) - \varphi(0)$ instead of φ, we may assume that

$$\varphi(0) = 0 = \min\{\varphi(u); u \in H\}.$$

Then, by definition of $\partial\varphi$, we have

$$\varphi(u) \leqslant (y, u) \text{ for all } [u, y] \in \partial\varphi$$

respectively

$$(y_0, u) \leqslant \varphi(u) \text{ for all } u \in H \text{ and } y_0 \in \partial\varphi(0).$$

By hypothesis, there exists $\rho > 0$ such that $\{y; |y| \leqslant \rho\} \subset \partial\varphi(0)$. On setting $y_0 = \rho u/|u|$ in the last inequality, we obtain

$$\rho|u| \leqslant \varphi(u) \quad \text{for all } u \in H$$

so that the former relation implies that

$$\rho|u| \leqslant |(\partial\varphi)^0 u|\ |u| \text{ for every } u \in D(\partial\varphi). \tag{2.106}$$

Now, let $u(t)$ be the solution of problem (2.69) provided by Theorem 2.3. We have to show that there exists $T>0$ such that $u(t)=0$ for any t in $[T, \infty[$. Suppose, however, that this is not the case. As $|u(t)|^2$ is convex and bounded as a function of t, we may conclude that it is monotonically non-increasing on $[0, \infty[$. Then we should have $u(t) \neq 0$ everywhere on $[0, \infty[$. Estimate (2.106) then yields

$$\left|\frac{d^2u}{dt^2}(t)\right| \geqslant \rho, \quad \text{a.e. on }]0, \infty[$$

which contradicts the fact that $t^{3/2}\dfrac{d^2u}{dt^2} \in L^2(0, \infty; H)$. This contradiction establishes the conclusion of our proposition.

Remark 1°. As mentioned before, Proposition 2.1 can be stated under the following equivalent form: *Given* $\varphi: H \to]-\infty, +\infty]$ *with* $0 \in \operatorname{int} \partial\varphi(0)$ *and* $x \in D(\varphi)$ *there is a unique solution with compact support of the minimization problem*

$$\min\left\{\int_0^\infty\left(\frac{1}{2}\left|\frac{dv}{dt}(t)\right|^2 + \varphi(v(t))\right)dt;\ \frac{dv}{dt} \in L^2(0, \infty; H),\ v(0) = x\right\}.$$

2°. It is easy to see that condition $0 \in \operatorname{int} \partial\varphi(0)$ is equivalent to the following one

$$\varphi(u) \geqslant \rho|u| \quad \text{for all } u \in H$$

where ρ is some positive constant.

2.4. The square root of a nonlinear maximal monotone operator

Let A be a maximal monotone set in $H \times H$ such that $0 \in R(A)$. From Theorem 2.3, for every $x \in \overline{D(A)}$, there is a unique continuous solution u_x: $[0, \infty[\to \overline{D(A)}$ for the problem

$$\frac{d^2u}{dt^2} \in Au, \quad \text{a.e. on }]0, \infty[\tag{2.107}$$

$$u(0) = x, \quad \sup_{t \geqslant 0} |u(t)| < \infty.$$

For $t \geqslant 0$ and $x \in \overline{D(A)}$ we define $S_{1/2}(t)x = u_x(t)$. By (2.73) we have

$$|S_{1/2}(t)x - S_{1/2}(t)y| \leqslant |x - y|, \ \forall x, y \in \overline{D(A)}, \ t \geqslant 0,$$

$S_{1/2}(0)x = x$ by definition and $\lim_{t \to 0+} S_{1/2}(t)x = x$ because u_x is continuous as a function of t. Moreover, the functions $S_{1/2}(t)\, S_{1/2}(s)x$ and $S_{1/2}(t+s)x$ are both solutions of Eq. (2.107) with $u(0) = S_{1/2}(s)x$ and hence coincide. Thus $S_{1/2}$ is a semigroup of nonlinear contractions on the closure $\overline{D(A)}$ of $D(A)$ (we recall that $\overline{D(A)}$ is a convex subset of H).

By Theorem 1.2, established in Chapter IV, there is a unique maximal monotone set $A_{1/2}$ such that $-A^0_{1/2}$ is the generator of $S_{1/2}$. As usual, $A^0_{1/2}$ denotes the *minimal section* of $A_{1/2}$, i.e.,

$$|A^0_{1/2}x| = \inf \{|y|;\, y \in A_{1/2}y\}.$$

By the result just quoted from Chapter IV, we have

$$D(A^0_{1/2}) = D(A_{1/2}), \ \overline{D(A_{1/2})} = \overline{D(A)}.$$

Moreover, if $x \in D(A)$, then in virtue of Theorem 2.3, $u_x(t) = S_{1/2}(t)$ is continuously differentiable as a function of t on $[0, \infty[$. Hence $x \in D(A_{1/2})$ and the following inclusion relations holds:

$$D(A) \subset D(A_{1/2}) \subset \overline{D(A)}.$$

If A is linear then according to a well-known result due to A. V. Balakrishnan [1], the operator $A_{1/2}$ introduced before coincides with the *square root* $A^{1/2}$ of A, i.e. $(A^{1/2})^2 = A$. For this reason the operator $A_{1/2}$ will be called *the square root* of A.

In general it is not clear whether there is any relationship between $(A_{1/2})^2$ and A. Nevertheless we shall see that the operator $A_{1/2}$ has several remarkable properties which are analogous to the corresponding one for square roots of linear operators. We begin with the convexity inequality.

Corollary 2.5. *Let A be maximal monotone in $H \times H$ such that $0 \in R(A)$ and let $A_{1/2}$ be its square root. Then*

$$|A^0_{1/2}x| \leqslant 2|A^0x|^{1/2}|x - y_0|^{1/2} \tag{2.108}$$

for any $x \in D(A)$ and every $y_0 \in A^{-1}0$.

Proof. Let $y_0 \in A^{-1}0$ be fixed. Replacing the operator A by $x \to A(x + y_0)$ if necessary, we can reduce it to the case where $y_0 = 0 \in A^{-1}0$. Let $S_{1/2}(t)$ be the semigroup generated by $A_{1/2}$.

From Theorem 2.4 (estimate (2.96)) it follows that

$$\left|\frac{d}{dt} S_{1/2}(t)x\right| \leqslant 2|A^0x|^{1/2}|x|^{1/2} \text{ everywhere on } [0, \infty[\text{ for each } x \in D(A).$$

This clearly implies estimate (2.108).

Remark. As we shall see later, we have a better estimate than (2.108) if the operator A is cyclically maximal monotone.

An important property of the square root of a linear maximal monotone operator is that it generates a differentiable (even analytic) semigroup of contractions (see e.g. S. G. Krein [1], Chapter III). It turns out that the semigroup $S_{1/2}$ generated by $A_{1/2}$ has comparable regularity properties, namely

THEOREM 2.5. *$S_{1/2}(t)\,\overline{D(A)} \subset D(A)$ for all $t > 0$. Moreover,*

(i) For every $x \in \overline{D(A)}$ the function $t \to S_{1/2}(t)x$ is continuously differentiable on $]0, \infty[$ and

$$\frac{d}{dt} S_{1/2}(t)x + A^0_{1/2}\, S_{1/2}(t)x = 0, \quad \forall\, t > 0. \tag{2.109}$$

(ii) For every $x \in \overline{D(A)}$ and $y \in A^{-1}(0)$,

$$\int_0^\infty t \left|\frac{d}{dt} S_{1/2}(t)x\right|^2 dt \leqslant |x - y|^2/2,$$

$$\left|\frac{d^+}{dt} S_{1/2}(t)x\right| \leqslant |x - y|/t \quad \text{for} \quad t > 0.$$

In addition, $\dfrac{d^2}{dt^2} S_{1/2}(t)x$ exists almost everywhere on $]0, \infty[$ and

$$\int_0^\infty t^3 \left|\frac{d^2}{dt^2} S_{1/2}(t)x^2\right|^2 dt \leqslant \frac{3}{2}|x - y|^2.$$

(iii) If $x \in D(A_{1/2})$, then the function $t \to S_{1/2}(t)x$ is continuously differentiable on $[0, \infty[$ and $t^{1/2} \dfrac{d^2}{dt^2} S_{1/2}(t)x \in L^2(0, \infty; H)$.

Proof. We will consider only the special case where $y = 0 \in A^{-1}0$. Just as in the proof of Corollary 2.5 the general result is obtained by appropriately translating the special case we shall consider.

Let x be arbitrary but fixed in $\overline{D(A)}$, and let $u(t) = S_{1/2}(t)x$. By Theorem 2.3, $t^{3/2} \dfrac{d^2u}{dt^2} \in L^2(0, \infty; H)$ and $u(t) \in D(A)$ a.e. on $]0, \infty[$. Next from the definition of $A_{1/2}$, we have

$$\frac{du}{dt}(t) + A^0_{1/2}u(t) = 0, \quad \text{a.e.} \quad \text{on }]0, \infty[.$$

Since $\frac{du}{dt}$ is continuous on $]0, \infty[$ and $A_{1/2}$ is closed (being maximal monotone) we may therefore conclude that this equality holds at every $t > 0$. The estimates given in part (ii) have been established in the proof of Theorem 2.3 (see Lemma 2.3).

Now suppose that $x \in D(A_{1/2})$. By definition, this implies that $\frac{d}{dt} S_{1/2}(t) = -A^0_{1/2} S_{1/2}(t)x$ is continuous from the right in $t = 0$. Therefore, $S_{1/2}(t)x \in C^1(0, \infty; H)$. To prove that $t^{1/2} \frac{d^2}{dt^2} S_{1/2}(t)x \in L^2(0, \infty; H)$ we shall use the same technique as in the proof of Lemma 2.3, but with some simplifications. We set $v(t) = u(t+h) - u(t)$ where $h > 0$. A being monotone, we have

$$\left(\frac{d^2v}{dt^2}, v\right) \geqslant 0, \quad \text{a.e. on }]0, \infty[$$

and therefore

$$\int_0^t s\left(\frac{d^2v}{ds^2}, v\right) ds \geqslant 0 \quad \text{for all} \quad t \geqslant 0.$$

Integrating by parts, we obtain

$$t\left(\frac{dv}{dt}(t), v(t)\right) - \int_0^t s\left|\frac{dv}{ds}\right|^2 ds \geqslant -\frac{1}{2}|v(0)|^2.$$

Hence

$$\int_0^t s\left|\frac{dv}{ds}\right|^2 ds \leqslant \frac{1}{2}|v(0)|^2 \quad \text{on} \quad [0, \infty[$$

because, as we have already observed, the function $t \to |v(t)|^2$ is nonincreasing. But

$$\lim_{h\to 0} \frac{1}{h} v(0) = \lim_{h\to 0} \frac{u(h) - x}{h} = -A^0_{1/2}x.$$

Hence

$$\int_0^t s\left|\frac{d^2u}{ds^2}\right|^2 ds \leqslant \frac{1}{2}|A^0_{1/2}x|^2 \quad \text{for all } t > 0.$$

In other words,

$$\int_0^\infty t\left|\frac{d^2}{dt^2} S_{1/2}(t)x\right|^2 \leqslant \frac{1}{2}|A^0_{1/2}x|^2.$$

Thus the proof of Theorem 2.4 is complete.

Remark. A significant consequence to be drawn from the above theorem is that the semigroup $S_{1/2}$ generated by $A_{1/2}$ has a *smoothing* effect on initial data (see Section 2 in Chapter IV and Theorem 2.6 in Chapter III for other relevant classes of nonlinear semigroups having this property). The outstanding new feature is the fact that this semigroup is differentiable everywhere on a positive half-axis.

The operator $A_{1/2}$ is hard to describe in elementary terms, even in a linear case. As we shall see later, however, there are some results of practical significance that can be deduced in the case in which A is cyclically monotone, i.e. is of the form $A = \partial\varphi$ where φ is a lower-semicontinuous convex function on H.

In order to see what results might be possible concerning the operator $A_{1/2}$, it is instructive to recall first the case where $A = \partial\varphi$ is linear. We know (see Proposition 2.7 in Chapter II) that a linear operator A from H to itself is cyclically maximal monotone if and only if it is self-adjoint. Moreover, the function φ is given up to an additive constant by

$$\varphi(u) = \begin{cases} \dfrac{1}{2}|A^{1/2}u|^2 & \text{if } u \in D(A^{1/2}) \\ +\infty & \text{otherwise} \end{cases}$$

where $A^{1/2}$ is the square root of the self-adjoint positive operator A. In other words

$$\frac{1}{2}|A^{1/2}u|^2 = \varphi(u) \quad \text{for} \quad u \in D(A^{1/2}) = D(\varphi).$$

This result leads one to conjecture that a similar theorem might be true for nonlinear cyclically maximal monotone operators. The conjecture turns out to be true, as the next theorem shows.

THEOREM 2.6. *Let φ be a proper and lower-semicontinuous convex function from H into $]-\infty, +\infty]$ such that $0 \in R(\partial\varphi)$. Let $A = \partial\varphi$. Then*

(i) $D(A_{1/2}) = D(\varphi)$ *and*

$$|A^0_{1/2}x|^2 = 2(\varphi(x) - \varphi_\infty) \text{ for every } x \in D(\varphi) \tag{2.110}$$

where $\varphi_\infty = \inf\{\varphi(u);\ u \in H\}$.

(ii) *There exists a lower-semicontinuous proper convex function $\psi : H \to$ $\to]-\infty, +\infty]$ such that*

$$A_{1/2} = \partial\psi.$$

Proof. Firstly, we note that condition $0 \in R(\partial\varphi)$ simply says that the function φ attains its minimum on H. Let y_0 be arbitrary in $A^{-1}0$. Then

$$\varphi(y_0) = \min\{\varphi(u);\ u \in H\} = \varphi_\infty.$$

Considering, if necessary, the function $u \to \varphi(u) - \varphi(y_0)$ instead of φ. we may assume without loss of generality that

$$\varphi(u) \geqslant \varphi(y_0) = 0 \text{ for all } u \in H. \tag{2.111}$$

Let $x \in \overline{D(A)}$ and let $u(t) = S_{1/2}(t)x$. From Theorem 2.3 we have

$$\frac{d^2u}{dt^2}(t) - \partial\varphi(u(t)) \ni 0, \quad \text{a.e.} \quad \text{on }]0, \infty[\tag{2.112}$$

and $t^{3/2} \dfrac{d^2u}{dt^2} \in L^2(0, \infty; H)$. This implies that (see Lemma 2.1 from Chapter IV) $t \to \varphi(u(t))$ is absolutely continuous on every compact interval of $]0, \infty[$ and

$$\left(\frac{du}{dt}(t), \partial\varphi(u(t))\right) = \frac{d}{dt}\varphi(u(t)), \quad \text{a.e.} \quad \text{on }]0, \infty[.$$

Now we multiply (2.112) by $\dfrac{du}{dt}$ and make use of the last relation to obtain

$$\frac{1}{2}\frac{d}{dt}\left|\frac{du}{dt}(t)\right|^2 - \frac{d}{dt}\varphi(u(t)) = 0, \quad \text{a.e.} \quad t > 0. \tag{2.113}$$

Since the function $\left|\dfrac{du}{dt}(t)\right|^2$ is monotonically decreasing in t, it follows from Eq. (2.113) that so is $\varphi(u(t))$. Moreover, by definition of $\partial\varphi$ we have

$$\varphi(u(t)) \leqslant \left(\frac{d^2u}{dt^2}(t), u(t) - y_0\right), \quad \text{a.e.} \quad \text{on }]0, \infty[$$

so that (2.111) implies that

$$\lim_{t\to\infty} \varphi(u(t)) = 0$$

(because $t^{1/2} \dfrac{d^2u}{dt^2} \in L^2(0, \infty; H)$ and $|u(t)|$ is bounded on $[0, \infty[$). Thus integrating (2.113) over $]t, \infty[$ and keeping in mind that $\lim\limits_{t\to 0} \dfrac{du}{dt}(t) = 0$, we find that

$$\left|\frac{du}{dt}(t)\right|^2 = 2\varphi(u(t)) \quad \text{on} \quad]0, \infty[. \tag{2.114}$$

Now we shall assume that $x \in D(A_{1/2})$. Then $\left|\frac{du}{dt}(t)\right| = \left|\frac{d}{dt} S_{1/2}(t)x\right| \leqslant |A^0_{1/2}x|$ everywhere on $[0, \infty[$. Since $\lim_{t\to 0} u(t) = x$ and φ is lower-semicontinuous on H, it follows from Eq. (2.114) that

$$|A^0_{1/2}x|^2 \geqslant 2\varphi(x). \tag{2.115}$$

We have shown, in particular, that $\varphi(x) < \infty$, i.e., $x \in D(\varphi)$. Therefore, $D(A_{1/2}) \subset D(\varphi)$.

Now let $x \in D(\varphi)$. We set $x_\lambda = (1+\lambda A)^{-1} x$ and define $u_\lambda(t) = S_{1/2}(t)x_\lambda$, where $\lambda > 0$. Since $x_\lambda \in D(A)$, it follows from Theorem 2.4 that $\frac{d^2u_\lambda}{dt^2} \in L^2(0, \infty; H)$ and $\frac{du_\lambda}{dt} \in L^2(0, \infty; H)$. Then relation (2.113) implies that $\frac{d}{dt}\varphi(u_\lambda) \in L^1(0, \infty)$ and

$$2\varphi(x_\lambda) = \left|\frac{du_\lambda}{dt}(0)\right|^2 = |A^0_{1/2}x_\lambda|^2. \tag{2.116}$$

We know (see Section 2.2 in Chapter II) that $\lim_{\lambda\to 0} x_\lambda = x$ and $\varphi(x_\lambda) \to \varphi(x)$ as $\lambda\to 0$. Hence $\lim_{\lambda\to 0} |A^0_{1/2}x_\lambda|^2 = 2\varphi(x)$. Since $A_{1/2}$ is maximal monotone we may conclude therefore (see Proposition 3.5 in Chapter II) that $x \in D(A_{1/2})$ and $|A^0_{1/2}x| \leqslant 2\varphi(x)$. Comparing this inequality with (2.115) we conclude that $|A^0_{1/2}x|^2 = 2\varphi(x)$ on $D(A_{1/2}) = D(\varphi)$ as desired.

From Theorem 2.3 in Chapter II, in order to prove part (ii) of the theorem, we have to verify that $A_{1/2}$ is cyclically monotone in $H\times H$. Firstly we shall prove that $A^0_{1/2}$ is cyclically monotone, i.e.,

$$(A^0_{1/2}x_0, x_1-x_0)+(A^0_{1/2}x_1, x_2-x_1)+ \dots +(A^0_{1/2}x_n, x_0-x_n)\leqslant 0, \tag{2.117}$$

for every finite sequence $\{x_i\} \subset D(A_{1/2})$. Let $u_n(t) = S_{1/2}(t)x_n$. Since A is cyclically monotone, we have

$$\left(\frac{d^2u_0}{dt^2}, u_1 - u_0\right) + \left(\frac{d^2u_1}{dt^2}, u_2 - u_1\right) + \dots + \left(\frac{d^2u_n}{dt^2}, u_0 - u_n\right) \leqslant 0,$$

a.e. $t > 0$.

Consequently

$$\frac{d}{dt}\left[\left(\frac{du_0}{dt}, u_1 - u_0\right) + \dots + \left(\frac{du_n}{dt}, u_0 - u_n\right)\right] - \left[\left(\frac{du_0}{dt}, \frac{du_1}{dt} - \frac{du_0}{dt}\right)\right.$$
$$\left. + \dots + \left(\frac{du_n}{dt}, \frac{du_0}{dt} - \frac{du_n}{dt}\right)\right] \leqslant 0, \quad \text{a.e. } t > 0.$$

Finally we obtain

$$\frac{\mathrm{d}}{\mathrm{d}t}\left[\left(\frac{\mathrm{d}u_0}{\mathrm{d}t}, u_1 - u_0\right) + \ldots + \left(\frac{\mathrm{d}u_n}{\mathrm{d}t}, u_0 - u_n\right)\right] \leqslant 0, \quad \text{a.e.} \quad t > 0.$$

Hence the function

$$t \to \left(\frac{\mathrm{d}u_0}{\mathrm{d}t}, u_1 - u_0\right) + \ldots + \left(\frac{\mathrm{d}u_n}{\mathrm{d}t}, u_0 - u_n\right)$$

is monotonically decreasing on $]0, \infty[$. Since $A^0_{1/2}x_i = -\dfrac{\mathrm{d}u_i}{\mathrm{d}t}(0)$ we obtain (2.117) by recalling that

$$\lim_{t\to\infty}\left[\left(\frac{\mathrm{d}u_0}{\mathrm{d}t}, u_1 - u_0\right) + \ldots + \left(\frac{\mathrm{d}u_n}{\mathrm{d}t}, u_0 - u_n\right)\right] = 0.$$

To conclude the proof we need the following general result.

LEMMA 2.4. *Let H be a real Hilbert space and let A be maximal monotone. Then A is cyclically monotone in $H \times H$ if and only if its minimal section A^0 is cyclically monotone.*

Proof. The "only if" part is obvious. We shall prove that if A^0 is cyclically monotone then so is A. Let $E(A^0; \overline{D(A)})$ be the set of all maximal monotone sets $\tilde{A}$ with domain in $\overline{D(A)}$ which extend A. In Chapter II (see Corollary 3.2) we have shown that $E(A^0; \overline{D(A)}) = A$. To conclude the proof it suffices to show that $E(A^0; \overline{D(A)})$ contains a cyclically monotone set. By Zorn's lemma there exists a cyclically maximal monotone set B such that $B \supset A^0$. Thus Theorem 2.3 from Chapter II implies that $B = \partial\varphi$ where φ is lower-semicontinuous and convex from H into $]-\infty, +\infty]$. Let $I_{\overline{D(A)}}$ be the indicator function of $\overline{D(A)}$, i.e., $I_{\overline{D(A)}}(x) = 0$ for $x \in \overline{D(A)}$ and $I_{\overline{D(A)}}(x) = +\infty$ if $x \,\bar{\in}\, \overline{D(A)}$. Define,

$$\psi = \varphi + I_{\overline{D(A)}}\,.$$

Then $\partial\psi \supset \partial\varphi + \partial I_{\overline{D(A)}} \supset B$ because $0 \in \partial I_{\overline{D(A)}}(x)$ for every $x \in \overline{D(A)}$. Moreover, $D(\partial\psi) \subset \overline{D(A)}$ since $D(\psi) = D(\varphi) \cap \overline{D(A)}$. Hence $\partial\psi \in E(A^0; \overline{D(A)})$. The proof is complete.

Before closing this paragraph, we shall state a theorem which is important in deciding whether a given cyclically maximal monotone operator is of type $A_{1/2}$ or not.

THEOREM 2.7. *Let B be a cyclically maximal monotone set in $H \times H$ such that $0 \in R(B)$. There exists a cyclically maximal monotone set A in $H \times H$ such that $B = A_{1/2}$, if and only if the function $u \to |B^0 u|^2$ is convex on H.*

Proof. Without any loss of generality we may assume that $0 \in B0$.

The "if" part. Let $\varphi(u) = \dfrac{1}{2} |B^0 u|^2$ be convex.

The maximality of B implies its demiclosedness in $H \times H$ so that φ is lower-semicontinuous from H into $[0, \infty]$. Let $A = \partial\varphi$. Thus, from Theorem 2.6, $D(A_{1/2}) = D(\varphi) = D(B)$ and

$$|B^0 u|^2 = |A^0_{1/2} u|^2 \quad \text{for all } u \in D(B). \tag{2.118}$$

Let $\varphi_{1/2}$ and $\psi_{1/2}$ be two lower-semicontinuous convex functions on H such that $B = \partial\varphi_{1/2}$ and $A_{1/2} = \partial\psi_{1/2}$. Since $0 \in B0$ and $0 \in A_{1/2}0$ we may choose $\varphi_{1/2}$ and $\psi_{1/2}$ satisfying

$$\varphi_{1/2}(u) \geqslant 0, \; \psi_{1/2}(u) \geqslant 0 \quad \text{for all } u \in H$$

and $\varphi_{1/2}(0) = \psi_{1/2}(0) = 0$.

From Theorem 2.1 in Chapter IV, for every $x \in D(\varphi_{1/2})$ there exists a function $u \in C(0, \infty; H)$ satisfying

$$\frac{du}{dt}(t) + (\partial\varphi_{1/2})^0 u(t) = 0, \quad \text{a.e. on }]0, \infty[$$

$$u(0) = x$$

and

$$\frac{du}{dt} \in L^2(0, \infty;\; H).$$

We have

$$\left|\frac{du}{dt}(t)\right|^2 + \frac{d}{dt}\varphi_{1/2}(u(t)) = 0, \quad \text{a.e. on }]0, \infty[. \tag{2.119}$$

This equality shows, in particular, that $\lim_{t\to\infty} \varphi_{1/2}(u(t))$ exists. On the other hand, by definition of $\partial\varphi_{1/2}$ we have

$$\varphi_{1/2}(u(t)) \leqslant ((\partial\varphi_{1/2})^0 u(t), u(t)), \quad \text{a.e. on }]0, \infty[$$

so that $\lim_{t\to\infty} \varphi_{1/2}(u(t)) = 0$ (because $(\partial\varphi_{1/2})^0 u \in L^2(0, T; H)$ and $|u(t)|$ is bounded on $[0, \infty[$). Thus, integrating Eq. (2.119) over $]0, \infty[$, yields

$$\varphi_{1/2}(x) = \int_0^\infty \left|\frac{du}{dt}\right|^2 dt.$$

On the other hand, the equality

$$\frac{\mathrm{d}}{\mathrm{d}t}\,\psi_{1/2}(u(t)) = \left(A^0_{1/2}u(t),\ \frac{\mathrm{d}u}{\mathrm{d}t}(t)\right), \quad \text{a.e. on }]0, \infty[$$

implies that $x \in D(\psi_{1/2})$ and

$$\psi_{1/2}(x) = \int_0^\infty (A^0_{1/2}u(t),\ B^0u(t))\,\mathrm{d}t \leqslant \int_0^\infty |B^0u|^2\,\mathrm{d}t.$$

Here we have used, in particular, the lower-semicontinuity of $\psi_{1/2}$. Thus we have proved that $D(\varphi_{1/2}) \subset D(\psi_{1/2})$ and

$$\psi_{1/2}(x) \leqslant \varphi_{1/2}(x) \quad \text{for} \quad x \in D(\varphi_{1/2}).$$

The same argument with $\psi_{1/2}$ instead of $\varphi_{1/2}$ leads to

$$\varphi_{1/2}(x) \leqslant \psi_{1/2}(x) \quad \text{for} \quad x \in D(\psi_{1/2}) \subset D(\varphi_{1/2}).$$

Hence $\varphi_{1/2} = \psi_{1/2}$. In other words, $B = A_{1/2}$.

The "only if" part is an immediate consequence of Theorem 2.6. The proof is complete.

Remark 1°. This theorem is analogous to the corresponding one for self-adjoint positive operators (see T. Kato [1], Section 2.7, Chapter VI).

2°. In the case of maximal monotone linear operators, the square root $A^{1/2}$ can be equivalently defined (see K. Yosida [1], Chapter IX) by explicit formula

$$A^{1/2}x = \frac{1}{\pi}\int_0^\infty \lambda^{-\frac{1}{2}}(\lambda + A)^{-1}Ax\,\mathrm{d}\lambda, \quad x \in D(A).$$

In particular, if A is self-adjoint, $A^{1/2}$ can be expressed as

$$A^{1/2} = \int_0^\infty \lambda^{1/2}\,\mathrm{d}P_\lambda \tag{2.120}$$

where $\{P_\lambda\}$ is the *spectral resolution of* A.

In the nonlinear case no such representation formula is known and this is one of the main limitations of the theory developed before. Nevertheless, there are some recent results due to E. Zarantonello [2] which permit in particular to define the square root for a certain class (unfortunately quite limited) of nonlinear cyclically maximal monotone operators by a formula of the type (2.120).

3°. We have incidentally shown in the proof of Theorem 2.7 (see also Section 2.1 in Chapter IV) that $D(\varphi)$ coincides with the "trace" class

$$\{u_0 = u(0);\ u \in W^{1,2}(0, 1; H);\ \partial\varphi^0(u) \in L^2(0, 1; H)\}.$$

More generally, if A is any maximal monotone set in $H \times H$, we can define $\mathscr{B}_{\alpha,p}(A) = \{u_0 = u(0);\ u(t)$ is absolutely continuous on $]0, 1]$, continuous on $[0, 1]$, $t^{1-\alpha-1/p}|A^0u(t)| \in L^p(0, 1)$; $t^{1-\alpha-1/p}\left|\frac{du}{dt}(t)\right| \in L^p(0, 1)\}$ where $0 < \alpha \leqslant 1$ and $1 \leqslant p < \infty$. It is easy to see that $\mathscr{B}_{\alpha,p}$ is an interpolation class with respect to Lipschitz (locally) operators, between $D(A)$ and $\overline{D(A)}$. For a detailed treatment of this subject we refer the reader to works by D. Brézis [1] and to a forthcoming paper by D. Brézis and H. Brézis.

As mentioned before, if A is cyclically maximal monotone, then $D(A_{1/2}) = \mathscr{B}_{2,\,1/2}(A)$.

Bibliographical notes

§ 1. Theorem 1.1 was originally established by J. L. Lions and W. Strauss [1] and it covers many important cases of hyperbolic equations. The simple proof given here is due to H. Brézis [3]. However, there are certain cases studied in the quoted paper by J.L. Lions and W. Strauss (see also Lions' book [1]) not included in this theorem. For these, one needs to combine the monotonicity method with compactness arguments. Theorem 1.2 is due to the author [2]. A theorem of the same type was previously established by H. Brézis [6]. The results presented in Section 1.4 are due to J. L. Lions [4]. For more details and references on this subject, we refer the reader to the lecture notes by J. L. Lions [6].

By quite different methods, results similar to those in Theorem 1.5 were established by K. Jörgens [1], F. E. Browder [4], J. L. Lions and W. Strauss [1], and many others. This kind of argument was recently used for a particular case of the equation by H. Brézis [7] we have considered.

§ 2. The results presented in Sections 2.1 and 2.2 were established in a paper by the author [8]. A related problem has been studied by the author in [10], [11], [12] in connection with optimal control problems in Hilbert spaces. A particular case of the problem (2.1), (2.2) not entirely covered by the results of this paragraph has been studied by N. Pavel [3]. Even specialized to ordinary differential equations, Theorem 2.1 differs from previous existence results for two-points boundary value problems by the unusually weak assumptions concerning the regularity of the functions defining the equation and the boundary conditions.

Theorems presented in Sections 2.3 and 2.4, together with their proofs, are essentially contained in the author's papers [3], [4]. However, Theorems 2.4 and 2.6 in the form in which they have been presented in this paragraph are sharpened forms, due to H. Brézis [9], of some theorems given by the author in the papers quoted.

An existence result being the same type as Theorem 2.3 was established by L. Veron [1], [2] recently for time-dependent equations having the form $p(t)\dfrac{d^2u}{dt^2}+q(t)\dfrac{du}{dt}\in Au(t)$. The paper by M. Biroli [4] contains some results in connection with the almost-periodic character of the solutions for Eq. (2.69). In the linear case this problem was previously studied by S. Zaidman [1], [2].

There are many interesting questions concerning how properties of A are reflected in properties of the square root $A_{1/2}$ and conversely. Some results on this point may be found in the works of H. Brézis [9] and Y. Konishi [8].

Bibliography

S. Agmon, A. Douglas, L. Nirenberg

1 *Estimates near boundary for solutions of elliptic partial differential equations satisfying general boundary conditions*, Comm. Pure. Appl. Math. 12 (1959), 623–727.

S. Aizicovici

1 *On a class of functional-differential equations*, Rend. Mat. (to appear).

L. Amerio, G. Prouse

1 *Almost periodic functions and functional analysis*, Van Nostrand, New York, 1971.

W. F. Ames

1 *Nonlinear partial differential equations in engineering*, Academic Press, New York – London 1965.

D. G. Aronson

1 *Regularity properties of flows through porous media*, SIAM J. Appl. Math., 17 (1969), 461–467.

2 *Regularity properties of flows through porous media, The interface*, Arch. Rational Mech. Anal., 37 (1970), 1–10.

M. Artola

1 *Sur les perturbations des équations d'évolution*, Ann Sci. Ecole Norm. Sup., 4 (1969), 137–253.

E. Asplund

1 *Averaged norms*, Israel J. Math., 5 (1967), 227–233.

2 *Topics in the theory of convex functions*, Proc. NATO Institute Venise (1968), Oderisi Gubio.

H. Attouch, A. Damlamian

1 *On multivalued evolution equations in Hilbert spaces*, Israel J. Math., 12 (1972), 373–390.

A. V. Balakrishnan

1 *Fractional powers of closed operators and the semigroups generated by them*, Pacific J. Math., 10 (1960), 419–437.

V. Barbu

1 *Dissipative sets and nonlinear perturbed equations in Banach spaces*, Ann. Scuola Norm. Sup. Pisa, XXVI (1972), 365–390.

2 *On the regularity of solutions of hyperbolic nonlinear equations*, Ann. Mat. Pura Appl., XCV (1973), 303–319.

3 *Sur un problème aux limites pour une classe d'équations différentielles nonlinéaires abstraites du deuxième ordre en t*, C. R. Acad. Sc. Paris, 274 (1972), 459–462.

4 *A class of boundary problems for second-order abstract differential equations*, J. Fac. Sci. Univ. Tokyo Sect. I, 19 (1972), 295–319.

5 *Regularity properties of some nonlinear evolution equations*, Rev. Roumaine Math. Pures Appl., 18, 10 (1973), 1503–1514.

6 *Continuous perturbations of nonlinear m-accretive operators in Banach spaces*, Boll. Un. Mat. Ital., 6 (1972), 270–278.

7 *Integro-differential equations in Hilbert spaces*, An. St. Univ. A. I. Cuza, XIX (1973), 365–383.

8 *Existence theorems for a class of two points boundary problems*, J. Differential Equations., 17 (1975), 236–257.

9 *Nonlinear Volterra equations in a Hilbert space*, SIAM J. Math. Anal., 6,4 (1975).

10 *Convex control problems of Bolza in Hilbert spaces*, SIAM J. Control, 13 (1975).

11 *On convex control problem of Bolza*, SIAM J. Control 13, 5 (1975).

12 *Convex control problems for linear differential systems of retarded type*, (to appear).

C. Bardos, H. Brézis

1 *Sur une classe de problème d'évolution nonlinéaires*, J. Differential Equations, 6 (1969), 343–395.

A. Belleni-Morante

1 *Trasporto di neutroni con sezioni d'urto dipendenti dalla temperatura* (to appear).

Belleni-Morante, L. Barone

1 *Nonlinear kinetic theory of vehicular traffic*, J. Math. Anal. Appl., 47 (1974), 444–457.

Ph. Bénilan

1 *Solutions intégrales d'équations d'évolution dans un espace de Banach*, C. R. Acad. Sc. Paris, 274 (1972), 47–50.

2 *Solutions faibles d'équations d'évolution dans un espace réflexif, Séminaire Deny sur les semi-groupes nonlinéaires*, Orsay, 1970–1971.

3 *Équations d'évolution dans un espace de Banach quelconque et applications, Thèse*, Orsay, 1972.

Ph. Bénilan, H. Brézis

1 *Solutions faibles d'équations d'évolution dans les espaces de Hilbert*, Ann. Inst. Fourier 22 (1972) 311–329.

L. Berkovitz, H. Pollard

1 *A nonclassical variational problem arising from an optimal filter problem*, Arch. Rational Mech. Anal., 26 (1967), 281–304.

M. Biroli

1 *Sur les solutions bornées ou presque-périodiques des équations d'évolution multivoques sur un espace de Hilbert*, Ricerche de Matematica, XXI (1972).

2 *Solutions bornées et presque-périodiques des équations et inéquations d'évolution*, Ann. Mat. Pura Appl. (IV), XCIII (1972), 1–79.

3 *Sul problema di Cauchy per una disequazione parabolica con convesso dipendente dal tempo*, Rend. Acad. Naz. Lincei VIII, 1–2 (1971), 1–7.

4 *Sur la solution bornée ou presque-périodique d'une équation d'évolution du deuxième ordre et du type elliptique*, Boll. U.M.I., 6 (1972), 229–241.

A. Bove, G. Da Prato, G. Fano

1 *An existence proof for the Hartree-Fock time-dependent problem with bounded two-body interaction* (to appear).

R. V. Bresan

1. *On an evolution equation in Banach spaces*, Rend. Sem. Mat. Padova, 51 (1974), 279–291.

H. Brézis

1 *Équations et inéquations nonlinéaires dans les espaces vectoriels en dualité*, Ann. Inst. Fourier, 18 (1968), 115–175.

2 *Inéquations variationnelles associées à des opérateurs d'évolution*, Proc. NATO Institute, Venise (1968), Oderisi Gubio.

3 *Semi-groupes nonlinéaires et applications, Symp. sur les problèmes d'évolution*, Istituto Naz. Alta Matematica, Roma (1970).

4 *On a problem of T. Kato*, Comm. Pure Appl. Math., 24 (1971), 1–6.
5 *Propriétés régularisantes de certains semi-groupes nonlinéaires*, Israel J. Math., 9 (1971), 513–534.
6 *Problèmes unilatéraux*, J. Math. Pures Appl., 51 (1972), 1–164.
7 *Monotonicity methods in Hilbert spaces and some applications to nonlinear part. diff. equations*. Contributions to Nonlinear Functional Analysis, E. Zarantonello (editor), Acad. Press (1971).
8 *Opérateurs maximaux monotones et semigroupes de contractions dans les espaces de Hilbert*, Math. Studies, 5, North Holland, 1973.
9 *Équations d'évolution du second ordre associées à des opérateurs monotones*, Israel J. Math., 12 (1972), 51–60.
10 *On some degenerate nonlinear parabolic equations*, Nonlinear Functional Analysis, 28–38, Proc. Symp. Pure Math., Vol. 18 AMS (1970).
11 *Un problème d'évolution avec contraintes unilatérales*, C.R. Acad. Sci. Paris, 274 (1972), 310–313.
12 *Solutions with compact support of variational inequalities*, (Russian), Uspehi Mat. Nauk. (1974).

H. Brézis, M. G. Crandall, A. Pazy
1 *Perturbations of nonlinear maximal monotone sets*, Comm. Pure Appl. Math., 23 (1970), 123–144.

H. Brézis, J. L. Lions
1 *Sur certains problèmes unilatéraux hyperboliques*, C.R. Acad. Sci. Paris, 264 (1967), 928–931.

H. Brézis, A. Pazy
1 *Semigroups of nonlinear contractions on convex sets*, J. Funct. Anal., 6 (1970), 367–383.
2 *Accretive sets and differential equations in Banach spaces*, Israel J. Math., 8 (1970), 367–383.

H. Brézis, G. Stampacchia
1 *Sur la régularité de la solution d'inéquations elliptiques*, Bull. Soc. Math. France, 96 (1968), 153–180.

H. Brézis, W. A. Strauss
1 *Semi-linear second-order elliptic equations in L^1*, J. Math. Soc. Japan 25 (1973), 565–590.

D. Brézis
1 *Classes d'interpolation associées à un opérateur monotone*, C.R. Acad. Sci. Paris, t. 276 (1973), 1553–1556, 1597–1600.

F. E. Browder
1 *Existence and uniqueness theorems for solutions of nonlinear boundary value problems*, Symposia in Appl. Math. Vol. 17 A.M.S. (1965), 24–49.
2 *Nonlinear elliptic boundary value problems*, Trans. Amer. Math. Soc., 117 (1965), 530–550.
3 *Problèmes nonlinéaires*, Les presses de l'Université de Montreal, 1966.
4 *On nonlinear wave equation*, Math. Zeitschr. 8 (1962), 249–264.
5 *Nonlinear initial value problems*, Ann. of Math., 82 (1965), 51–87.
6 *Nonlinear maximal monotone operators in Banach spaces*, Math. Annalen, 177 (1968), 283–301.
7 *Nonlinear operators and nonlinear equations of evolution in Banach spaces*, Nonlinear Functional Analysis, Symposia in Pure Math., Vol. 18, Part. 2.
8 *Periodic solutions of nonlinear equations of evolution in infinite dimensional spaces*, Lect. Diff. Eqs., Van Nostrand, New York, 1969, 71–96.

B. Calvert
1 *Nonlinear evolution equations in Banach lattices*, Bull. Amer. Soc., 76 (1970), 345–350.

B. Calvert, K. Gustafson
1 *Multiplicative perturbation of nonlinear m-accretive operators*, J. Functional Analysis, 10 (1972), 149–158.

Bibliography

A. Cellina

1 *On the local existence of solutions of ordinary differential equations*, Bull. Acad. Pol. Sci., XX, 4 (1972), 293–296.

P. Chernoff

1 *Note on product formulae for operators semi-groups*, J. Functional Analysis, 2 (1968), 238–242.

I. Cioränescu

1 *Aplicaţia de dualitate în analiza funcţională neliniară*, Editura Academiei, Bucureşti, 1974.

B. D. Coleman, M. E. Gurtin

1 *Equi presence and constitutive equations for rigid heat conductors*, Zeit. fur. Angew. Math. und. Phys., 18 (1967), 199–208.

A. Corduneanu

1 *La somme de deux génerateurs infinitésimals*, An. Şt. Univ. "Al. I. Cuza" Iaşi, fasc. II (1972), 825–829.

2 *A note on minimum property of R(A) for monotone mappings in Hilbert spaces*, Atti Acad. Naz. Lincei, LIII (1972), 56–59.

C. Corduneanu

1 *Almost Periodic Functions*, Interscience tracts in pure and applied mathematics, J. Wiley & Sons, New York — London — Sydney — Toronto, 1968.

2 *Integral Equations and Feedback Systems*, Academic Press, New York, 1973.

M. G. Crandall

1 *Differential equations on convex sets*, J. Math. Soc. Japan, 22 (1970), 443–455.

2 *Semigroups of nonlinear transformations in Banach spaces*, Contributions to Nonlinear Functional Analysis, E. Zarantonello (editor), Academic Press, New York, 1971, 157–179.

3 *A generalized domain for semigroup generators*, Proc. Amer. Math. Soc., 37 (1973), 434–440.

4 *The semigroup approach to first-order quasi-linear equations in several space variables*, Israel J. Math., 12 (1972), 108–132.

M. G. Crandall, T. Liggett

1 *Generations of semi-groups of nonlinear transformations on general Banach spaces*, Amer. J. Math., 93 (1971), 265–298.

M. G. Crandall, A. Pazy

1 *Semi-groups of nonlinear contractions and dissipative sets*, J. Funct. Anal., 3 (1969), 376–418.

2 *On accretive sets in Banach spaces*, J. Funct. Anal., 5 (1970), 204–217.

3 *Nonlinear evolution equations in Banach spaces*, Israel J. Math., 11 (1972), 57–94.

R. Curtain, A. J. Pritchard

1 *The infinite dimensional Riccati equations*, J. Math. Anal. Appl., 4 (1974), 43–58.

C. Dafermos

1 *On abstract Volterra equations with applications to linear viscoelasticity*, J. Diff. Equations, 7 (1970), 554–569.

G. Da Prato

1 *Somme d'applications nonlinéaires dans des cônes et équations d'évolution dans des espaces d'opérateurs*, J. Math. Pures et Appl., 49 (1970), 289–348.

2 *Somme d'applications non-linéaires*, Symposia Mathematica, Academic Press, 1971, p. 233–268.

3 *Quelques résultats d'existence, unicité et régularité pour un problème de la théorie de contrôle*, Ann. Sci. École Norm. Sup. T. 3 (1973).

1 *Équations d'évolution dans des algèbres d'opérateurs et applications à des équations quasi-linéaires*, J. Math. Pures Appl., 48 (1969), 59–107.

H. Debrunner, P. Flor

1 *Ein Erweiterungssatz für monotone Mogen*, Archiv. Math., 15 (1964), 445–447.

J. Dorroh
1 *A non-linear Hille-Yosida-Philips theorem*, J. Funct. Anal., 3 (1969), 345–353.

J. A. Dubinski
1 *Weak convergence for nonlinear elliptic and parabolic equations* (Russian), Mat. Sbornik, 67 (1965), 609–642.
2 *Quasilinear elliptic and parabolic equations of arbitrary order*, Uspehi Mat. Nauk., 23 (1968), 45–90.

N. Dunford, J. Schwartz
1 *Linear operators*, Part I, Interscience, New York, 1958.
2 *Linear operators*, Part II, Interscience, New York, London 1963.

G. Duvaut, J. L. Lions
1 *Sur les inéquations en mécanique et en physique*, Dunod, 1972.

R. E. Edwards
1 *Functional Analysis*, Holt, Rinehart and Winston, New York, 1965.

G. Fichera
1 *Problemi elastostatici con vincoli unilaterali*, Memorie dell'Acad Naz dei Lincei, S VIII, 7 (1964), 91–140.

P. M. Fitzpatrick
1 *Surjectivity results for nonlinear mappings from a Banach space to its dual*, Math. Annallen, 204 (1973), 177–188.

A. Friedman
1 *Generalized heat transfer between solids and gases under nonlinear boundary conditions*, J. Math. Mech., 8 (1959), 161–184.
2 *Partial differential equations of parabolic type*, Prentice Hall, 1964.
3 *The Stefan problem in several space variables*, Trans. Amer. Math. Soc., 132 (1968), 51–87.

S. Fucik, J. Nečas, J. Soucek, V. Soucek
1 *Spectral analysis of nonlinear operators*, Lecture Notes in Mathematics 346, Springer Verlag, 1973.

H. Gajewski, K. Zacharias
1 *Über eine Klasse nichtlinearer Differentialgleichungen im Hilbert-Raum*, Match. Nach. Band., 57 (1973), 127–140.
2 *Über eine Klasse nichtlinearer Differentialgleichungen im Hilbertraum*, J. Math. Anal. Appl., 44 (1973), 71–87.

J. P. Gossez
1 *Opérateurs monotones non-linéaires dans les espaces de Banach nonréflexifs*, J. Math. Anal. Appl., 34, 2 (1971), 371–395.

J. M. Greenberg, R. C. Mac Camy and V. J. Mizel
1 *On the existence, uniqueness and stability of solutions of the equation* $\sigma'(u_x)u_{xx}+\lambda u_{xtx}=$ $=\rho_0 u_{tt}$, J. Math. Mech., 17 (1969), 707–728.

K. Gröger
1 *Regularitätsaussagen fur Evolution mit stark monotonen Operatoren* (to appear).

A. Halanay
1 *Differential Equations: Stability, Oscilations, Time Lags*, Academic Press, New York and London, 1965.
2 *On the asymptotic behavior of the solutions of an integrodifferential equations*, J. Math. Anal. Appl., 10 (1965), 319–324.

K. B. Hannsgen
1 *On a nonlinear Volterra equation*, Mich. Math. J., 16 (1969), 365–377.

E. Hille, R. S. Philips
1 *Functional analysis and semigroups*, Amer. Math. Soc. Coll. Pub., 31 (1957).

L. Hörmander
1 *Linear Partial Differential Operators*, Springer Verlag, 1963.

M. Iannelli

1 *Non-linear semigroups on cones of a non reflexive Banach space*, Bol. Un. Mat. Ital., 3 (1970), 412–419.

2 *Note on some nonlinear semigroups*, Boll. Un. Mat. Ital., 6 (1970), 1015–1025.

K. Jörgens

1 *Das Afangswertproblem in Grossen für eine klasse nichtlinearen wellengleichungen*, Math Z., 77 (1961), 295–308.

R. I. Kachurovski

1 *On monotone operators and convex functionals* (Russian), Uspehi Mat. Nauk SSSR, 15 (1960), 213–215.

2 *Nonlinear monotone operators in Banach spaces* (Russian). Uspehi Mat. Nauk SSSR, 23 (1968), 121–168.

T. Kato

1 *Perturbation theory for linear operators*, Springer Verlag, Berlin–Heidelberg–New York, 1966.

2 *Nonlinear semigroups and evolution equations*, J. Math. Soc. Japan, 19 (1967), 508–520.

3 *Accretive operators and nonlinear evolution equations in Banach spaces*, Nonlinear Functional Analysis, Proc. Symp. Pure Math., 18, Amer. Math. Soc. (1970), 138–161.

4 *Differentiability of nonlinear semigroups*, Global Analysis, Proc. Symp. Pure Math. Amer. Math. Soc. (1970).

Y. Komura

1 *Nonlinear semi-groups in Hilbert spaces*, J. Math. Soc. Japan, 19 (1967), 493–507.

2 *Differentiability of nonlinear semi-groups*, J. Math. Soc. Japan, 21 (1969), 375–402.

3 *Nonlinear semi-groups in Hilbert spaces*, Proc. Int. Conf. Funct. Anal. Tokyo (1969).

Y. Konishi

1 *A remark on perturbation of m-accretive operators in Banach spaces*, Proc. Japan Acad., 47 (1971), 452–455.

2 *Some examples of nonlinear semigroups in Banach lattices*, J. Fac. Sci. Univ. Tokyo, 18 (1972), 537–543.

3 *On the nonlinear semi-groups associated with* $u_t = \Delta\beta(u)$ *and* $\rho(u_t) = \Delta(u)$, J. Math. Soc. Japan, 25, 4 (1973), 622–628.

4 *On* $u_t = u_{xx} - F(u_x)$ *and the differentiability of the nonlinear semigroups associated with it*, Proc. Japan Acad., 48 (1972), 281–286.

5 *Une méthode de résolution d'une équation d'évolution nonlinéaire dégénérée*, J. Fac. Sci. Tokyo, Sect I A, 19 (1972), 243–255.

6 *Une remarque sur la perturbation d'opérateurs m-accrétifs dans un espace de Banach*, Proc. Japan Acad., 48 (1972), 157–160.

7 *Semi-linear Poisson's Equations*, Proc. Japan Acad., 49 (1973), 100–105.

8 *Compacité des résolvantes des opérateurs maximaux cycliquement monotones*, Proc. Japan Acad., 49 (1973), 303–305.

G. Köthe

1 *Topological vector spaces I*, Springer Verlag, Berlin–Heidelberg–New York, 1969.

S. G. Krein

1 *Linear differential equations in Banach spaces* (Russian), Ed. Nauka, Moskva, 1967.

J. J. Levin

1 *On a nonlinear Volterra equation*, Japan-United States Seminar on Ordinary Differential Equations and Functional Equations, Springer Lecture Notes, 243 (1971), 66–75.

J. J. Levin, J. Nohel

1 *On a system of integro-differential equations occurring in reactor dynamics*, J. Math. Mech., 9 (1960), 347–368.

2 *Note on a nonlinear Volterra equation*, Proc. Amer. Math. Soc., 14 (1963), 924–929.

3 *Perturbations of a nonlinear Volterra equation*, Proc. Amer. Math. Soc., 14 (1963), 924–929.

J. L. Lions

1 *Quelques méthodes de résolution des problèmes aux limites non-linéaires*, Paris, Dunod, Gauthier-Villars, 1969.

2 *Sur le contrôle optimal de systèmes gouvernés par des équations aux dérivées partielles*, Dunod, Gauthier-Villars, Paris, 1968.
3 *Sur quelques nouveaux exemples de problèmes unilateraux*, J. Fac. Sci. Univ. Tokyo, Sect. 1 A, 17 (1970), 1–9.
4 *Quelques problèmes de la théorie des équations non-linéaires d'évolution*, Problems in Non-Linear Analysis, C.I.M.E., Varenna, 1970, Ed. Cremonese (1971), 191–342.
5 *Partial differential inequalities* (Russian), Uspehi Mat. Nauk, 26 (1971), 202–263.
6 *Perturbations singuliers dans les problèmes aux limites et en contrôle optimal*, Lecture Notes in Mathematics, 323, Springer Verlag, 1973.

J. L. Lions, E. Magenes
1 *Problèmes aux limites nonhomogènes et applications*, Vol. 1, Dunod, Paris, 1968.

J. L. Lions, W. A. Strauss
1 *Some nonlinear evolution equations*, Bull. Soc. Math. France, 93 (1965), 43–96.

S. O. Londen
1 *On a nonlinear Volterra integral equation*, J. Differential Eqs., 14 (1973), 106–120.
2 *Volterra equations on a Hilbert space*, Report A, 43 (1973), Helsinki University of Technology, Institute of Mathematics.

D. L. Lovelady
1 *A functional differential equation in a Banach space*, Funk. Ekvac., 1973.
2 *A necessary and sufficient condition for exponentially bounded existence and uniqueness* (to appear).

D. L. Lovelady, R. H. Martin
1 *A global existence theorem for a nonautonomous differential equation in a Banach space*, Proc. Amer. Math. Soc., 35 (1972), 445–449.

R. C. MacCamy
1 *Approximations for a class of functional differential equations* (to appear).
2 *Asymptotic stability for a class of functional differential equations* (to appear).
3 *Nonlinear Volterra equations on a Hilbert space*, Report 73–14, Dep. of Math., Carnegie–Mellon Univ., 1973.

R. C. MacCamy, J. S. Wong
1 *Stability theorems for some functional differential equations*, Trans. Amer. Math. Soc. 164 (1972), 1–37.

W. R. Mann, F. Wolf
1 *Heat transfer between solids and gases under nonlinear boundary conditions*, Quarterly Appl. Math., 9 (1951), 163–184.

Gh. Marinescu
1 *Tratat de analiză funcţională*, II, Ed. Academiei, Bucureşti, 1972.

R. H. Martin Jr.
1 *A global existence theorem for autonomous differential equations in a Banach space*, Proc. Amer. Math. Soc., 26 (1970), 307–314.
2 *Differential equations on closed subsets of a Banach space*, Trans. Amer. Math. Soc., 179 (1973), 399–414.
3 *On the asymptotic behaviour of autonomous differential equations* (to appear).
4 *Generating an evolution system in a class of uniformly convex Banach spaces*, J. Functional Analysis, 11 (1972), 62–76.
5 *A theorem on critical points and global asymptotic stability*, J. Math. Anal. Appl., (1971), 124–130.

J. Mermin
1 *Accretive operators and nonlinear semigroups*, Thesis, University of California, Berkeley, 1968.
2 *On exponential limit formula and nonlinear semigroups*, Trans. Amer. Math. Soc., 150 (1970), 469–476.

R. K. Miller
1 *Nonlinear Volterra Integral Equations*, Benjamin, New York, 1971.

G. J. MINTY

1 *Monotone (nonlinear) operators in Hilbert space*, Duke Math. J., 29 (1962), 341–346.

2 *On the generalization of a direct method of the calculus of variations*, Bull. Amer. Math. Soc., 73 (1967), 315–321.

3 *On the extension of Lipschitz-Hölder continuous and monotone functions*, Bull. Amer. Math. Soc., 76 (1970), 334–339.

J. J. MOREAU

1 *Proximité et dualité dans un espace hilbertien*, Bull. Soc. Math. France, 93 (1965), 273–299.

2 *Fonctionnelles convexes*, mimeographed lecture notes, Séminaire sur les équations aux dérivées partielles, Collège de France, 1966–1967.

3 *Problème d'évolution associé à une convexe mobile d'un espace hilbertien*, C.R. Acad. Sci. Paris, 276 (1973).

4 *Rafle par un convexe variable*, Séminaire d'analyse convexe, Montpelier, 1972, exposé n° 3.

5 *Rétraction d'une multiapplication*, Séminaire d'analyse convexe, Montpelier, 1972 exposé n° 13.

U. MOSCO

1 *Perturbation of variational inequality* Proc. Amer. Math. Soc. Symp. Pure Mat., XVIII (1970), 182–194.

I. MIYADERA

1 *On the convergence of nonlinear semigroups*, Tohoku Math. J., 21 (1969), 221–236; 403–412.

2 *Some remarks on semi-groups of nonlinear operators*, Tohoku Math. J., 23 (1971), 245–258.

I. MIYADERA — S. OHARU

1 *Approximations of semi-groups of nonlinear operators*, Tohoku Math. J., 22 (1970), 24–47.

J. NEUBERGER

1 *Existence of a spectrum for nonlinear transformations*, Pacific J. Math., 31 (1969), 157–159.

M. NICOLESCU

1 *Analiza Matematică*, II, III, Ed. tehnică, Bucureşti, 1960.

S. ÔHARU

1 *Note on the representation of semigroups of nonlinear operators*, Proc. Japan Acad., 42 (1966), 1149–1154.

O. A. OLEINIK, A. S. KALASHNIKOV, CHZOU YUI-LIN

1 *The Cauchy problem and boundary value problems for equations of the type of non-stationary filtration*, Izv. Akad. Nauk SSSR Ser. Mat., 22 (1958), 667–704.

S. OUCHI

1 *On some degenerate system of parabolic semi-linear equations*, J. Fac. Sci.Univ. Tokyo Sect. I A, 18 (1972), 487–498.

C. V. PAO

1. *On nonlinear neutron transport equation*, J. Math. Anal. Appl., 42 (1973), 578–593.

D. PASCALI

1 *Operatori neliniari, Editura Academiei, Bucureşti, 1974*.

N. PAVEL

1 *Approximate solutions of Cauchy problem for some differential equations in Banach* Funkcial, Ekvac. 17 (1974), 85–94.

2 *Integral solutions for nonlinear evolution equations* (to appear).

3 *Nonlinear boundary value problems for second-order differential equations*, J. Math. Anal. Appl. (to appear).

4 *Nonlinear evolution equations associated to accretive operators on general Banach spaces*, mimeographed lecture notes, 132 p., Iași, 1974.

A. Pazy

1 *Semigroups of nonlinear contractions in Hilbert spaces*, C.I.M.E. Varenna, 1970, Ed. Cremonese (1971), 343–430.

2 *Asymptotic behaviour of contractions in Hilbert spaces*, Israel J. Math., 2 (1971), 235–240.

L. A. Peletier

1 *Asymptotic behaviour of solutions of the porous media equation*, SIAM J. Appl. Math., 21 (1971), 542–551.

J. C. Peralba

1 *Un problème d'évolution relatif à un opérateur sous-différentiel dépendent du temps*, C.R. Acad. Sci. Paris, 275 (1972), 93–96.

2 *Équations d'évolution dans un espace de Hilbert associées à des opérateurs sous-différentiels*, *Thèse*, Université des Sciences et Techniques du Languedoc, 1973.

M. Pierre

1 *Perturbations localement Lipschitziennes et continues d'opérateurs m-accretifs* (to appear).

A. J. Pritchard

1 *The linear, quadratic problems for systems described by evolution equations*, Report 10, Univ. of Warwick, Control Centre.

H. M. Riedl, G. F. Webb

1 *Relative boundedness conditions and the perturbation of nonlinear operators* (to appear).

R. T. Rockafellar

1 *Characterization of the subdifferential of convex functions*, Pacific. J. Math., 17 (1966), 497–510.

2 *Convex functions, monotone operators and variational inequalities*, Proc. NATO Institute, Venice (1968), Oderisi Gubio.

3 *Local boundedness of nonlinear monotone operators*, Michigan Math. J., 16 (1970), 397–407.

4 *On the maximality of sums of nonlinear monotone operators*, Trans. Amer. Math. Soc., 163 (1971).

5 *Convex Analysis*, Princeton Univ. Press, 1969.

6 *On the maximal monotonicity of subdifferential mappings*, Pacific. Math. J., 33 (1970), 209–216.

7 *Existence and duality theorems for convex problems of Bolza*, Trans. Amer. Math. Soc., 159 (1971), 1–40.

8 *Conjugate convex functions in optimal control and the calculus of variations*, J. Math. Anal. Appl., 32 (1970), 174–222.

K. Sato

1 *On the generators of non-negative contraction semi-groups in Banach lattices*, J. Math. Soc. Japan, 20 (1968), 431–436.

2 *A note on nonlinear dispersive operators*, J. Faculty Sci. Univ. Tokyo, 18 (1972), 456–473.

L. Schwartz

1 *Théorie des distributions*, Herman, Paris, 1967.

2 *Théorie des distributions à valeurs vectorielles*, Ann. Institut Fourier, VII, VIII (1958).

I. Segal

1 *Nonlinear semigroups*, Ann. of Math., 78 (1963), 339–364.

M. Slemrod

1 *An application of maximal dissipative sets in control theory*, J. Math. Anal. Appl., 46 (1974), 369–387.

G. Stampacchia

1 *Variational inequalities*, Proceedings of a NATO Adv. Study. Inst., Venice, 1968, Oderisi, Gubio.

I. Straskraba, O. Vejvoda

1 *Periodic solutions to abstract differential equations*, Czech. Math. J., 23 (1973), 635–669.

W. A. Strauss

1 *Evolution equations nonlinear in the time derivative*, J. Math. Mech., 15 (1966), 49–82.

2 *The initial value problem for certain nonlinear evolution equations*, Amer. J. Math., 89 (1967), 249–259.

L. Tartar

1 *L'équation de Riccati en dimension infinie* (to appear).

R. Temam

1 *Sur l'équation de Riccati associée à des opérateurs non bornés, en dimension infinie*, J. Funct. Analysis, 7 (1971), 85–116.

M. M. Vainberg

1 *A new principle in the theory of nonlinear equations*, Uspehi Mat. Nauk SSSR, 15 (1960), 243–244.

O. Vejvoda

1 *Nonlinear boundary value problems for differential equations*, Proceedings of the Conference on Differential Equations, Prague, 1962 (New York, Academic Press 1963), 119–215.

L. Veron

1 *Problèmes d'évolution du second ordre associés à des opérateurs monotones*, C.R. Acad. Sc. Paris, 278 (1974), 1099–2101.

J. Watanabe

1 *Semi-groups of nonlinear operators on closed convex sets*, Proc. Japan, Acad. Sci., 45 l(1969), 219–223.

2 *On nonlinear semi-groups generated by cyclically dissipative sets*, J. Fac. Sci. Tokyo, 18 (1971), 122–137.

3 *On certain nonlinear evolution equations*, J. Math. Soc. Japan, 25 (1973), 446–463.

G. F. Webb

1 *Continuous nonlinear perturbations of linear accretive operators*, J. Functional Analysis, 10 (1972), 191–203.

2 *Nonlinear perturbations of linear accretive operators in Banach spaces*, Israel J. Math, 12 (1972), 237–248.

3 *Autonomous nonlinear functional differential equations and nonlinear semigroups*, J. Math. Anal. Appl., 46 (1974), 1–12.

4 *Accretive operators and existence for nonlinear functional differential equations*, J. Differential Equations, 14 (1973), 57–69.

5 *Finite difference methods and nonlinear semigroups* (to appear).

6 *Abstracts scattering for nonlinear evolution equations*, Math. System Theory (to appear).

K. Yosida

1 *Functional Analysis*, Springer Verlag, Berlin–Heidelberg–New York, 1969.

S. Zaidman

1 *Almost-periodicity for some differential equations in Hilbert spaces*, Atti Accad. Naz. Lincei (8), 37, 5 (1964), 253–257.

2 *Soluzioni limitate e quasi periodiche della equazione di Poisson*, Ann. Mat. Pura Appl. 4), 64 (1964), 365–405.

E. H. Zarantonello

1 *Solving functional equations by contractive averaging*, Math. Research. Center Report 160, Madison, Wisc. 1960.

2 *Projections on Convex sets in Hilbert space and spectral theory*, Contributions to Noninear Functional Analysis, Academic Press 1971, New York and London, 237–424.

Subject index